WORDS ON WINGS

By the same author:

Blokes and Sheds

Inventions from the Shed

Mates

Back in the Shed

WORDS ON WINGS

AN ANTHOLOGY OF NEW ZEALANDERS IN FLIGHT

EDITED BY
JIM HOPKINS

HarperCollins*Publishers*

To my father. Those who knew him will know why I stand in awe of his life; for all sorts of reasons but not least because, in 1953, while he was Vicar at St Mary's in Addington, it fell to him to conduct the funeral service for a solitary patient at Sunnyside Hospital — Richard Pearse.

National Library of New Zealand Cataloguing-in-Publication Data

Words on wings : an anthology of New Zealanders in flight / Jim Hopkins, editor.
ISBN 1-86950-464-X
1. Aeronautics—New Zealand—Anecdotes. 2. Aeronautics—New Zealand—
History. I. Hopkins, Jim, 1946–
629.130993-dc 21

First published 2004
HarperCollins*Publishers (New Zealand) Limited*
P.O. Box 1, Auckland

ISBN 1 86950 464 X

Cover design and illustration by Murray Dewhurst
Typesetting by IslandBridge

Printed by Griffin Press, Australia, on 80 gsm Bulky Book Ivory

Contents

Part five: Later days

Introduction

This is the age of invisible inventions, an age when we are being transformed by what is out of sight. It's the age of algorithms and engineered genes; wisps and fragments that are the real vehicle of humanity's next journey. And we never see them.

Of course, we can see and touch the larger objects in which such potent creations are embedded. But it's hard to imagine a bug-resistant turnip as an object of awe and wonder, no matter what it might mean for food supplies in far-off parts. And equally hard to imagine excited crowds flocking to a grassy paddock to witness the arrival of the latest laptop or PlayStation. For laptops and turnips (and much more besides) may reveal great ingenuity at work but they don't inspire us. We might welcome (or fear) what they do but we're not stirred by what they are.

That's how it is in this age of ours. As the speed of change increases, so our capacity for excitement seems to shrink. When novelty is a commonplace, it loses its power to amaze.

And when adventure becomes a virtual reality, it will lose something. Best you should brace yourself, for that day's coming — and quickly. Pretty soon, we will be able to 'experience' the most fantastic adventures in a helmet. And when that happens, some nameless pioneer will have the dubious distinction of becoming the first human being to die flying a Sopwith Camel in the upholstered comfort of their favourite armchair.

Mind you, the really big story will be the lawsuit afterwards. There'll be much debate about the size of the health warnings on the software, allegations of corporate indifference, calls for more regulations — and an inquiry, you name it. The great army of meddlers will have a virtual field day.

It wasn't like that when the original Sopwith Camels took to the air. By all accounts, they were temperamental wee beasts, unstable, tricky to fly and therefore almost certain to wear a lawsuit in these litigious times.

Indeed, the same goes for most of the earliest aircraft. Like computers, these frail and complex craft had a disconcerting tendency to crash, but with consequences far more serious than inconvenience.

Yet there was no great clamour to leave them on the ground till every last defect had been rectified. The fraught and anxious timidity that's so much a part of our times isn't evident in the comments of the public or the columns of the press. Instead, there was amazement, and not a little awe. Risk was something to embrace, not avoid. Cotton wool was something to put in your ears to deaden the sound of clattering engines, not something to wrap yourself in so you could cushion the impact of experience.

They may seem like curios now, quaint and awkward relics, but in their day the Wright Flyers, the Bleriots, the Caudrons and others were the essence of the new, the sharpest point on the cutting edge. They were taking everybody, not just the pilots, on a welcome adventure in a new dimension.

There is much about that era, its lifestyles, clothing, attitudes and values, which may seem comic to the knowing modern mind, but there was surely something in the air back then that we no longer have. The crowds and pilots of the first years of the twentieth century, all the builders and bystanders, were part of a grand adventure, a most visible adventure and they shared an excitement we can barely imagine.

Today, when most experience air travel only as a banality, a means of getting from A to B in cramped discomfort, few of those who flash through the thin, freezing air at 30,000 feet even bother to look out the window. And fewer still remember the times when crowds flocked to Wigram or Mangere or the shores of Auckland Harbour to watch — yes, just watch — an aeroplane land.

From memory, it was Bill Gates who said aviation was the Internet of another era, and that's right. It was the first fast and efficient way to link people and places all around the world.

So the arrival of a pioneer aeroplane was a grand moment, a new measure of human achievement and possibility. And people came to celebrate, whether it was the *Southern Cross* or Jean Batten's Gull or the big flying boat, *Aotearoa*, or Scotland's Caudron or the Creamoata de Havilland; whether it was *Manurewa* or *Brittania* or Clouston's Comet or the big, white, four-engined, delta-winged Vulcan that clipped the runway when they opened Wellington's windy airport — having previously flown through a clear blue sky straight over Christchurch South Intermediate, gleefully watched by an owlish lad lying on the grass and squinting into the sun.

In truth, it's always been that way. I've always been fascinated by aircraft and flight. As a boy, I was amazed that weapons of war like the de Havilland Hornet or the Hawker Hunter could look so … beautiful. (Those who know will understand!) I used to think we could see some essence of nations expressed in designs: Germanic logic, French folly, American power, the British affection for Heath Robinson.

I joined the Air Training Corps and drilled for hours in a Wigram hanger, hoping I might score a flight in a Vampire. No such luck. I wanted to be a Spitfire pilot, till a friendly air force fellow pointed out you actually needed 20/20 vision. Cancel that.

But I still accumulated great piles of *R.A.F. Flying Reviews* (long since out of publication) and made every available Airfix model — 3s 11d for the small ones in plastic bags at the Selwyn Street general store.

Finally, I got to see (and hear) the fabled Spit at Wanaka and this year, at Omaka, also watched the Fokker Triplanes, the Bristols and Nieuports and Camels of a time long ago. Then flew back home, low, below the peaks, through the Molesworth in Pionair's DC 3.

Ahh, yes! If you seek it, the excitement's still there. On the margins at least — doing aerobatics in a Tiger Moth high over Croydon airfield near Gore; perched precariously at the front of a microlight flying low over the Waikato countryside or taking the controls of a Cessna heading up the valley between Lake Wakatipu and Te Anau with the peaks on both sides towering over the wingtips.

I hope some of that excitement, indeed much of it, lives in these pages. Not all, obviously, because this isn't a history; it consciously avoids disaster and, for all sorts of reasons, some great stories are missing. But I hope those that are here share some part of the excitement and peril of pioneer pilots in peace and war.

As the world celebrates the Wright stuff and our first century of powered flight, it's good to remember there were heroes here too, brave people, men and women who ignored the risks and amazed those less fortunate on the ground when they took their very visible inventions up into the long white clouds.

Acknowledgements

This was certainly not a solo flight and it would be wrong to take-off without acknowledging the rest of the crew, beginning with all the authors and publishers who have generously allowed their writings to be included in these pages.

More specifically, profound thanks to researcher Judi Ball for her diligent aerial reconnaissance. And to that most conscientious of editors, Sue Page for skilfully navigating through great blizzards of paper. Thanks also to sage and steady Air Traffic Controller, Lorain Day who ensured a smooth landing despite the dodgy cross winds.

Finally, thanks to Tess and Tom for (silently) tolerating great stacks of books and piles of papers in a house that's far too small to be an office as well.

Contributors

The publishers are grateful to the following authors, publishers and copyright holders for their permission to reproduce copyright material:

Part one

Ross Ewing and Ross Macpherson, *The History of New Zealand Aviation*, (Heinemann Publishers (NZ) Ltd, 1986); Professor Baldwin's Ascent: A Sensational Feat, *Otago Daily Times*, Tuesday 22 January 1889; Attempted Parachute Descent: Narrow Escape, *Auckland Weekly News*, 10 March 1894; Leo White, *Wingspread*, (The Unity Press, 1941); David Mulgan, *The Kiwi's First Wings*, (The Wingfield Press, 1960); Roy Sinclair, *New Zealand Aviation Yarns*, (Grantham House Publishing, 1994); Aeroplane flies from Timaru to Christchurch, *Press* 7 March 1914.

Part two

The Kia Ora Coo-ee, The Magazine for the ANZACS in the Middle East, 1918; A. R. Kingsford, *Night Raiders of the Air* (John Hamilton Ltd, 1988); Lord Douglas of Kirtleside, *Years of Combat: The First Volume of the Auto-biography of Sholto Douglas*, (Collins, London, 1963); G. H. Cunningham, *Mac's Memoirs: The Flying Life of Squadron-Leader McGregor*, (A H & A W Reed Ltd); A. R. Kingsford, *Night Raiders of the Air* (John Hamilton Ltd, 1988); A. R. Grimwood, *Flights and Fancies*, 1943.

Part three

David Yerex, *Yerex of TACA*, (Ampersand Publishing Associates Ltd, 1985); G. H. Cunningham, *Mac's Memoirs: The Flying Life of Squadron-Leader McGregor*, (A H & A W Reed Ltd); E. F. Harvie, *George Bolt — Pioneer Aviator*, (A H & A W Reed Ltd,1974); *The Winged Express*, Adventure of Flying Height and its Reactions Safe, Swift and Absorbing Travel, *Otago Daily Times* 1 February 1936; R. Campbell Begg, *The Secret of the Knife*, (Jarrold & Sons Ltd); Arthur Bradshaw, *Flying by Bradshaw: Memoirs of a*

Pioneer Pilot 1933–1975, (Proctor Publications Nelson); John Stannage, *Smithy*, (Oxford University Press); Jean Batten, *My Life*, (George G Harrap & Co Ltd); John Stannage, *High Adventure*; E. F. Harvie, *Venture the Far Horizon*, (Whitcombe and Tombs Limited); Air Commodore A. E. Clouston, *The Dangerous Skies*, (Cassell & Company Limited, 1954).

Part four

Squadron Leader I. A. Ewen, 'Pilot Error' from *Wings of War: Airmen of All Nations Tell Their Stories 1939–1945*, (Hutchinson, London, 1983); Edited by J. D. McDonald, *The Pitcher and the Well* (Pauls Book Arcade, 1961); Maurice McGreal, *A Noble Chance: One Pilot's Life* (Wellington, 1994); Noel Monks, *Fighter Squadrons: The Epic Story of Two Hurricane Squadrons in France*, (Angus and Robertson, 1941); Alan C. Deere, *Nine Lives*, (Hodder & Stoughton, London, 1959); Squadron Leader Bob Spurdle DFC and Bar, *The Blue Arena*, (William Kimber & Co Ltd, London, 1986); Desmond Scott, *Typhoon Pilot*, (Century Hutchinson Ltd, 1989); Wing Commander H. L. Thompson, *New Zealanders with the Royal Air Force Volume II: Europe Theatre, January 1943–May 1945*, (War History Branch, Department of Internal Affairs, Wellington, 1956); Desmond Scott, *One More Hour*, (Century Hutchinson Ltd, 1989); Tony Williams, *Anzacs: Stories from New Zealanders at War*, (Hodder Moa Beckett, 2000); Maurice McGreal, *A Noble Chance: One Pilot's Life*, (Wellington, 1994); Edited by J. D. McDonald, *The Pitcher and the Well*, (Pauls Book Arcade, 1961); Squadron Leader Bob Spurdle, *The Blue Arena*, (William Kimber & Co Ltd, London, 1986); Wing Commander H. L. Thompson, *New Zealanders with the Royal Air Force Volume II: Europe Theatre, January 1943–May 1945*, (War History Branch, Department of Internal Affairs, Wellington, 1956); Brian Joyce, *Great New Zealand Adventures*, (A H & A W Reed Ltd, 1983); James Sanders, *Venturer Courageous: Group Captain Leonard Trent V.C., D.F.C., A Biography*, (Hutchinson Group (N.Z.) Ltd, Auckland, 1983); Desmond Scott, *Typhoon Pilot*; Desmond Scott, *One More Hour*, (Century Hutchinson Ltd, 1989); Desmond Scott, *One More Hour*, (Century Hutchinson Ltd, 1989); Flight Lieutenant Bryan E. Young, *Beckoning Skies: A Pilot's Story*, (Reed Publishing); Desmond Scott, *One More Hour*, (Century Hutchinson Ltd, 1989); Gerald Bowman, *Jump for It! Stories of the Caterpillar Club*, (Evans Brothers Limited, 1955); A. M. Feast, *They Got Back: The Best Escape Stories from the R.A.F. Flying Review'*, (Herbert Jenkins Ltd, London, 1961); J. Norby King,

Green Kiwi versus German Eagle: The Journal of a New Zealand Spitfire Pilot, (J. Norby King (self-published), 1991); Squadron Leader Bob Spurdle DFC and Bar, *The Blue Arena*, (William Kimber & Co Ltd, London, 1986); Miles King, *The Sky My Canvas: An Autobiography*, (Mallinson Rendel Publishers Ltd, 1985); James Bertram, *The Shadow of a War: A New Zealander in the Far East 1939–1946*, (Whitcombe & Tombs Ltd, 1947); Squadron Leader Bob Spurdle DFC and Bar, *The Blue Arena*, (William Kimber & Co Ltd, London, 1986); Desmond Scott, *One More Hour*, (Century Hutchinson Ltd, 1989).

Part five

Maurice McGreal, *A Noble Chance: One Pilot's Life* (Wellington, 1994); Noel Holmes, *To Fly a Desk*, (Reed Publishing (NZ) Ltd, 1982); Roy Sinclair, *New Zealand Aviation Yarns*, (Grantham House Publishing, 1994); David Yerex, *Yerex of TACA*, (Ampersand Publishing Associates Ltd, 1985); Roger Crow, *Flying Low*, (Hodder & Stoughton Ltd, Auckland, 1978); Miles King, *The Sky My Canvas: An Autobiography*, (Mallinson Rendel Publishers Ltd, 1985); John King, *Famous New Zealand Aviators*, (Grantham House, 1998); Ian H. Driscoll, *Flightpath South Pacific*, (Whitcombe & Tombs); Paul Beauchamp-Legg, *West Coast Memories, Volume One*, (Paul Beauchamp-Legg, Picton, 1994); Brian Waugh, *Turbulent Years: A Commercial Pilot's Story*, (Hazard Press, Christchurch, 1990); Captain Fred Ladd, MBE with Ross Annabell, *A Shower of Spray and We're Away*, (A H & A W Reed Ltd, 1971); Ross Ewing, *Catalina Dreaming*, (David Ling Publishing Limited, 1996); Roy Sinclair, *Journeying with Aviators in New Zealand*, (Random House New Zealand Ltd, Auckland, 1998); Hanafi Hayes, *Hayes over New Zealand*, (Methuen Publications NZ Ltd, Auckland, 1981); Mike Bennet, *The Venison Hunters*, (A H & A W Reed Ltd, 1979).

While every attempt has been made to contact copyright holders and secure permission to reproduce copyright material, it has not always been possible to do so. Copyright holders are invited to contact the publisher.

PART ONE

PEARSE
AND THE
PIONEERS

Long before Abel Tasman hove to, there were man-made objects up in the long white clouds. Some were small but some were large, with wingspans of four or five metres. Some were bird-shaped, some were circular or triangular, others soared as diamonds or as crosses. But whatever their shape and size, they were all kites (manu tukutuku), flown by children and adults alike.

The anthropologist Elsdon Best notes that kites were frequently 'made and flown by the Maori during their social gatherings. At such times many persons would bring kites, each person flying his own.' He also describes how 'young folks occasionally played a joke on their elders by flying a kite at night, and attaching it to a firebrand. They would then run, in great apparent excitement, to their elders and inform them that a fire demon was darting about in the heavens.'

Yet there was more to kites than merriment. They were also used to send signals and flown before a battle to win the support of Tu (the God of War). There are even stories (some say legend, some say fact) of kites being used to 'fly' a lightweight volunteer over the walls of an enemy's pa in order to open the gates and permit an attack.

Not surprisingly, having spent six months in their leaky boats, getting airborne was a notion that also appealed to some of New Zealand's first British settlers. One such visionary soul was Sir Julius Vogel, who served as prime minister from 1873 to 1876. Nine years later he published *Anno Domini 2000 or a Woman's Destiny* in which his heroine, Lady Taieri, dashes across the Tasman in an 'air cruiser'. This was an airship whose engine gained its power from 'the atomic structure of substances'.

Vogel's magnificent machine was 'beautifully constructed of pure aluminium' and featured 'everything conducive to the comfort of the passengers'. It flew about 'fifty feet above the sea, and, without any strain on the machinery, made easily a hundred miles an hour'.

A trifle slow, for sure, but still better than cattle class and DVT. Then again, the 'air cruiser' was a work in fiction, a flight of fancy. There were those who realised how hard it would be for anyone to put a real cruiser into the air ...

The History of New Zealand Aviation
Ross Ewing and Ross Macpherson

The *New Zealand Graphic* commented in 1882 that 'unquestionably the greatest mechanical problem unsolved is that of aerial navigation. When one considers the far-reaching effect which its solution would have on the civilisation of the age, and weighs the benefits which would accrue to mankind, the wonder is that the matter has not been, ere this, more seriously studied under government supervision. Various governments and societies have been making limited experiments, but the problem has never been attacked in any vigorous way; nor has there been any attempt to bring together the minds which in various quarters of the globe have individually been engaged on the problems involved.'

New Zealanders had been addressing the problem, even before the subject was aired in the pages of the *Graphic*. Government involvement, however, was to be conspicuously absent until the 1920s.

The first known recorded ideas on 'aerial navigation' in New Zealand came from the pen of a Christchurch doctor in July 1868. Dr A. C. Barker, well known as 'Barker of Cathedral Square', was a man of wide-ranging interests, his place in history assured by his photographic studies of early Canterbury. In July 1868 he wrote to his brother Matthias in England and described a then rather radical idea for achieving flight.

'My notion of flying would be to raise the collapsed machine somewhat in the way of a rocket. Then, as the ascending power becomes exhausted, the wings should expand to a certain fixed degree and position which would enable the machine to be propelled by the power of gravity along an inclined plane, until the proximity of the earth required fresh explosive power to raise the machine to a summit of a fresh inclined plane, the level of the wings to be modified by simple machinery so as to alter the obliquity of the plane at will, and by opening a representation of the broad tail of the bird to oppose a sufficient surface to moderate the descending impetus, when you alight on the ground. I doubt its being possible to gain the ascending power by means of a pulsation of wings, as that would require so great a weight of machinery as would defeat its own design. I think therefore attention should be given to some of these new explosives compounds, if any are known which have a sustained action, like the mixture used for propelling rockets, only it would require to be more portable and with a gradually increasing rapidity of explosion. My absence from England prevents me knowing anything of the discoveries which are daily being made there, but I hear there is an aeronautic society established.'

The doctor's concept of an aircraft equipped with wings that could alter their configuration during different stages of flight was remarkably far-sighted.

A year later, in neighbouring Otago Province, another New Zealander made an astonishing revelation to the world at large on the subject of aerial transport.

One Samuel Goldston (given his location on the banks of the Arrow River in Central Otago, the name may well have been a pen name) wrote to the editor of the *Otago Witness* with some news of far-reaching implications. The issue for 10 July 1869 carried his letter:

'Sir, With no small consternation I observed in a late issue of your paper that the executive is again occupied with a railway scheme. Not only is the way they have taken up the question to be condemned, and ought to be put a stop to, but the idea itself may now be looked upon as antiquated, since something better, cheaper, and also more expeditious, can be provided. Sir, I can without boasting state, that my efforts to construct a flying machine are about to be crowned with success.

'The construction of the machine is so cheap, and the plan so simple, that I may well be excused from going into the particulars thus publicly. Naturally averse to boasting, I may briefly inform your readers and the world at large, that the machine is fast approaching completion, and that in the course of the next month this grand problem will be solved by me.'

Unfortunately, nothing more appears to be recorded in the pages of the *Otago Witness* as to the achievements of Samuel Goldston, the 'modest' inventor from the banks of the Arrow.

Dr Barker wrote, in April 1872, another letter to his brother on the subject of aeronautics. He described an aircraft with short, fixed wings and a rotor (a configuration similar to that first tested on the British-built Fairey Jet Rotodyne of 1954).

'I see by the "Cornhill" that the aeronautics society is bestirring itself and is adopting a plan which appears from the account given of it to be nearly that I have long thought of and spoken about here viz. a circulating disc above fixed wings — my notion is a very wide disc made of a wheel like that of the old ventilator but the complete circle consisting of two or three vanes only which should be made to revolve with great rapidity perhaps by means of an engine worked by dynamite or one of the new explosives.'

Dr Barker died the following year — not, however, as a result of testing a flying machine powered by explosives.

Close at hand, at Waikerakikari on Banks Peninsula, lived a colourful eccentric called Harry Head. Born in 1835 in Wiltshire, England, as Harry Head Alexander, he received a good education by the standards of the day, becoming an accomplished mathematician. Head was somewhat of a character. He is said to have travelled around the world five times, staying for a period with Indians in the USA, and he spent time goldmining in Central Otago. One of his feats was a journey on foot from Auckland to Wellington and then on down through the South Island to Christchurch during the 1860s.

Head's able mind dwelt on many things, including the possibilities of flight. H. C. Jacobson, in his 1917 edition of *Tales of Banks Peninsula*, wrote that 'one of his strangest notions was that, with properly manufactured appliances, human beings would be able to fly'. Jacobson noted, 'nearly 50 years ago Harry Head broke his arm in trying one of his roughly made aeroplanes.' This would make Harry Head, *circa* 1869, the first person to experiment with practical aviation in New Zealand. One presumes that his efforts were centred around a roughly constructed hang-glider; a view supported by an 1872 entry in his diary. The entry theorises on the angle of an 'aeroplane' necessary to 'get the most lifting force with the least forward resistance'.

Harry Head's broken arm is the first known example of an aviation accident in New Zealand. The Hermit of Waikerakikari is believed to have died in Melbourne, Australia, during the 1920s.

New Zealand, despite its geographical isolation from the main centres of scientific endeavour, occasionally made it to the forefront of international aeronautical advances during the nineteenth century. The 1879 edition of the *Encyclopaedia Brittanica* devoted 14 pages to the subject of flight, with some decidedly novel descriptions of the latest in aerial machinery, and closed with a reference to the most recent experiments in aerial transit — by a New Zealander.

Henry Skey, a draughtsman and meteorologist living in Dunedin, had advanced his ideas for what he described as a tension wheel to the Otago branch of the New Zealand Institute on 9 October 1877. The design, resembling a twin-rotored helicopter, incorporated two contra-rotating discs to provide lift and stability for a small cabin mounted beneath. Seated in the cabin, the pilot would provide both foot and arm power to both rotate coaxial drives to the rotors and also propel a 'screw propeller' for horizontal movement. The configuration outlined was similar to that later adopted with some success by numerous twentieth-century helicopter pioneers and by manufacturers such as Gyrodyne, Kamov and Focke-Wulf.

Henry Skey's experimentation with inclined surfaces was of sufficient status to be acknowledged by Octave Chanute, an American engineer, in his 1894 classic *Progress in Flying Machines*. It was this book that was to assist the Wright brothers in beginning their experiments.

Mention must be made of a Mr Nicolson and of Arthur Marychurch. Nicolson is reputed to have been involved with the construction of a flying machine at Fern Bush near Lochiel in Southland during the 1880s. Again, in a manner similar to the entry in the *Otago Witness* of 1869 for Samuel Goldston, a tantalising snippet appeared in the pages of the *Auckland Weekly News* for 2 September 1893 about the activities of Marychurch and his flying machine at 'Waoto'.

'He has for some time past been experimenting in this matter and at last is reported to have succeeded in inventing a machine which will actually ascend into the air, rising, falling, and turning at the will of the navigators, and is not affected by a head wind.'

What manner of flying machine was in existence at 'Waoto' (probably Waotu, on the Waikato River near Lake Arapuni)? As with Samuel Goldston, the modest optimist of some 24 years earlier, nothing further has been traced of Arthur Marychurch and his flying machine.

Not yet, anyway. But if Mr Marychurch is a mystery, others are not. Our earliest successful 'aeronauts' copied the Montgolfier Brothers, who'd sent two brave Parisians aloft in a balloon back in 1783. The first to ascend down under was a good-looking, lithe-limbed fellow who called himself Professor Thomas S. Baldwin. Not content with the excitements of life as a tightrope walker (which he'd been), Prof. Baldwin decided to go up in the world – and did, in Dunedin, in 1889 ...

Professor Baldwin's Ascent: A Sensational Feat
Otago Daily Times, Tuesday 22 January 1889

Another large crowd assembled on and round about the Caledonian ground by 7 o'clock yesterday evening to see Professor Baldwin take his daring flight through the air. Within the enclosure there were not nearly so many people as on Saturday, but the hill at the rear of Smith and Fotheringham's brickworks and the Town Belt at Montecillo were packed with sightseers, who had a good view of the exhibition without going through the idle ceremony of paying a shilling. As the advertised time for the ascent approached considerable doubt was entertained among the public as to whether the balloon would get safely away after all, rather a brisk breeze springing up just before sundown. However, Professor Baldwin continued actively superintending the inflation of his balloon, which was rather dangerously agitated by the gusts of wind every now and then. At a few minutes before 7 Professor Baldwin mounted a form and, as before, made a short preliminary speech to the spectators. He is a well-built, lithe-limbed American, with dark complexion and moustache, good-looking, and with considerable alertness and resolution in his manner. That he is a man of wonderful pluck and iron nerve, his aerial feats amply testify. Standing up to address his patrons, attired in the orthodox silk hat and black frock coat, he looks scarcely like a man on the eve of taking such a startling journey. He might be intending to sell some town allotments, or say a few words on the political situation. A little later, divested of hat and coat, quick yet cool amid all the bustle attending his departure, he is seen at his best. What the professor has now to say is brief and to the point. He explains Saturday's failure in a frank and manly way. The pressure of so high a wind on the frail fabric of the balloon was not to be withstood, but had it not been for the purely accidental bursting, he himself would have been willing to make the ascent. He could control his balloon and his parachute once he got fairly away, but he could not control the elements. In spite of the wind then blowing, he would endeavour to make the ascent that evening at 7 o'clock sharp — i.e., in 10 minutes' time — and he begged them all to stand back and keep quiet while the attempt was made. There would be danger again of the balloon bursting in that wind, directly it was raised off the ground, and the air pressure got underneath it; but if such an accident did happen it would not be his fault. If he could only get up he would guarantee to come down right enough. He regretted to see a statement in that evening's paper to the effect that he had purposely ripped the balloon up on Saturday. He was not standing within yards of it at the time, and it was certainly no advantage to him at his first exhibition in a new country to tamper with the feelings of the public. If he had to stay here all the summer he would give them an ascent as promised, and he could assure them he would rather lose a leg than miss the ascent that evening. This short speech was well received by the people, and Mr Baldwin then hurried away and began to make final preparations for his excursion, in which he was assisted by his manager, Mr Farini. The balloon was raised well off the ground, being held captive by several men, and although it swayed rather violently in the breeze the fabric kept together on this occasion. Everything being nearly completed, Professor Baldwin, who is now bareheaded and clad in a dark close-fitting vest, runs across to the bench near at hand and gives his wife a hasty parting kiss. There is nothing whatever of the theatrical element about this ceremony, which is quickly and unostentatiously performed, and is not even observed by the majority of the spectators. Confident as the aeronaut is in the efficacy of his invention, he is probably too shrewd a man not to recognise that the 'wisest schemes of mice and men gang aft aglee.' He has all the assurance of safety that personal attention to his apparatus and splendid coolness and nerve can give him, but there are chances against him too. Some blunder on the part of an attendant, or some unforeseen hitch at the last moment, may wreck him before he is sufficiently clear of the earth to rely upon his parachute, — or what if away in the clouds some little thing — some very little thing — should go amiss with the parachute itself? Professor Baldwin, no doubt, does not believe in this latter contingency, and would bet long odds against the parachute ever failing him. It is to be hoped it never will, and that the adage about the pitcher and the well will not be verified in the case of this daring man. His leave taking over, the professor bends down and disappears for some minutes within the folds of some silky looking drapery, which is held for him by Mr Farini. This mass of limp-looking cloth is the wonderful parachute, and it may easily be guessed what Professor Baldwin is doing inside it. He is adjusting the hoop which, when the machine is expanded, will form the orifice at the top, and this orifice through which the air escapes in his descent is perhaps the most important feature about Mr Baldwin's invention. He emerges presently, and then the folded parachute is drawn up to the netting which hangs loose around the neck of the balloon. It can be seen that depending from the parachute are a number of long ropes attached to a stout hoop, which is presently passed over the aeronaut's head. In descending he will hang by both hands to this hoop.

There is a great shouting of orders now, and the excitement among the spectators is very great., 'Lift her up,' cries Professor Baldwin, 'but hold her,' and as the struggling balloon rides a few yards above the ground he is seen to have taken his position immediately below her, and to be surrounded by a confusing array of ropes. An excited shout by Mr Farini to some assistant to 'Leave go of that rope' shows that it is a critical moment, and then, before the spectators well realise it, balloon and balloonist are away. She mounts swiftly and smoothly like a bird released, the professor sitting apparently upon some small bar with outstretched hands, in much the attitude of a driver handling a team of horses. Spontaneous cheering and applause break from the crowd at the ascent, but it is only a matter of seconds before the bold aeronaut is out of ear shot. The ascent is made from the leeward side of the stand, and the wind being from the north-east the balloon is driven at once in the direction of Caversham. In consequence of this wind which is taking him rapidly away from the spectators, Professor Baldwin does not go to anything like the height he has sometimes reached. He goes so high, however, that he and his balloon look very small objects indeed against the clear sky. About 1000ft would perhaps be the height, and it has taken an incredibly short space of time for him to reach it. Before his movements become indistinguishable with the naked eye he has been seen to extend one leg and pass his foot into a loop of rope that is hanging within reach. Whether or not this is part of the preparations for casting loose can only be guessed. Suddenly there is unmistakable movement in the diminutive figure aloft, and the next instant the folded parachute and its inventor have left the balloon which turns upside down and floats aimlessly about in the empyrean for awhile. The parachute retains its limp appearance, and at the end of the long ropes that depend from it is the figure of the falling balloonist. He is holding on with his arms raised above his head, and his whole form is perfectly rigid; feet together and frame erect. He comes down in that fashion as straight as a stone and in a standing posture for nearly half his journey, and then the onlookers draw a sudden breath of relief, for the air has caught the parachute, and it has expanded into umbrella shape. The aeronaut's fall is instantly checked, and from that point he descends steadily with a gentle swaying motion that soon brings him apparently among the housetops of South Dunedin. Here he swings himself into a sitting posture, evidently steering the parachute towards a safe alighting place, and finally comes easily to earth in a vacant section off the Cargill road, near the Railway Workshops Hotel.

Ten minutes later the professor was again at the Caledonian ground, and, accompanied by Mr Farini, appeared in the front of the stand, receiving quite an ovation. He then gave a short address as announced, claiming (of course with perfect truth) to have made the first descent of the kind that had ever been attempted in New Zealand. The parachute, of which he was the originator, required, he explained, two feet of surface to every pound weight of the object attached to it. The orifice at the top was 18in or 20in in diameter, and this, by allowing the compressed air in the parachute to escape, formed a kind of column of air, down which he slid. As regarded the long drop before the parachute expanded, that was merely a bit of sensationalism he introduced. How soon the parachute expanded depended upon the size of the hoop he placed in the orifice at the top. He could, if he desired it, make the parachute expand directly after leaving the balloon. Mr Farini, who followed with a few words, added some further information as to the way in which Professor Baldwin had perfected his invention, and remarked that having solved the difficulty, there had yet remained the necessity of finding a plucky fellow to jump from the balloon and test the truth of the theory. That man they had found in Professor Baldwin. — (Loud applause.)

Soon the crowds were clapping other balloonists, including the world's 'only living lady aerialist', Miss Millie Viola. In 1892 she came from Melbourne to Invercargill, intent on heading into Southland's skies beneath her hot-air balloon. Alas, her first attempt failed because, as the *Southland News* reported, 'the radiation of the heat is too rapid to allow of a sufficient gathering of caloric in the huge bag to undertake the weighty responsibility of lifting Miss Viola'. Fortunately, the caloric did accept its responsibilities some weeks later and carried her about 300 metres (or 1000 feet) up in the air.

Some early aeronauts not only went up, they also came down — by parachute, to further thrill the crowds. But it was never easy attempting such feats, as Miss Leila Adair discovered in Auckland in 1894 ...

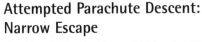

Attempted Parachute Descent: Narrow Escape
Auckland Weekly News, 10 March 1894

On Saturday, March 3, Miss Leila Adair, a lady aeronaut, was announced to make a balloon ascension from the Takapuna Racecourse, and afterwards made a descent by means of a parachute. To witness the feat large numbers journeyed to the North Shore, though only about 500 or 600 paid for admission. The Flagstaff Hill, however, was crowded with people, while even as far as Mount Eden hundreds took up positions. Some time was spent in mending a rent in the balloon, but about four o'clock the process of inflation commenced. This process consisted of filling the balloon with hot hair from an improvised furnace, the fire being assiduously fed with kerosene. While this was going on speculation was rife as to whether Miss Adair would be able to make her descent to *terra firma*, the direction and force of the wind holding out a prospect of her being carried into the harbour. About five all was ready: the balloon seemed full, the parachute was attached to the bottom, Miss Adair clasped a life belt around her, in view of harbour eventualities, and then seating herself on the bar of a trapeze which was suspended from the bottom of the balloon, gave the signal to let go. Somewhat slowly she rose in the air, drifting close to the flagstaff topping the totalisator house. 'Good-bye,' she cheerily called out as the balloon caught the wind and rose higher. But when it had attained a height of about 60 feet or so it was seen to descend; its sustaining power was not sufficient. Miss Adair threw herself upon her back as she came to the ground, but being unable to disentangle herself from the trapeze, was dragged a distance of some forty or fifty yards across some rough ground, and at least two gullies. A rush of spectators at once set in directly it was seen the balloon was falling, and it was eventually seized and held. Miss Adair escaped all injury, save a few scratches on the shoulder. It was a narrow escape, though, for had she shown less presence of mind she might have been very badly hurt.

Leila Adair was lucky but, five years later, another balloonist was not. On 2 November 1899 crowds gathered at Lancaster Park in Christchurch to watch a great ascent. According to the *Press*, the aeronaut was determined this climb 'should surpass anything he had previously given. As he himself expressed it, *Christchurch is the ideal place for a balloonist, and if the wind does not carry me too far, I am going to try and break my record of 20,000 feet.*

There was no sturdy wicker basket for the passenger. Rather, he would sit on a trapeze beneath the balloon – possibly doing some acrobatics as he climbed. When the time was right, he would pull a lanyard, release his parachute from its clasp on the side of the great balloon and float back down to the waiting throng below.

He'd done it before – many times, for this Auckland-born pioneer was serious about ballooning. David Mahoney had trained in England (the first New Zealand balloonist to do so); he'd given displays in Auckland and Wellington as well as Christchurch. And, after this his final ascent, he was off to South Africa to serve as a military aeronaut with a regiment fighting in the Boer War.

So there was much excitement at Lancaster Park in the minutes before the climb. A band played *Rule Brittania* as the balloonist made his final preparations 'without a thought of accident or peril'. It seemed certain that a new record was about to be set by the young Mr Mahoney.

 **Wingspread**

Leo White

To the public, however, he appeared as 'Captain Lorraine.' About a score of assistants helped to fill the bag with gas from the Christchurch Gas Works, while young 'Lorraine' chatted with his wife and a group of notables, including the Mayor. He already had several ascents at Lancaster Park to his credit, and this was to be a benefit performance. It was a gay scene.

'Let her go, gentlemen,' called Lorraine, with a cheery wave.

The balloon shot skyward. Then a cry of horror rose from the crowd. The parachute had fouled, coming adrift and billowing out *beneath* the rapidly rising bag. As it ascended the young aeronaut climbed the netting, trying to tip the balloon and make it come to earth as a parachute itself. The balloon and its doomed human cargo disappeared over the Port Hills. Soon afterwards a signalman at Port Levy, ten miles from Lyttelton, was surprised to see it descend into the water. 'Lorraine' struck out for the shore, but disappeared before help could reach him. The big balloon rapidly sank.

* * *

A name to be conjured with in those early days was that of Noah Jonassen. Self-styled 'The Aerial King,' he was billed that way from Northland to Otago, and gave exhibitions throughout the country. The tights he wore showed off his fine physique, and with his 'Captain's' peaked cap he cut quite a figure. Guaranteeing an exhibition under any circumstances, he performed at anything from humble sports meetings to special demonstrations. Witness to his boast were the number of bones he broke. In his first rush upward he vowed his big hot-air balloon was faster than a rocket. He was a true lover of the air, though, and he used to say that to soothe jangled nerves there was nothing like floating above the clouds, the ropes and bag working to a music only the balloonist knew.

One little foible Jonassen had. He always pinned an aneroid to his natty pink tights to check his height. His most sensational exhibition was at Gisborne. At 600 feet the bag burst. Smoke poured from the rapidly expanding hole in the canvas. Bag and man rushed to earth before a petrified crowd, but Noah Jonassen suffered no more than a couple of leg fractures! These proved only temporary obstacles, and before long he announced that he would attempt a landing in the Waikato River. When this stunt was tried he missed the river narrowly, was nearly dashed against a stone wall, and finally found a happy haven in a plum tree.

At Oamaru he landed unintentionally in the harbour, but it was only two feet deep where he splashed to rest. His balloon came down on a nearby wharf. Even in those days he was a fine parachutist, though, and no risk frightened him. Nearly a quarter of a century later, when he made his first descent from an aeroplane in 1930 at Rotorua, he broke yet more bones coming down in scrub.

Noah Jonassen and his wife lived for many years at Thames, and both recalled the early days of aviation in Christchurch, where they were enthusiastic members of a gliding club in 1912.

* * *

In Invercargill in 1907 the locals came to the conclusion that they must have a lunatic in the town in a man named Murie, a well-known cycle dealer. Murie had made a trip overseas and was impressed with balloons he had seen. On his return he announced that he was going to build one. To taunts that he was crazy he made the pithy reply that it was his money and his balloon. That seemed to be that. Nobody could do much about it. When he hired the Town Hall and started laboriously to pump the big bag up by hand, though, it became the joke of the day. The balloon filled the hall to capacity. When the place was needed for a touring theatrical company the bag had to be taken outside.

Then one day he made the great experiment and took his balloon up in the air. He landed in the thick estuary mud behind the town. Deflating his balloon, he stored it in the loft above his shop, but before long spontaneous combustion ignited the canvas. A practical man, he decided while repairing it to alter it to carry passengers. With a pair of shears he cut away the damaged canvas and his dream-child eventually emerged as a pear-shaped novelty instead of as a sphere.

General excitement grew as the bag was taken to the Gas Company to be filled. It became greater when one of Murie's assistants sniffed too much gas crawling beneath the filling balloon and had to be rushed to hospital. The honour of being the first aerial passenger in New Zealand fell to the lot of a Miss Ball, a young woman from Murie's office, who jumped at the chance to ride in the basket while the balloon was shifted from the Gas Company's premises to Queen's Park, where the ascent was to be made. Holding the guy ropes, helpers guided the balloon over the railway line, but then came the problem of getting it past the telegraph wires. The only way was to pass the ropes over them, and thus Miss Ball got her first 'flight.'

Many Southlanders made their first 'flights' at the Park at a shilling a time, the balloon rising and being hauled down with a windlass. One of the helpers was Bert Mercer, who later wrote his name large in the history of New Zealand aviation.

Murie, ambitious as ever, dreamed of aerial travel. One day, with a favourable wind, he went aloft and sailed inland for three and a-half hours, covering 20 miles as the crow flies. This crowning achievement nearly ended in unexpected tragedy. A farmer, tilling his land, suddenly looked up and saw the floating apparition. 'Give me my gun,' he yelled.

Fortunately for Murie and his adventure he was out of range before the amazed farmer could get a sight on him. Had he done so he might have ranked as New Zealand's first anti-aircraft gunner.

And what of Murie? He left his aerial exploits in New Zealand far behind him, and up to the outbreak of World War II was managing a gold mine in South Africa.

* * *

In 1910 there appeared a New Zealand issue of 'The Cloudland Times,' which was described as 'a journal of ballooning devoted to the interests of the Beebe Balloon Company.' It was of course, a propaganda stunt, and with an eye to creating an effect 'the highest' circulation on earth was claimed for it. Beebe was, apparently, an American. His 'act' included two giant balloons, named (with an eye to publicity) 'King Edward VII' and 'President Roosevelt.' His performers were two daring parachutists, Albert Eastwood, of Brisbane and Christopher Sebphe, a Spaniard. A display of fireworks and a local band were thrown in for good measure. Up to 1914 Beebe's Balloon Company performed at Lancaster Park, Christchurch, and at the Auckland Exhibition.

Undoubtedly Beebe was a showman; but he had the goods. His parachutists were exponents of the triple parachute descent. As the balloon rapidly rose they performed the most amazing feats on the one trapeze.

At a given signal they would drop away. Spinning down, they would first loose a red 'chute, then a white, finally landing from a big blue one. Sebphe, a small man with iron muscles, had to his credit a drop of 8,000 feet and a quadruple descent. Eastwood was more lackadaisical but he, too, was an expert.

Filling the bag was a widely advertised feature of the show, and invariably had its effect in inducing numbers of dead-heads to gather outside the grounds. The top of the bag was hauled up a pole and willing spectators helped to hold out the guy ropes and guide the balloon neck over the funnel leading from a fire in the ground. Right inside the bag was an attendant with a bucket. If the canvas caught alight it was his job to douche it with water. Beebe, complete in peaked cap, strutted importantly around while his two performers stood idly by in red tights, the cynosure of all eyes.

'Stand by to cover the funnel,' roared Beebe as the supporting poles fell away, leaving the bag swaying easily. Buffalo Bill style, he carried a big pistol.

'Are you ready, Mr. Sebphe?'

'Yes,' would come the reply.

'Are you ready, Mr. Eastwood?'

'Yes,' rather languidly.

The intrepid performers replied with never a quiver.

'Then let her go!'

Beebe's great voice led to the climax, a blank fired from his pistol. Out crawled the attendant, black with soot. Up shot the balloon as the cold air found its way beneath it; a momentary hush over the crowd, and two red-clad figures … swinging … swaying. Though then unheard-of, 'the man on the flying trapeze' might have learnt a wrinkle or two from that pair. They had no nets to catch them. Each moment they were carried higher. As they dwindled from sight and all eyes strained upwards, bang went the pistol again. Beebe judged the psychological moment nicely. First one figure would drop; then the other. Spinning earthward they came at a crazy speed.

Yelling itself hoarse, the crowd roared again as first one parachute and then another opened and was cut away. White 'chutes replaced the red, and finally, with blue ones billowing out, the artists found their feet … bowing to the madly enthusiastic crowd.

It would be a great and marvellous thing if the same enthusiasm had greeted our next pioneer. But it did not. Instead, his name has become synonymous with frustrated invention and thwarted dreams.

Born in 1877, Richard Pearse was one of nine children whose parents, Digory and Sarah, bought land in 1895 at Waitohi (near Temuka).

As a lad, young Richard was a very able tennis player and also a keen musician, continuing to play the cello throughout his life. But he didn't shine at school, although he did show an early enthusiasm for engineering. Indeed, he had hopes of furthering this interest at Canterbury University College but, sadly, the cost of funding an older brother's medical degree meant his father vetoed that idea.

So Richard found another kind of classroom — his shed — where he started the tinkering and experiments that ultimately earned him the nickname 'Mad' Pearse. In 1902 he was granted a patent for a new type of bicycle he'd designed and built in splendid isolation, then proudly ridden to Temuka on sale day. The bike was a genuinely innovative machine, with new pedal and gear systems as well as a back wheel rim brake and automatically inflating tires.

In an article published in 1903, the local paper, *The Temuka Leader*, waxed lyrical about the Pearse cycle, particularly the crank-based driving gear, which 'it is claimed ... makes riding easier as there is no waste of energy, the rider being able to keep a steady strain on the machine the whole time.' Yet, despite such enthusiasm, the bike didn't go into production and the bamboo-framed prototype no longer exists.

Not to worry. The young inventor had set his sights on another, even more ambitious goal — he had decided to build a flying machine. Most New Zealanders (and an increasing number of people around the world) know a little of his solitary experiments.

Well, solitary to a point, because he did have access to reports in *The Scientific American* and he also had some help with his engines from another extraordinary South Canterbury inventor, Cecil Wood, a Timaru engineer who built New Zealand's first locally designed motor bike and motor car.

But, for the most part, Pearse worked alone, using bamboo and scrap metals he found on the farm, including cast-iron irrigation pipes, which he modified to use as the cylinders of his aircraft's engine. (At least one of his neighbours wasn't impressed when his cows were spooked by the alarming noise of an early engine test.) Ultimately, Pearse was ready to trial his high-wing monoplane design — and that's where the debate begins. Because he worked alone, because so little was recorded, because the accounts of his endeavours are anecdotal, there's still no agreement about when Pearse took to the air and if he was in control of his craft when he did so.

Some insist he was. They're adamant he was the first to fly a powered aircraft. Others are more circumspect. Even the inscription on the base of the memorial erected to commemorate his achievments in the green rolling hills of Waitohi goes no further than stating that his was 'the first powered flight to be made by a British citizen in a heavier than air machine.'

Yet that too is an assertion challenged by aviation historians who argue there is no clear evidence that Richard Pearse ever achieved a controlled flight in one of his flying machines.

Perhaps an American author came as close to truth as anyone when he simply said of Richard Pearse: 'He was in the wrong place at the Wright time.'

Apposite as that is, let's also offer a more detailed assessment and there's no-one better qualified to provide it than Gordon Ogilvie, whose biography, *The Riddle of Richard Pearse*, was first published in 1973 and is now in its fourth edition. In 1982, Gordon Ogilvie also wrote about Pearse for the fiftieth anniversary edition of the magazine *New Zealand Wings* and, two decades later, he still stands by the conclusions he reached in that thoughtful and measured article.

Richard Pearse of Waitohi — time for some stocktaking
Gordon Ogilvie

The Richard Pearse debate shows no signs of tapering off. At intervals a fresh protagonist leaps into the fray and imagines he is disposing of the issue once and for all. But it is never that easy.

Viewpoints cover the whole gamut of possibilities: from the complete debunking of Pearse and what some are now pleased to call 'the Pearse cult', to confident claims that Pearse not only 'flew before the Wright brothers' but did so in 1902. Though I do not claim to have any monopoly of wisdom and truth, I have been involved in Pearse research longer than anyone else and would like to come to a few tentative conclusions.

I first encountered the Pearse Story while teaching at Pleasant Point District High School, 1964–7. Waitohi, where Pearse spent his first 34 years, was one of the school's contributing areas. I included a chapter on Pearse in my history of the district, 'Moonshine County' (1971), and followed this up with 'The Riddle of Richard Pearse' (1973), a book which remains the only full length investigation of this lonely genius.

Since then I have kept in close touch with the Pearse problem, corresponded with hundreds of friendly and hostile enquirers from all over the English speaking world, got involved in numberless arguments with extremists on both flanks, and have still continued to hunt for new insights into the dilemmas posed. No new testimony has come to light since my biography was published. The riddle remains.

It is significant, I think, that the most radical viewpoints from either edge of the battlefield emanate from Auckland. As with most problems, the further you are from the scene of inquiry, the simpler the issues seem to be. I have the advantage of knowing the Waitohi district intimately, and of having met all the key witnesses who were still alive in the 1960s. From this much closer range I would hesitate to be dogmatic about anything relating to Pearse, particularly the dating of Pearse's first aeronautical experiments and the degree to which they were successful. The evidence is irretrievably contradictory and arguable. The main difficulties?

In the first place, did Richard Pearse ever fly at all? My view is that he did not, in any acceptable sense, if you include control and duration in your definition. Pearse said in his two press letters of 1915 and 1928 that he did not fly and it is profitless trying to contradict Pearse himself. The claim made for a 'terrace flight' is not sufficiently corroborated, it cannot be dated with any certainty, and looks more like a powered glide even if it did take place.

The 'flights' that numbers of elderly witnesses described in newspaper reports and affidavits collected by George Bolt and others 1954–1972 are best described as 'powered take offs', 'tentative flights' or 'long hops'. Most of them seem to have been from 10 to 50 metres in distance and several ended in crashes.

But it is equally futile to persist, as some 'Pearse-bashers' do, that the inventor did not even get off the ground. Pearse in his 1928 letter to the

Christchurch 'Star' makes it clear that he did take off in his first aircraft, but did not regard this as flying. 'At the trials it would start to rise off the ground when a speed of twenty miles an hour was attained. This speed was not sufficient to work the rudders so, on account of its huge size (700 square feet of wing area) and low speed, it was uncontrollable, and would spin round broadside on directly it left the ground. So I never flew with my first experimental plane'.

'Two newspaper reports from 1909 also refer to Pearse making short hops, and numbers of witnesses later attested to seeing Pearse off the ground on a variety of occasions. You can call this 'flying' if you wish, but it was evidently not successful enough to satisfy Pearse. Yet that he got off the ground should not be doubted.

The dating of Pearse's first flight attempts is the other area of vigorous dispute. Here the betting ranges all the way from 1902 to 1909. Let us consult Pearse first. In his 1915 Dunedin 'Evening Star' letter he writes. 'After Lanley's failure in 1903 I was still of the opinion that aerial navigation was possible, and I started out to solve the problem about March 1904.'

But the answer may not be this simple. Elsewhere in the letter Pearse says 'I only built one aeroplane which was designed before anyone made a flight ...' As he is presumably referring to the Wright brothers who first flew on December 17th 1903, it appears that he may have been working on his aircraft well before March 1904. Also, according to the Sunnyside Hospital files, Pearse told a doctor there in 1951 that he had built an engine during the Boer War.

In addition to this, the more you examine the substantial body of Pearse's 'flight testimony', much of which can be dated, the more convinced you have to be that Pearse was carrying out flight experiments in 1903 and possibly even earlier. Two April Fool's Day jokes (from quite different sources) plus other testimony and circumstantial evidence suggest that Pearse may in fact have made his first adequately witnessed flight attempt on March 31st 1903. This was the much chronicled take-off down the Main Waitohi Road from the school cross-roads, which ended in disaster on his own gorse frontage.

Testimony relating to this flight attempt as well as others which followed it can be end-dated by a variety of methods. For example a witness tells you he saw Pearse in action shortly before the witness moved to another farm elsewhere in the country. Lands and Survey records will show you exactly when the transaction took place. Or another witness may have seen Pearse trying to fly soon after her marriage and before the birth of her first child. Marriage and birth records can help pinpoint the time. Or several witnesses may have sworn that a particular teacher at Upper Waitohi School let her class out early to see Pearse experimenting one day. Education Board records will show when that teacher left the school. Other witnesses recalled that Pearse's aircraft was left stranded on a hedge after another flight attempt and covered by the 'great snow' for some days. Meteorological records can indicate the exact day of this snow storm — July 31st 1903.

A considerable amount of such testimony from a wide variety of old-timers tracked down in various parts of the country (with no chance of collaboration — and some of them didn't even like the taciturn Pearse), points to experimentation by the inventor in 1903. In his farm paddocks as well as on nearby roads. Two or three witnesses had actually left the district by 1904 and could not have seen what they did if the date had been later than 1903. Britain's foremost aviation historian Charles Gibbs-Smith, was sceptical of Pearse from the very first and we exchanged a number of animated letters both before and after the publication of my Pearse biography. Gibbs-Smith would never be budged from the view that Pearse could not have attempted to fly before March 1904 at the earliest. He argued quite defensibly that Pearse's own dating must be preferred to any other possibility.

Yet Gibbs-Smith did concede at the finish, in the light of this other evidence and conflicting statements from Pearse himself, that the inventor might have been 'doing something' with an aeroplane in 1903 — using a model or carrying out un-manned test flights for example though not piloting it himself (no evidence has ever been found of these two test methods.) Another possibility with Pearse's March 1904 dating is that some years after the event he might have got the date wrong himself. (He seems not to have kept a diary or log book and was probably working from memory). Or Pearse may have discounted his early experiments as he discounted the Wright brothers' preliminary efforts in the same letter. Or his use of the expression setting out to 'solve' the problem of aerial navigation may suggest a later stage in the process, not his initial experimentation.

Whichever way you look at the evidence it is difficult to be categorical about either 1904 or 1903. Certainly 1904 is safer but Pearse seems to have been carrying out experiments of some sort in 1903 — designing and building his machine and testing its original two cylinder engine, to say the least I think we must continue to keep our options open.

Hard-line sceptics are not satisfied by 1903 or 1904 for there is no contemporary documentation for either year — newspaper reports, diary entries, letters or anything comparable. The only newspaper descriptions of Pearse at work date from 1909. This anomaly is easily accounted for. Pearse wishing no one to get the advantage of him shunned publicity of any kind in the early years of his experimentation. He once ordered a 'Temuka Leader' reporter off his farm it is said. But by 1909, powered flight was an accomplished fact in other parts of the world and he had nothing much to lose from disclosure. The Temuka paper published two reports on Pearse in November 1909.

These reports describe a late modification of Pearse's Waitohi mono-plane a four cylinder oval-winged version, saying Pearse had 'for some years past been working in secret in an endeavour to perfect a flying machine'. The paper also notes: 'He has improved on previous perfor-mances every time and in his latest effort he flew about 25 yards.'

In the following month an 'Otago Witness' reporter, probably alerted by the 'Temuka Leader' items, filed a more detailed account of Pearse's aircraft saying it had made a short jump recently and noting how Pearse had 'toiled for five long years in putting together this intricate con-struction'. This indicates 1904 for a starting date. However the machine the reporter describes seems markedly different from the one earlier witnesses describe, and does not greatly resemble the version Pearse patented in 1906. This has led Geoff Rodliffe to speculate that Pearse built two quite different aircraft and that the first of these — a smaller machine — was the more successful. However Pearse, as we have already seen, claims to have built only one plane at Waitohi.

So the 1909 newspaper reports are the only contemporary evidence of Pearse getting off the ground at all. If this is the only form of evidence you are prepared to accept, Richard Pearse's achievement is still impressive enough. He is still the first New Zealander to achieve a tentative powered flight in a heavier-than-air machine — indeed the first person in the

Southern Hemisphere. His powered take-offs of October–November 1909 would place him well among the front-runners in the British Commonwealth, where (apart from Horatio Phillip's 'hop flight' at Streatham in 1907) no one else was to have any success until 1909 either. James McCurdy's flight in Canada on February 23rd 1909 was the first in the Empire. J.T.C. Moore-Brabazon made the first true flights in Britain by a British-born Briton in April-May 1909 and A.V. Roe made a powered hop in July 1909. Richard Pearse even if you take into account nothing but his 1909 take-offs would be next in line. This alone is an honour not to be sneezed at.

If March 1904 is your preference for Pearse's first flight experiments, he then lies seventh in the field of powered take-offs after Du Temple, Mozhaiski, Ader, Maxim Langley and Jatho. (Incidentally Du Temple, Maxim and Langley managed only a metre or two each). If March 1903 is your decision he lies fifth after Maxim. He may have beaten the Wright brothers into the air, if the 1903 observations are valid, but I am not prepared to argue that he 'flew' in any strict sense. With either 1904 or 1903 Pearse would be, by a wide margin, the first British citizen to achieve powered take-off. This once more, is honour enough, even if he did not beat anyone else to 'fly'. None of these dates, 1903, 1904 or 1909 — is absolutely foolproof. The best one can do with the information available is to say that Pearse was probably conducting flight experiments as early as 1903–4 and had certainly achieved a number of powered take-offs by the end of 1909.

It is a pity in a way that almost the entire emphasis with Pearse has been from the very outset on whether or not he 'flew' or 'beat the Wright brothers'. In 'The Riddle of Richard Pearse' I tried also to show how visionary and original his thinking was with both his Waitohi aircraft and his later Utility Plane, constructed in Christchurch. His other inventions also show considerable ingenuity, particularly his vertical-drive bicycle, patented in 1902. I believe that more attention should now be paid to the inventions themselves and less time spent on largely inconclusive 'flight' arguments.

Above all else with Pearse, one must constantly marvel at the persistence and the courage of this man in almost complete isolation with little encouragement, no contact with others of like talent, few resources and only the most basic workshop facilities, this solitary back-country farmer came within an ace of writing his name in the aviation record books. The fact that despite his undoubted brilliance, Richard Pearse died in obscurity and had no influence at all on the course of aviation history, should remain a matter for sober reflection.

Richard Pearse was a reserved and unassuming gentleman. He did not ask to become a figure of controversy, and he deserves to be treated with sympathy and respect. The least we can do is to handle the evidence of his experimentation with care, thoroughness and some sense of proportion. It is immediately evident that most of the extreme viewpoints for and against Pearse are based on an incomplete examination of the evidence: sometimes even, I'm afraid to say, an amateurish or even negligent handling of the facts and testimony.

It is high time that all parties in the debate took stock once more.

So Gordon Ogilvie's conclusion is clear. While he's certain that Pearse did get airborne he doesn't think the lonely inventor made a true flight. 'His efforts are more properly described as powered take-offs, tentative flights, long-hops, lift-offs or flight attempts. Not flying. Sometimes I have to feel just plain sorry for Richard Pearse ... he would have recoiled in dismay from the hullabaloo his name has recently generated.' The notion circulated by over-zealous enthusiasts that Pearse flew before the Wright Brothers has for years created acrimony and distracted the public from more important issues concerning this far-sighted pioneer inventor.

'The proposition is, I believe, a dead duck.'

That's a view shared by the much-published English aviation historian, Philip Jarrett: 'Pearse has become a victim of those who seek to elevate his accomplishments to unrealistic heights. They are not interested in the truth, they are aeronautical evangelists seeking to promote their own nation as the scene of the world's first powered *flight*. Part of their strategy is to devalue the definition of flight so that any uncontrolled hop off the ground qualifies.'

Another historian, Maurice Hendry, is much more blunt. In 2003, writing in the New Zealand aviation magazine *Pacific Wings*, he described Richard Pearse as someone 'so intent on doing everything by himself ... that he defeated his own purpose and never achieved anything of lasting value. He was our Fred Dagg of the air. There have been several would-be claimants to the "first flight title", but the list is of minor interest only' mainly because none of these efforts 'led to controlled sustained flights as did the Wrights. Therefore ... the Wrights are uniquely distinguished from all others.'

But let the last word go to Pearse himself. In his 1915 letter to the Dunedin *Evening Star* he wrote: 'The honor [sic] of inventing the aeroplane cannot be assigned wholly to one man; like most other inventions, it is the product of many minds. After all, there is nothing that succeeds like success, and for this reason pre-eminence will undoubtedly be given to the Wright Brothers, of America, when the history of the aeroplane is written, as they were the first to actually make successful flights with a motor-driven aeroplane ... [but] ... I may say that my object is to show that New Zealand brains anticipated the essential features of the aeroplane. If I have claimed anything unduly, I want to know it, as I am open to correction. All my experimenting in aerial navigation was pioneer work, and when a history of the pioneers is being written I hold that I am within my rights in asserting my claims.'

There were others, of course. In 1909 an Invercargil engineer (and former champion cyclist) started building a tubular steel monoplane. Herbert Pither also designed his own engine and, after some months of testing, flew a kilometre along the Oreti Beach on 4 July 1910. As a comprehensive display in the Invercargill Museum notes, this was the world's first sustained flight by a metal-framed aircraft. And things were happening in the North Island too ...

Wingspread
Leo White

While ballooning was still in its heyday the aeroplane age had its birth in the Hawkes Bay district. News of powered flying machines overseas stimulated an irrepressible spirit which was to place New Zealand to the fore in air-mindedness. But for a twist of fortune, many pioneer New Zealand followers of aviation might have been heading huge combines to-day. Bertram Ogilvie, of Napier, was one of them.

Over a number of years Ogilvie and a band of enthusiasts worked long night hours, testing the results of their efforts in the greatest secrecy.

In 1907 Ogilvie hit on the idea of flaps, or ailerons, to supersede the then popular wing-warping method of aeroplane control. He was an employee of Hawkins and Rome, well-known Hastings engineers, and Hawkins, a military man, encouraged Ogilvie and expressed faith in the work he was pioneering. Finance was arranged among Hawkes Bay well-wishers and revolutionary ideas were conceived and tested, improved upon or discarded. With Ogilvie were five other enthusiasts, Bob Goodger and his son, H. Suckling, Lerew and J. Munro.

Visitors were discouraged. The midnight oil burned for months on end. Out of it all came a framework of cycle tubes, white pine wing frames, and fabric drawn taut, dope being undreamed-of for covering fabric surfaces. Three such machines were actually completed between 1907 and 1910. Ingenuity came into play to test the idea of aileron control. Four foot wings were built with miniature ailerons and towed against the wind. Cords attached to the flaps were pulled ... and proved Ogilvie's idea to be a winner.

Launching his aeroplane into space was a matter of trial and error. Only a Heath Robinson could have illustrated the efforts those pioneers made, but to them the result was heartening. A 20-foot high framework ramp was built 80 to 100 feet in length. Towed to the top, Ogilvie would clench his teeth and take his place behind the weird 10 h.p. V motor coupled to a propellor that looked like a double paddle for a canoe. Down the ramp the contraption roared, shaking loose the broomstick struts, while the vibration would spring every screw in the machine. In those secret dawn hours the craft would bound into the air and land with a resounding thud. Then everybody would get to work again with new ideas. The wings of the first machine had flat surfaces, but on a later model they were cambered. Yacht rigging screws did duty as turnbuckles. The flaps, or ailerons, hung by piano hinges. There was a box kite tail. One propeller, of hickory, had an eight-foot diameter. The motor was incapable of turning it until it had been severely whittled down. A later one — the best of the lot — was of metal, and could be adjusted. Eventually it was agreed that the power unit was useless. The hinged wing control, however, was enthusiastically endorsed, and at a cost of £800 or so world rights were taken out.

It was at that time that Lord Kitchener was in New Zealand. Hawkins' description awoke his interest, and he visited the secret testing ground, promising the assistance of the War Office if the invention was taken to England. The syndicate chose Hawkins and Ogilvie to make the trip, and an excited crowd farewelled them when they sailed from Napier for Great Britain. On the voyage Ogilvie built a beautiful model triplane. This was later made into practical reality by the famous Handley Page Company, and Ogilvie joined that select little band of early flying members of the Royal Aero Club.

Financial and other trouble, however, caused the syndicate to wind up in 1912, but influential aviation interests continued working on hinged control, and to-day the whole flying world knows what ailerons mean in a modern aeroplane.

During 1909 a Wellington photographer, Arthur Schaef, commenced to build an aeroplane. What he lacked in time and money he made up for in enthusiasm. He became associated with Percy Fisher, one of the foremost New Zealand engineers of his day. The illustrated efforts of overseas pioneers formed Schaef's only guide, but fifteen months after starting work he announced to an incredulous public that he had built a flying machine that was neither a biplane nor a monoplane. It was christened 'Schaef's New Zealand Vogel,' and was taken to Hutt Park by horse transport.

The J.A.P. engine was found to be unsuitable, and it was rebuilt. Over twenty castings were made before Schaef and Fisher could get four cylinders without flaws. The motor itself was then looked upon as an engineering triumph, and was expected to develop over thirty horsepower. Cedar and ash were chosen for the laminated propellor.

'As soon as I have made tests,' said Schaef, 'I hope to be able to carry passengers, and realise my ambition to take photographs from the air.'

Little did he dream to what a high state of proficiency aerial photography would be brought within the short space of thirty years. Infra red plates, enabling photography through haze, and aerial photographic surveys would have sounded like something out of Jules Verne or a Wellsian phantasy.

Schaef's New Zealand Vogel rather resembled a big dragonfly, but he gave short shrift to Hutt sightseers, particularly when a rival photographer set up his camera to record the scene. If any photographs were to be taken, bristled Schaef to his rival, pushing the cap to the back of his head, he himself would take them. No professional 'shadow catcher' was going to beat him for the few shillings he hoped to get back for all his efforts.

The Hutt trials proved unsuitable, and location was moved to Lyall Bay. There on the foreshore, not far from where Rongotai Aerodrome is to-day, an old tram did duty as headquarters. Early morning low tides were awaited, and in mid January, 1911, the Vogel roared along the sands with the excited Schaef at the controls. The exhaust outlet came directly into his face and covered his head with a nauseating oily film. For a brief space there was daylight under the wheels. Then, owing to the limited runway, Schaef skidded to a standstill.

A thrilled and vociferous band of onlookers surged forward. Schaef, his long, fair hair covered with sand and streaked with oil, groped his way from the machine to grasp welcoming hands.

'I hardly knew I was up,' he cried, his voice pitched with excitement.

In February and March he made further attempts, his best effort being a straight dash of about 150 feet at varying altitudes of a few feet above the wet sands. Various propellors were tried. Some of them made the machine so unruly that Schaef metaphorically had to dig his heels in to keep his seat. It was a mechanical bucking broncho, but mishaps seemed only to make him more determined. There were rumours spread that the machine had not actually flown. Schaef's comeback was a large advertisement in the Wellington press signed by well known citizens who had witnessed the spectacle. The aeroplane was placed on view at Adams', in Manners Street, Wellington, sixpence being charged for admission and lectures given. As an added attraction the motor was started for people to see the propellor buzz round. The threshing prop often nearly blew out the light, but that was a mere bagatelle.

Schaef was next attracted by the Canterbury Plains as a trial ground. He had hoped to put the machine through its paces in secrecy at Hagley Park, Christchurch, but he had reckoned without the youngsters of 1911. It was during the school holidays, and a local spy had been detailed to report when Schaef had his contraption ready. When the day dawned, Hagley Park simply swarmed with boys on bicycles, and women pushing perambulators.

The more Schaef taxied, the more the cyclists pedalled in front of him. Then just as his frayed temper commenced to bubble over, the crankshaft broke … and that put an end to the trials. Far from beaten, however, Schaef immediately announced that he would get a new engine from England. Then he shipped his machine back to Wellington.

Another 18 months went by. A new type of aircraft made its appearance, with a 30 h.p. Anzani engine. At once Schaef's ingenuity became apparent, for the machine was now fitted with floats. Once he had cooled off after the Hagley Park incident, Schaef treated boys on bikes and women pushing prams as he would have a scientific problem to be overcome. They could hardly walk the water to get in his way!

The Vogel II took to the water at Evans Bay, Wellington, as New Zealand's first amphibian aircraft. But something was wrong. Schaef taxied round and round the bay, but that was all that happened. Not if he had taxied around the Pacific could the amphibian have become unstuck, for he had unwittingly uncovered that potent snag, drag. His top speed was no more than 10 m.p.h. Furthermore, the choppy water clipped the tips of the propellor blades. Schaef came back to earth, and wheels again replaced the floats.

Schaef's versatility was demonstrated by the way in which he often started the motor himself and then jumped into the seat as the aircraft gathered way along the sloping sands of Lyall Bay. Side winds not infrequently caused the machine to dive into the sea, and one day, as he was about to make a respectable flight he suddenly dived into the water just as he was being photographed.

Arthur Schaef and his new mechanic, a man named Swan, were never daunted until one day in March, 1914, a blow-lamp caught some inflammable material and the machine and shed were destroyed. Schaef was away at the time, but this really disheartened him and he did not try again.

Not long before he died in 1940 I sat in his shop in Vivian Street, Wellington, while he dived into dusty books of old photographs.

'No, White, my boy,' he said, 'the only tangible thing I have left is that old step-ladder there. I made it from a few spars, and it has helped me to obtain elevation in making a few shillings at my real profession, photography.'

Down in Christchurch, Sir Henry Wigram wasn't much interested in cameras. Rather, he was focused on flight and in September 1916 he launched New Zealand's second flying school in a paddock that became an air base that took his name. Eleven months earlier, on 2 October 1915, the New Zealand Flying School had opened its doors at Kohimarama in Auckland. This northern school was started by the Walsh brothers, Leo and Vivian, who'd earlier formed the Aero Club of New Zealand. But beyond such institutional achievements, history grants these two an even greater distinction.

The Kiwi's First Wings
David Mulgan

The first successful aeroplane flight ever to be made in New Zealand took place at Glenora Park, Papakura, on the morning of February 5, 1911. It was the climax to weeks of painstaking preparations and almost daily taxiing trials until the self-taught, 23-year-old aviator, Vivian Walsh, as well as his brother and members of the syndicate, felt they had come to know the *Manurewa* well enough to take an upward plunge with her.

After a preliminary run over the ground, she rose gracefully to a height of 60 feet, flew about 400 yards, and then made a safe landing. A few local residents as well as a proud father, Mr Austin Walsh, and two equally proud sisters were present. News of the flight soon spread and it was not long before newspapermen and photographers were on to the scent.

'The syndicate deserve all the praise that can be bestowed on them for their enterprise, and their action in keeping everything quiet until success was assured is to be commended,' said the *Auckland Star* that night.

It was soon found that the *Manurewa* had a good reserve of power as the engine was capable of 1200 r.p.m., but only 900 were needed for flight.

Sir Joseph Ward left for England to attend the Imperial Conference a week or so after this epic and the syndicate hoped that he would see the plane in flight before leaving. The Walshs wrote to him saying that, should aeronautical matters come up for discussion in London, he could claim credit for New Zealand being in the van of the Australasian colonies in aviation construction and the ability to produce aerial machines locally.

Also, as aviation was coming forward so prominently in defence considerations, the Government should interest itself in this important national question and the syndicate offered to place themselves at the country's disposal should it be decided to add flying machines to the defence forces.

The syndicate was prepared to construct a military type of biplane to carry pilot and observer, of the Howard Wright type similar to the *Manurewa*, the letter added. The syndicate also offered to give a thorough practical course of instruction to an officer appointed by the Government. The Prime Minister, however, had a crowded programme before sailing and was unable to pay a second visit to Papakura....

A number of circular flights as well as aerial excursions over short distances were made by Vivian in the remaining weeks of February. Towards the end of the following month a public exhibition was planned to which the Mayor, councillors, and school children of Papakura were invited but a mishap occurred in the early morning during trials. The machine began to take off along the field, hopping off the ground at intervals and taking leaps of from 20 to 30 feet until she rose gracefully into the air.

About 40 feet up a pocket of wind caught and tilted the plane until she was nearly vertical. With her tail down she dropped to the ground. Fortunately the shock was taken up to some extent by the skids and the machine righted. The mishap, which resulted in a few broken streamers and struts in the under section, was regarded as a blessing in disguise as it showed up a structural weakness. This was duly remedied with the assistance of the latest Howard Wright plans which had then arrived from England.

About this time the *Auckland Star* was moved to remark, 'Another astronomer announces that after six years' arduous work he has completed a map of the moon, and no doubt the Aero Club of New Zealand will be properly grateful to him; but until the *Manurewa* is ready for a longer flight than across the Papakura showgrounds, such a chart is not, perhaps, so essential to her equipment as a knowledge as to where petrol may be obtained in Onehunga, or which is the cheapest repair shop in Newmarket. If any ambitious hydrographers really want a job, why don't they tackle the Th   Kings, and fix them so they won't shift their moorings again.'

The syndicate planned exhibition flights in various centres with the idea of demonstrating that a modern flying machine could be built by New Zealanders and was preparing to take orders for the construction of any type of machine. The selling price of a machine of the *Manurewa* type was to have been £1500 with free tuition, and a flying school was envisaged.

There were several minor mishaps in the first weeks of the *Manurewa*'s career and the syndicate was kept busy patiently repairing the damage each time. The handicaps that beset the Walsh brothers in their pioneer flying were very pointedly summarized in a [sic] editorial in the *New Zealand Herald* on April 12.

'There is something about the utter boldness of the conquest that man has made over the air that makes for long hesitation', the editorial said. 'There are for certain two aeroplanes in New Zealand that can fly ... their pilots are eager enough, they are confident enough, they know that they can fly over the wash-house chimney if they want to, and go as far as from here to the store and back. But they don't do it.... There must be something rather appalling in the consciousness that one is doing something exceedingly picturesque, that no one else in the country can do, and that appalment hangs its weight on the front of the elevator planes and brings the machine to earth again. The aviators in New Zealand seem to be badly in need of two things — practice and company. To placard a bicycle with notices of one's intention to start in the cycle trade, and then commence learning to ride before the populace is not a good way to learn; the beginner is apt to get "rattled". But company is perhaps as necessary as practice. Here, with one aeroplane to the province, the pilot has only his own experience to go on; he has not the advantages of airmen at Home, who can watch each other falling down and catching themselves, and learn by others' examples what tricks man can achieve to fool the unsteady breezes.'

An exhibition flight by the *Manurewa* from the Auckland Domain cricket ground was the syndicate's next objective and this was scheduled for April 22. Extensive advertising preceded the event and personal invitations were also sent to people. The City Council granted permission to use the ground at the sole risk of the syndicate and authorized admission charges not exceeding one shilling to the ground   ee   n extra shilling for every horse or vehicle, and an additional   shilling for admission to the pavilion. A large marquee was actually erected in the Domain to house the machine.

A major calamity, however, was to thwart these expectations. While on a practice flight at Glenora Park, Vivian tried taking off from another part of the field. Before the plane left the ground one of the skids caught in an unsuspected mound of earth. There was a sudden jerk, then a summersault and the *Manurewa* became a total wreck.

Leo Walsh, who was watching, heard the crash and saw the wreck swallowed up in a cloud of dust. He reached the scene in time to see his brother crawl out, badly shaken but unhurt except for a minor knee injury. When he had had time to recover from the shock he realized what a miraculous escape he had had. The engine bolts had all been broken by the impact and the engine had been held a few inches above the pilot's head only by a stray wire that had become twisted round the propellor.

The tail elevator and engine were the only parts intact after the wreck, and the sago, which helped make the flying machine was now to contribute further to its undoing. The wreckage was left in a field overnight but the fact that calves were grazing there was overlooked. To them sago-covered linen was like manna from Heaven and next morning it was found that they had smashed most of what was left of the wooden framework in the process of 'grazing' on the fabric.

Aviation was very much a private preserve in those early days — the government was literally all at sea. It wanted dreadnoughts, not aeroplanes, and didn't know quite what to do in 1913 when it was unexpectedly presented with a Bleriot monoplane — a gift from the Imperial Air Fleet Committee in England. The Bleriot (named *Britannia*) was New Zealand's first military aircraft and, like many others since, quickly became controversial. Principally because the pilot preferred a pretty passenger to a pompous one ...

New Zealand Aviation Yarns
Roy Sinclair

In England, the aeroplane had been named *Britannia* in a special ceremony at Hendon on 22 May 1913. A former New Zealand prime minister, Sir Joseph Ward, was then taken for a flight before the Blériot was dismantled and sent to become New Zealand's first military aircraft. But its use here was to be short-lived, as was New Zealand's first official government pilot appointed to fly the Blériot. He was Second Lieutenant Joseph Hammond of the Royal Flying Corps. Born in Feilding on 19 July 1886, he obtained his pilot's 'ticket' in France on 4 October 1910.

Having taken the Blériot on two successful flights from the Epsom Showgrounds on 17 January 1914 and, on the following day experienced an aborted take-off with a terrified journalist passenger, Hammond was ready to fly a number of eager 'notables' over Auckland. But he offered the first official flight, on 24 January, to Miss Esme McLalland, a glamorous showgirl from the visiting Royal Pantomime Company. She was delighted with her half-hour joy ride but the 'notables' left earthbound did not share in her ecstasy. Hammond was immediately sacked and the Blériot placed in storage in Wellington, returning to England at the outbreak of the First World War.

Not to be outdone by an unsporting officialdom, Joe Hammond also returned to England — and his RFC uniform. In 1916, while flying an experimental Robey Peters Fighting Machine, he is said to have landed it skilfully on the roof of a lunatic asylum near Lincoln after something seriously malfunctioned. The aeroplane quickly caught fire, but not before an unharmed Hammond got out and calmly walked to the other end of the capacious roof. There he sat smoking a cigarette and watching the fire brigade's gallant attempt to rescue a supposedly scorching pilot.

Captain Hammond was killed on 22 September 1918 while serving as an exchange flying instructor at Indianapolis in the United States. Nothing more was heard of the Blériot *Britannia* but a magnificent replica — made in Dunedin by David Comrie — can be seen in the Royal New Zealand Air Force Museum at Wigram.

It was Alan Whicker — once (in)famous for his television exposés — who called New Zealanders 'the flyingest people in the world'. The fact that he'd been hired by Air New Zealand to front a series of ads may have encouraged this generous aeronautical endorsement, but it's also true our fondness for things that fly is a logical response to the tyranny of distance.

The first pilot to span the gap between here and there was 22-year-old James William Humphrys Scotland. On 5 January 1914, having earned his British Aviator's Certificate in the U.K., he came back to New Zealand, bringing with him a brand new Caudron. By February, he was in the deep south where his 'fascinating piece of modern mechanism' with its 'powerful' Anzani motor and 'minimum of gravity attraction' quickly became the talk of the town. Just before 7.00 p.m. on the 20th, he took off from the New Show Grounds 'as gracefully as a swallow' and headed for Gore. He arrived there at 7.38 p.m.

This pioneering cross-country flight between New Zealand towns was quickly followed by a more ambitious journey — from Timaru to Christchurch, with an aerial mail drop over Temuka en route.

Aeroplane flies from Timaru to Christchurch
The Press, 7 March 1914

Mr J. W. H. Scotland, the aviator, yesterday made aviation history for Canterbury by putting up the finest and longest flight yet performed in New Zealand. This was no less a feat than flying from Timaru to Christchurch, a distance of roughly one hundred miles by the railway line (which Mr Scotland followed throughout) between 8.30 a.m. and 5 p.m.

This is a protracted time, but it was occasioned by a delay at Orari of some six and a half hours. When Mr Scotland did leave Orari, however, he made rapid progress towards the city. Rising into the air at 3.20 he passed over Ashburton precisely as the clock struck four, and was sighted from Christchurch coming rapidly from the west-south-west at ten minutes to five.

He came up at a great pace, and successfully landed at five minutes past the hour.

Mr Scotland's story

Mr Scotland rose at Timaru at 6 a.m., when the conditions were very favourable, but as some adjustments required to be made in the controls, he descended again, and restarted at 8.35, with the engine running well.

When passing over Temuka, Mr Scotland dropped a parcel for Mr Andrews from a height of 2500 feet. About this time, however, the conditions were getting very bumpy — the plane was being lifted up and dropped about it an uncomfortable fashion — and after rising to between 4000 and 5000 feet, Mr Scotland vol-planed down and landed in Mr Bell's paddock at Orari because of a little engine trouble, due to over-lubrication.

For the last 300 feet the propeller stopped, which made the landing all the more creditable. A mechanic was wired for from Timaru. He came up by motorcar, and had the aeroplane ready for flight again by about 3.20.

Very bumpy conditions were encountered soon after the restart, and after a height of 6000 feet had been reached near Ashburton, Mr Scotland came to a lower level of 1000 feet, and enjoyed more equable conditions.

The gusts on the high levels were very disconcerting, and had the machine tested to its utmost several times.

'If it had not been for the extreme stability of my machine,' said Mr Scotland on this point, 'I would have come to grief several times. The hills seemed to form funnels and produce very heavy gusts, which subjected the machine to a severe strain.'

As Mr Scotland followed the railway line, he passed a number of trains. The first express leaves Timaru at 1.30, and Mr Scotland left Orari, twenty miles this side, at 3.20, yet at Bankside, twelve miles outside Christchurch, the aviator sailed past the express at a height of a couple of thousand feet, and had safely landed at Addington fully five minutes before the express reached Christchurch.

The passengers on the express had the aviator in sight for the best part of thirty miles, and could see him dodging about amongst the clouds at an altitude of about 5000 feet. Sometimes he would disappear from view behind clouds, but when he descended lower, Mr Scotland found the air very clear.

A trying journey

'Yes, the journey has been a very trying one,' said Mr Scotland. 'The air was bumpy all the way, especially on the higher altitudes. I seemed to be striking into a head wind, and was continually pushing down the lever to keep the plane down. This has tired my whole arm, and taxed my wrist very much. I am glad the trip is over.'

Mr Scotland is a small but wiry young man, with a keen sharp face. When he landed, his eyes were very red with the exposure to the wind, but a bath soon put him to rights. In flying he wears a comfortably padded helmet, with, of course, suitable warm clothing. The air in the early part of the journey was extremely cold, and when he descended at Orari the aviator said he was almost frozen.

The most intense excitement in the flight was manifested all along the route, and the aviator could see groups of people, like small specks, away below, running together in farm and hamlet to see him pass.

Arrangements for today

Today the aeroplane will be on exhibition only in the morning at the Show Grounds, but in the afternoon at 3.30, exhibition flights will be made. These flights will be in the direction of the west, and away from town.

Probably the most interesting features of aeroplane work from a spectacular point of view, are the actions of rising and alighting. The vol-plane down, the dip, glide, and hover just before striking the ground; these recall the action of a pigeon in judging its landing on a roof.

PART TWO

WORLD WAR ONE

For most New Zealanders today, the monstrous conflicts of World War One have been distilled into a single potent image. Ninety years after the first shots were fired, we remember only Gallipoli and the Anzacs who battled Johnny Turk on the bare hills of the Dardanelles.

Today, those Anzacs are officially seen as brave men betrayed; the victims of imperial delusion and useless generals whose courage and character still made them the emblem of a nation. It's a seductive notion, but one which serves new agendas better than it serves the facts.

In truth, New Zealand then was still part of the British Empire and proud to be so. When newspapers wrote of Home it came with a capital letter, as in this January 1914 *New Zealand Herald* article about how to train pilots and provide airships for military use: 'If it were the intention of the Government to establish an aviation corps in the Dominion, there existed but little necessity to send men Home to be trained . . . at present there are any amount of young fellows falling over each other for an opportunity to join an aviation corps . . .'

And many did. Nearly 1000 New Zealanders flew with the Royal Flying Corps (RFC) or the Royal Naval Air Service — which merged on 1 April 1918 to become the Royal Air Force.

Some of those pioneer pilots also played a crucial role in the RAF during World War Two. One such was Keith Park, whose 11 Group squadrons defended London and the south of England during the Battle of Britain. (Later, he commanded the air defence forces on the Mediterranean island of Malta, a vital staging post for the 8th Army in Egypt.)

Another equally notable World War One veteran was Arthur 'Maori' Coningham, described in a *New Zealand Herald* obituary as 'one of the great Allied commanders in the Second World War'. Coningham led the Desert Air Force in North Africa and subsequently the Tactical Air Force in Europe. After his death in an aircraft accident in 1948, he was also described as 'a brilliant strategist' by *The Times*, and *Aeroplane* magazine called him 'one of the war's great men'. Before and during the battle of El Alamein, Coningham developed tactics and methods that enabled air and land forces to work as an integrated team. Montgomery praised him for this, saying 'The Desert Air Force and the Eighth Army are one . . . if you knit together the power of the Army on land and the power of the Air in the sky, then nothing will stand against you and you will never lose a battle.' Coningham is certainly a genuine New Zealand hero yet, despite his tactical influence (as evident in Iraq) he's now unknown and unsung in his own country.

Like Coningham and Park, most of New Zealand's World War One aviators fought on the Western Front. Some flew reconnaissance aircraft such as the RE 8 or bombers but the majority fought in fighter squadrons operating nimble biplanes like the Sopwith Camel and S.E. 5a.

Eleven New Zealanders became 'aces', pilots who shot down five or more enemy machines. Major Keith Caldwell M.C. (who commanded 74 'Tiger' Squadron) was the most lethal New Zealand pilot, officially credited with 25 German aircraft. Nicknamed 'Grid' (after his habit of calling all aircraft grids), Keith Caldwell was a great fighter and a dashing patrol leader. According to Wing Commander Ira Jones 'he expected every one of us to fight like hell'.

Renowned for telling his pilots to 'kill the sods, the Hunnerinoes, at all costs', Grid Caldwell had little regard for tactics or orders. In his book *Tiger Squadron*, Jones writes 'Where the enemy is, there goes Grid. Without any disrespect to all the wonderful fellows I have met so far in this war, I feel I must hand the palm to Grid for individual valour. It is prodigious and impossible to describe. All our squadron patrols which he leads are nightmares. He frightens us as much as we frighten the Huns.'

On one occasion, Grid collided with one of his own pilots while both were attacking a German aircraft. While his compatriot made a successful landing, the whole squadron saw Grid go spinning down. That night they held a wake and all got drunk. About midnight Grid walked in ... and it turned into a celebration. As described in *War Birds: Diary of an Unknown Aviator*, 'he was all bloody and his clothes were torn to pieces. He had set his tail stabilizer and gotten out of his seat and crawled out on the wing and gotten the plane out of the spin. His aileron control was jammed and part of his wing tip was gone but he balanced it down and landed it on his side of the trenches by reaching in and pulling the stick back before he hit. The plane turned over and threw him into a clump of bushes. It had taken him ever since to get back, as he crashed about thirty miles away. He resumed command and took charge of the drinking and when the squadron went out on dawn patrol he led it. Then he went to the hospital.'

Captain Clive Collett was another 'ace', but according to John Thomson in *Warrior Nation: New Zealanders at the Front*, 'his greatest claim to fame is his daring test drop by parachute from an aircraft — the first official RFC jump, made in January 1917. At the time controversy raged over whether parachutes were a sensible safety back-up for pilots or an inducement to cowardice: the official British Air Board view favoured the latter, despite a parliamentary report indicating there had been 800 fatal accidents to trainee pilots that year, and the Germans were believed to be testing parachutes. Collett died in an aircraft accident in Scotland in the same year.'

Let's also remember Lieutenant Alfred de Barthe Brandon, one of the first RFC pilots to shoot down a Zeppelin. From 1915 to 1917, these huge airships made numerous raids on London and other cities, sparking panic and outrage, much to the delight of one German newspaper editor, who wrote: 'London, the heart which pumps lifeblood into the arteries of the degenerate huckster nation, has been mauled and mutilated with bombs by brave German fighting men in German airships.' Needless to say, the Zeppelins' victims were equally jubilant when one was shot down, hence this enthusiastic description of Lieutenant de Barthe Brandon's 31 March 1916 attack on airship L 33 in the darkness over the East End of London in F. V. Monk and H.T. Winter's *Great Exploits in the Air*.

> 'And now rushing upward was a young lieutenant of the No. 39 Squadron ... who had already figured in many an aerial fight. By the time he got within range of the enemy ship it had been hit but not disabled by shrapnel. With fine judgement and magnificent dash he manoeuvred for position, then blazed away. The airship faltered, fell. And righted itself. Lieut. Brandon followed it down and riddled it with bullets.'

Just six years earlier, in 1910, the British Secretary of State for War had boldly declared: 'We do not consider that aeroplanes will be of any possible use for war purposes.' And yet, when World War One finally ended on 11 November 1918, some 54,879 aircraft had been used by British forces during 'the great struggle'.

Even so, there were those for whom the 'great struggle' was simply getting out of an aeroplane. One Kiwi soldier explained why in the Anzacs' Middle East magazine, *The Kia Ora Coo-ee* ...

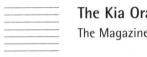

The Kia Ora Coo-ee
The Magazine for the ANZACS in the Middle East. 1918

My First Flight

It was at a pantomime that I caught the flying fever. A charming girl sang a charming song, 'Up In My Aeroplane'; and perhaps it was violet eyes and dimpled chin that really made me long to become an airman. Anyway, when the Great War started, I enlisted with visions of the Flying Corps; but — I became a Camelier.

Well, riding a 'hooshter' was a step upwards, as it were; and at length Fate tossed me into an A.F.C. camp. When the chance came to realise the hope of half a lifetime, I found my enthusiasm chilling a bit. You see, I had been an eye witness of some terrible smashes; and, somehow, flying didn't appear to be so divine now.

I donned the leather helmet and adjusted the goggles, clinched my teeth, and tried to look unconcerned, while I listened to advice from half a dozen grinning pilots.

'Mighty gusty,' says one, addressing the pilot who was taking me up. 'I wouldn't go high, Jack; it was just such a day as this when poor Bob Danton crashed.'

'The old 'plane looks pretty ricketty,' remarked another airman, to me. 'If you get the tip that she is going to crash, hop out on to the wing; you'll have a chance — Buckley's — of escaping with only a couple of broken limbs.'

I felt like doing a bolt, but could think of no reasonable reason for postponing my first flight. So I climbed into the front seat of the 'plane, fastened the strap, and gripped the sides of the 'car' till my knuckles became white.

'Petrol on, switch off, suck in, sir,' cried the mechanic, giving the propeller a twirl ... 'Contact.'

'All clear, contact,' shouted the pilot, and the machine started to quiver like a captive moth; presently she rushed over the ground, while the wind roared past and twanged the wing wires to madness. Then the 'plane rose with the grace of a swallow, and up, up we soared towards heaven. I never loved the earth so much as at that moment of leaving it. At an altitude of about 5000 feet, the pilot kept her straight; and though I felt sick and giddy, my nerves steadied a bit. In fact, I began to enjoy this thrilling aerial journey. I ventured to look over the side, but drew back my head in a flash. It was raining, and big drops stung my face as if it had stopped a charge of tin tacks. I was still feeling sore about the matter when the 'plane started to heel over and move in circles. This manoeuvre put the wind up me. I seemed to be spinning, a million miles above earth, on the crown of a giant top. Crumbs! it was awful; but worse was to follow.

Put not your trust in pilots. My skipper had vowed, on earth, that he wouldn't put up any stunts on me; but now that we were hurtling through space, he began to play dice with Death. We looped the loop, did side slips, and finished up with a spinning nose dive. I can't describe my sensations during these appalling events. All I remember is, that I clung desperately to whatever I could grip, shut my eyes, and waited to be dashed to bits.

When I opened my eyes again, the 'bus' was resting on a grassy place near the aerodrome, while a group of pilots and mechanics stood around, vainly striving to look sympathetic. Somebody helped me out — how good it was to feel the solid earth — and I crept away to my tent, cured forever of the flying fever: a camel's hump is high enough for me.

Palestine. 'Corporal Geebung.'

Maybe so, but others couldn't resist the newfangled flying machines, no matter how strange they seemed. Here's A.R. Kingsford, who left New Zealand a soldier but came home an airman.

Night Raiders of the Air
A. R. Kingsford

At an aerodrome, on the corner of the desert, not far from Heliopolis, I saw an aeroplane for the first time, and a queer looking object it was; a network of bamboo with a couple of men in front of the engine, the whole suspended between what was apparently the planes.

This queer object was running along the ground; Rooty, my pal, and I, were all eyes.

'So that's an aeroplane,' said Rooty.

'Gee, it's a queer looking affair,' I grunted.

'What's wrong with it?' said a voice behind us. Turning, we saw a real live airman; at least, we concluded so, for he was all dressed up fit to kill. On his head was a huge helmet affair, to serve as a bumper in case of crashes, he explained. A light oilskin suit arrangement covered the remainder of his anatomy, to say nothing of the gloves, with gauntlets nearly up to his elbows. Around his waist was a belt with a big pistol tucked in it and numerous cartridges. He was certainly an airman, for he informed us that he had just returned from bombing the enemy.

The bamboo affair was a Maurice Farman Shorthorn, so the airman told us. Rooty showed his ignorance by enquiring if it really did fly. Of course it would fly; it had been known to reach a height of five thousand feet, could carry two men at once, and bombs, which were thrown overboard in the hopes of hitting the enemy or something belonging to him.

'What about the armoury?' says Rooty, pointing to his belt.

'You mean the pistol?' queried the airman.

'Yep!'

'Ah! to pot the enemy airman with,' he replied.

'Lor! some shooter this feller,' thought I, 'this must be some sport, potting at one another in the air!'

This sport, as I considered it, impressed me, I would certainly have to give it a pop at once. Yes, an airman I would be, some big game hunting this, great fun dropping bombs overboard and potting at some other bird in the air with a pistol.

Next morning I appeared at the Orderly Room with the request for an Army form X.Y.Z. Duly filled, I handed it to the Adjutant, who, with an enquiring look, asked,

'Tired of life?'

I assured him 'No.'

'Well, what the hell do you want to transfer to the Flying Corps for?'

'I don't know, but I want to fly one of those bamboo affairs,' muttered I.

At any rate, the long and short of it was that there was nothing doing. No, I was a Lance-Jack, and I was told straight out that Britain would lose the War if I transferred. He couldn't lose any of his trained men and that was all about it.

'Get out of here and do some work,' he growled. I about turned, cursed and got out, determined to try again at an early date. I'd got the bug. Eventually I joined the Flying Corps, but much happened before that....

'I wish to transfer to the Flying Corps, Sir; is there any chance?'

Funnily enough, he asked me the same question as I'd had put to me by the Adjutant, 'What, are you tired of life?'

'No,' I told him, 'I just have a kink for it.'

'Well,' he said, 'we don't wish to lose any of our N.C.O.'s, but, if you're keen, there's a chance. Two hundred vacancies to the Flying Corps have just been allotted to the Anzac Division.'

This was great news, and at this moment the Adjutant butted in and I had to take a back seat. I pictured a golden opportunity lost, but as the General was leaving, he called me over.

'Still want that transfer, Corporal?' he asked.

'Too right, Sir,' I replied.

'Well, renew your application in the morning.'

I was too flabbergasted to thank him, but the application went in, and within a week I was ordered to appear before the Interviewing Officer of the Flying Corps at Corbie. It certainly looked as though my luck was in.

Major Boyd, the Interviewing Officer, was a gruff Scotchman and an old pilot. I stalked into the tent in my turn and he eyed me over critically. I saluted as I'd never saluted before, for, as a rule, we weren't too flash at that business, but I knew what appearance meant. I'd got my chance and I wasn't losing it.

He questioned me closely:

'Can you play football?'

'Can you ride?'

'Can you swim?'

'Can you row?'

'Have you ever been under shell fire?'

To all of which I said yes. He then questioned me about my weight. I was wearing a fur-lined British warm, which made me look heavier than I was. Apparently he doubted me and sent me with an orderly to the station to be weighed. On our return, he dismissed me, with no satisfaction, just said I'd hear one way or the other in due course. I didn't feel too happy after the interview, as I seemed to be possessed with the idea that he had taken a dislike to me. However, I had nothing to do but to wait and see.

On the way back to Amiens, an ambulance passed me and the driver offered me a lift, which was readily accepted. We took a rather round-about route and were passing the famous aerodrome at Bertangles when an officer hailed us. He was a flying chappie, decorated with wings and the M.C. He told us there had been a crash, the pilot was dead and he wanted us to take the body in to the hospital. How cheerful! and I'd just been doing my damnedest to get into this outfit. We pulled into the aerodrome and there, on the other side of the landing ground, lay all that was left of a perfectly good aeroplane. It had been on fire and was still smoking, and we saw that a gang of men were struggling with something. I hopped out of the ambulance to give a hand and found it was the dead pilot they were lifting. Poor devil, he was charred beyond recognition, and on looking closer, proved a truly terrible sight. This young dare-devil, who was no more, had, not two hours before, left that very spot on a reconnaissance over the enemy lines, and in the pink of condition. Five Huns had pounced on him, he'd put up a great fight against tremendous odds. After shooting one of them down, he himself was hit, and with his machine riddled with bullets, managed to reach his aerodrome, only to crash. The engine caught fire and the flames did the rest. Some grit that kid had.

Several times during the next few days, my thoughts returned to this incident and I found myself debating the subject. 'Don't be a fool,' I thought, 'get out of it while you can.' Both the General and the Adjutant thought me an ass and had asked me if I was tired of life. They ought to know. Yes, this flying was a mug's game and damned dangerous too, I could see that. I'd back out of it, and yet, no, I couldn't do that, it was yellow, or looked like it. I'd certainly got the wind up a bit, but in any case, I thought, one might as well be killed that way as any other. I reckoned that Fritz would get some of us with those bombs of his; he'd tried pretty hard and our luck couldn't always be in. Eventually I dismissed the subject and events took their course.

A fortnight after this incident, I received orders to report to the headquarters of the Flying Corps at St. Pol, away up by Abbeville....

One glorious morning, Lieut. Stewart said he'd take some of us down to Shoreham. He led the way, Brooky was on his right, I was on the left, while behind us came Watson and a chappie named Ball. Soon after starting, Brooky and Watson dropped out with dud engines, so the three of us had to go on without them. It was delightful going until nearing the coast, when we ran into a thick fog. Stewart knew where the aerodrome was, we didn't. However, I saw he had throttled back and was diving down through the mist, so I followed, with Ball behind me. I was soon in a hopeless maze and couldn't see a couple of yards. It was my first experience of the kind and I didn't know whether I was upside down or what; I did know that I had no desire to hit the ground, so I pulled the old bus up, put the throttle full over, and flew straight on, travelling for some distance, still in a thick mist. I hadn't the least idea where I was, or in what direction I was flying.

Suddenly the mist disappeared and I found myself flying over the sea, with nothing but water in every direction.

'Hell!' I thought, 'where am I now?'

Landing on the briny ocean was no good to me, so I turned her round and flew back into the fog, and after a few minutes decided to come down a bit. I accordingly throttled back and stuck her nose down, and found the mist had cleared somewhat. Then a bridge loomed up and I was apparently flying straight for it, on went the engine again, back went the stick, and up into the mist once more. I must have flown round and round, for the next time I stuck her nose down, a row of bathing sheds seemed to be coming at me, and I couldn't have been twenty feet above the beach. I zoomed over them and into the mist, trying once more to locate this infernal aerodrome. I remembered that an instructor had once told me I'd run out of benzine some day; perhaps this was going to be the day; goodness knows how long I'd been cruising round, and there were no compasses or instruments much on this old bus, or I might have done better. However, I gave it another go, and through the mist spotted what appeared to be a big, open space, where I eventually decided to land. The wheels touched — it was a ploughed field — we ran a few yards and ended in a most graceful somersault, the machine lying there on her back, with me underneath. Oil, or something sticky, started trickling on my face, stacks of it, until eventually I managed to crawl from underneath, just as some women, who were hoeing in a nearby field, came running up. They seemed very disappointed, for one exclaimed:

'Oh! he's not killed,' and the tone of her voice prompted me to say:

'No, I'm sorry to disappoint you; would you like me to do it again?' …

My old pal, Rooty, turned up at the aerodrome one day and requested a flip. He was game all right, for I hadn't done twenty hours solo then. The instructor said he'd shut his eyes, however, and away we went.

I showed him Richmond and Kew from the air and Rooty took a few photos, until in his excitement he dropped his camera overboard. He yelled out to me:

'What about stopping, I've dropped it.'

But just then something else happened, the engine cut out, so I yelled back to him:

'The damned engine heard you, it's stopped.'

I tried a few of the gadgets, but nothing happened, so the only alternative was to look out for a forced landing, which eventually took place in a wheat field, nose up. Rooty was shot out, followed by the benzine tin, which served him for a seat, and which cracked him on the head, while I had incidentally knocked the under-carriage off.

'Do you always land like this?' was Rooty's first remark, to which I replied:

'Occasionally, but at any rate, you said you wanted to stop, so you can look for your camera.'

I 'phoned the aerodrome, and was told to stand by, while a curious crowd seemed to come from nowhere. I decided that my passenger had better disappear, in case of a strafe; I'd just about get court-martialled for having a passenger aboard.

To Rooty's disgust, I told him he'd have to walk, but his reply doesn't bear repeating. At any rate, I reported that I'd gone up solo and had forgotten the ballast for the front seat. Someone called me a damn fool and that was that. Rooty had his flip and I had crash number two; bad habit this crashing.…

Now to combat, and the first Victoria Cross ever awarded to an airman. Some claim that airman as a New Zealander — he's certainly recognised in the Air Force Museum at Wigram. But others merely assert a strong New Zealand link, pointing out that William Rhodes-Moorhouse was actually born in England. Nevertheless, his mother was a New Zealander (the daughter of a wealthy pioneer land-owner and a Maori woman) who went to England to successfully contest her father's will before the Privy Council. In 1915 her son was flying with the RFC in France.

Years of Combat: The First Volume of the Autobiography of Sholto Douglas
Lord Douglas of Kirtleside G.C.B., M.C., D.F.C.

By late April we were receiving more and more detailed orders to bomb targets behind the enemy lines. I was flying quite a lot by then with a pilot by the name of W. B. Rhodes-Moorhouse. A small, slight, sandy-haired man of extraordinary vitality, he was one of the pioneers of flying in England, and he had already been a pilot for two years before he qualified for his certificate as early as 1911. He was a member of the Royal Flying Corps Special Reserve.

During the afternoon of the 26th of April the squadron received instructions to bomb Courtrai, which was well behind the German lines, in order to disrupt the movement of enemy troops to the area of the Ypres Salient. Rhodes-Moorhouse was detailed to the task, and he realised that with his B.E.2a he would only just be able to carry a bomb of one hundred pounds. A load as heavy as that meant dispensing with the observer. I begged him to take me with him and we had quite an argument about it, but he firmly refused on the grounds that he simply would not be able to manage with my extra weight on board.

Instructed to use his own discretion about the height from which he should bomb, Rhodes-Moorhouse made sure of his attack on the railway line west of the station at Courtrai by flying at a height of only three hundred feet. He was subjected to very heavy rifle and machine-gun fire, particularly, in the case of the latter, from the belfry of the church in Courtrai, which was almost at the level at which he was flying; and immediately after successfully bombing his objective he was hit in the abdomen, and then again in the thigh and in one hand. He managed to fly the thirty-five miles back to our airfield, never at any time more than a few hundred feet from the ground. He landed and when we helped him out of his aircraft and we saw the full extent of his wounds we were amazed that the poor chap had ever managed to get back. I also noticed that half a dozen bullet holes had appeared in the observer's seat, and it was quite obvious that I would not have survived if I had been with him.

Before he was taken to hospital Rhodes-Moorhouse insisted on making a full report on the flight and on the results that he had achieved. But he died the next day, and the memory of his funeral is a particularly sad one for he was a most likeable and gallant man. Just before he died he had read to him a message from the Commander-in-Chief which stated: 'But for the pressure of urgent work, the Field-Marshal Commanding-in-Chief would have visited 2nd Lieut. Moorhouse himself to express his admiration of his courage and the way in which he carried out his duties yesterday.' Following that came the announcement that he had been awarded the Victoria Cross. It was the first to be won by an airman.

Although born in England, Rhodes-Moorhouse had strong family ties with New Zealand, and the view was expressed recently by Linda Rhodes-Moorhouse, his widow, that his 'courage and endurance' were possibly the result of an infusion of the blood of the vigorous fighters of that country. There might be something in that because in addition to the honour of sharing in that first V.C. of the air, New Zealand also produced the only man to win the Victoria Cross twice in the Second World War.

William Rhodes-Moorhouse must have sensed the danger of that Courtrai rail junction raid, because the night before he wrote a letter to his son ...

No. 2 Sqd
Royal Flying Corps
Expeditionary Force
France
1915

My dear Sonny,
This is the first and last letter I shall ever write to you. You are now just

over a year old and the dearest happy little chap I have ever seen. And thank god you don't realise the awful war that is going on now.

Years hence you will be shown this letter and your Dad's photograph by your dear Mother who has always been the sweetest and dearest wife a man could possibly have. Your dear Mother and I have never had a quarrel or a misunderstanding of any sort ever, and it is for you to fill up the gap by exerting all your energies to be a great comfort to her who has been so sweet to your Dad and yourself. Tomorrow I am going out at a very early hour, long, long before you will be out of bed, on an expedition which, if this letter reaches home, I shall be dead. If things are as they are at present you will be fairly well off. It has always been my desire and wish that my son be an engineer, I don't mean by profession, but I know how useful a knowledge of machinery in all forms is and after all, the world is getting more mechanical day by day. But, my dear boy, it is for each of us to make his own way in this world and far be it from me to say now 'be an engineer'. Your tastes may be quite different. But whatever you choose for your profession, bear in mind that a fair knowledge of engineering will always be infinitely useful to you. If you want any advice on any subject, don't forget any subject, always ask your dear Mother. Your two godfathers, Eric Innes and Donovan Rancliffe, are both dear friends of mine and I know both will take a lively interest in your welfare. But, old boy, always if in doubt, go to your mother, doesn't matter what it is, always. Don't be afraid to tell her everything. I always told your Mother everything. You do the same. Parnham, I hope, will be yours in time to come. Always keep up your position as a land-owner and a gentleman and make your friends at school and the 'Varsity friends, true friends (not hangers on) that will be friends to you not for what they can get but friends because they like you as a good fellow, and above all my dear boy be always loving and sweet to your darling Mother who has gone through so much for me and for you. I am looking at your sweet little face now and God bless you my dear boy.

Goodbye Sonny
Your loving father,
Will Rhodes-Moorhouse
Merille, France

My dearest Linda,

Goodbye darling. God bless you always. I am off on a trip from which I don't expect to return but which will I hope shorten this war a bit. I shall probably be blown up by my own bomb, or if not, killed by rifle fire. Well, God bless you my darling. I have no regrets. We have never had a quarrel and dear old Sonny will help to fill the gap.

Goodbye my Darling pet. God keep you always dear.

From Will

... and in a footnote to history, 26 years after this letter was written, William's son was shot down and killed in a dogfight during the Battle of Britain.

Undoubtedly one of New Zealand's most colourful World War One pilots was Squadron Leader Malcolm 'Mac' McGregor, better known to his compatriots as 'Mad Mac'. Born in Mangamako in1896, Mac was 'upright in carriage, with a markedly strong face, prominent jaw, fair hair, and those unusual pale eyes given to people who do not recognise defeat nor know fear'. In 1917, whilst with 54 Squadron in France, Mac was injured when his engine failed shortly after take-off and he had to make a forced landing. After recuperating, he spent time as a ferry pilot in Britain, flying a wide range of types. His biographer, G.H. Cunningham, notes that Mac's letters 'contain entertaining accounts of his experiences with these machines, the following being a typical sample'.

Mac's Memoirs: The Flying Life of Squadron-Leader McGregor
G. H. Cunningham

'Of course there is only one type of machine really worth flying, and that is the scout, as the others practically fly themselves. The latest arrival here (October, 1917) is the Sopwith Camel, so-named because of the hump formed by the two guns. This is a marvellous bus, with a wonderful performance, though difficult to fly. It rolls and spins just like a top and beats the "Pup" hollow for quickness in manoeuvring. She is rather difficult to loop, however, as she gets on the top and spins around on her back, which makes one wonder if the safety belt is really meant to hold a fourteen stone New Zealander.'

Mac spent much of his spare time at Biggin Hill in improving his aerobatics, soon becoming recognised as one of the outstanding stunt pilots in England. His aerobatics frequently led him into collision with the authorities, who held that stunting near the ground should be frowned upon. It is not surprising to learn, therefore, that he was shortly 'placed on the mat' and as punishment given six days' Orderly Officer.

It appeared that he 'borrowed' the pet 'plane of his Commanding Officer, whom he thought was on leave in London. After cruising around above the aerodrome for a while he decided to seek other amusement. Sighting a football match in progress, he dived down and so disconcerted the players that they fled from the field. ('The teams scattered like a mob of frightened merino wethers.') He next swooped over a double-decker omnibus passing along a nearby road, nearly striking the top rail with the landing wheels, and frightened the passengers off the top deck. Finally he steeplechased around a few feet above the hedges, dived down a narrow gully, zoomed above the aerodrome and landed.

As Mac taxied on to the tarmac his Commanding Officer walked out of the office and informed him, in correct army fashion, that he was for it. After the fur had ceased to fly, the C.O. offered Mac a Flight (and his captaincy) in a squadron of Home Defence Sopwith Scouts. Mac declined the offer, despite the promotion; for he was anxious to return to France, and had applied for a position as a fighting pilot in No. 91 Squadron, then being formed.

A few days later he staged a spectacular flight. Outside one boundary of the aerodrome lay a deep and narrow gully in which several pilots had crashed, following trouble with their rotary engines. These choked readily when taking off unless the petrol mixture was most carefully adjusted. A special control was provided for this purpose. After leaving the ground Mac so increased his mixture that the engine began to backfire loudly. He then shut off the petrol and dived into the gully. When out of sight of the aerodrome he opened up the engine again, passed up a side gully and by flying low managed to return and land unobserved.

Members of the squadron thought that he had crashed, and a crash in this gully meant serious injury or worse. He was just in time, therefore, to see the staff, complete with ambulance, speeding to the spot to collect the pieces. Shortly they returned, to find Mac smoking a complacent cigarette. The Commanding Officer and other pilots hastened forward to congratulate him on his narrow escape. They seemed so genuinely pleased to learn that Mac was unhurt that he considered it advisable to keep to himself the fact that the whole performance was an elaborately staged leg-pull. A wireless officer who was in the machine as a passenger, and had been boasting that no mere pilot could frighten *him*, applied within the hour for a week's sick leave!

While at this aerodrome Mac developed a spectacular feat for which he later became renowned. This was to loop an aeroplane off the ground, commencing each loop when within two feet of the earth. Owing to the fine judgement required, this proved to be the most spectacular of his repertoire of aerobatics. It was therefore used by him on all possible occasions until —

'This is rather squashed at present as the C.O. threatened to stop me flying for 14 days if I did it again. However, this is not the only aerodrome in the world.'

...

A recrudescence of low flying was bound to occur, with the result that on March 8th, 1918, Mac was again 'placed on the mat.' His 'crimes' on this occasion were (1) chasing a train, (2) trying to demolish the wind vane on a church spire with his propeller, (3) landing on a racecourse, and (4) stunting low over a town. He was court-martialled and, after a fair hearing, in his own words — 'Got off with a severe reprimand, plonk!' ...

At the conclusion of a flight in which he had been testing a junior pilot, Mac flew the 'plane over Windsor Castle:

'It is a fine old place, with lovely grounds. We floated around the old place for quite a while, having a good old look, much to the annoyance of a fat policeman who thought probably we had come to bomb the place, or something; and after pulling out his notebook tried to get our number.'

The following afternoon Mac and this chap collected an Avro from Brooklands and flew it to Hounslow. On the way back they flew low over the River Thames:

> 'It was a glorious evening and there were crowds punting and rowing. We flew along a few feet above their heads, taking the bends in great style and turning back to see any extra pretty girls. Our wheels even touched the water in places.'

Machine gun tests from the air were conducted above the Staines reservoir, in the centre of which a target had been rigged. In the process of testing the Vickers guns on a Sopwith Dolphin, when diving on this target Mac shot off his propeller. This proved to be a fortunate accident, as the experience gained in fighting the terrific vibration which ensued, was to save his life, as will appear.

He had an even more exciting experience when testing the guns of an S.E.5a on the same target. The Lewis gun mounted on top of the centre section was apparently not properly secured. The terrific wind pressure generated by the speed of his dive forced it back into the cockpit so that the butt struck Mac on top of his head and partially stunned him. When he again became conscious he found the 'plane to be so low that when it was pulled out of the dive the wheels barely cleared the surface of the water. Mac returned to the 'drome in a dazed condition, and suffering from a violent headache, which was scarcely surprising! …

In May 1918 Mac returned to the Western Front as a first lieutenant in the newly formed 85 Squadron, RAF.

The position on the Western Front when 'Eighty-five' arrived in France was much more hopeful than during the gloomy days following the big German advance in March. For not only had this advance been checked, but the enemy was being driven slowly back at most points. The last desperate efforts of the Germans were made on May 27th, when they launched a formidable attack against the French Sixth Army in the Aisne, and on July 15th attempted a drive south-west and east of Rheims.

Germany's 'Black Day' arrived on July 18th, when Foch launched his deadly counter-attack ...

The first to score was Mac himself. On May 29th, near Armentieres, he shot down a two-seater, which was seen to crash by five other pilots. Thus he had the distinction of 'blooding' a squadron which was to become one of the most famous fighting units in the Royal Air Force, and of having his name appear first on the score board.

The following day Mac forced a second enemy machine down out of control; but as no one actually saw it crash, he did not bother to put it in his combat report.

On June 1st, near La Gorgue, he shot down two Pfalz single seaters, one of which fell upside down for some considerable distance.

In his letters Mac complained of the scarcity of enemy machines on this part of the front:

'This is a rotten part of the line, and the sooner we shift the better. The Huns around here take more stalking than an old stag. You cannot get within miles of them. For instance, last night (June 7th) we went 20 miles into Hunland, and still they went further in.

'A week ago five of us surprised six of them and before they knew who were after them we had shot down four, two of whom, by some chance, I happened to get. If it is like this all along the front, the sooner the Kaiser puts in a bigger contract for machines the better. We have spent three perfectly good days without shooting at anything.

'I wish Ewen could hear one of my old guns, with a new gadget on it, rattling off its hymn of hate, doing 1000 rounds a minute. I am anxious to find a Hun to test it on, but they are rather hard to find.'

...

On June 14th the squadron moved to an old French aerodrome at St. Omer, about five miles behind the lines. Hangars and Nissen huts were well placed on the edge of a small wood bordering the aerodrome....

Two days after the squadron had taken up its new quarters, Mac had an exciting experience and nearly ended the day as a prisoner of war. After his Flight had completed the afternoon patrol, he remained behind with a view to seeking Germans flying over the lines. Observing a burst of our 'Archie' in its vicinity, he spotted an enemy machine a few thousand feet above him. He tried to climb to the same level, but the enemy pilot saw him and hurried for home. Mac followed, some distance below, trying to sink him by stalling up and firing short bursts at intervals. Unsuccessful in this, Mac turned back and a little later saw four other enemy aeroplanes some distance below him. He dived on one, when all four evidently decided to call it a day and dived steeply away. Mac followed and, travelling at over 180 miles an hour, was firing a good burst into the machine he had selected, when he shot away one blade of his propeller. The terrific vibration which ensued, caused by the uneven swing of the propeller, and peaking of the engine (the revolutions jumping from 2,000 to nearly 5,000) nearly destroyed the machine. Screws fell out of the fascia board, flying wires slackened, and everything that was not split-pinned came adrift. Fortunately, his previous experience at Hounslow had shown him how to stop the vibration. He therefore switched off the engine, and stalled up until the propeller ceased to revolve.

Gliding slowly back to our lines, he was amused to see three of the Germans flying parallel with him, just out of range (the fourth, whom he had engaged, was by this time nearing Berlin). They made no attempt to attack, although it would have been a simple matter to have shot Mac down, as he was unable to travel at more than 70 miles an hour. Had he exceeded this speed, the propeller would have spun and recommenced the vibration. The Germans doubtless suspected that he was playing some trick to entice them within easy range. Mac managed to get over the lines, some 20 miles from his aerodrome, being 'Archied' vigorously the while, and — 'Plonked the old machine down without any damage to myself but plenty to the bus.' ...

On July 15th Mac secured his tenth victory. According to the official record he attacked near Armentieres an enemy two-seater, killing the observer. 'The machine was afterwards seen wrecked on the ground, but Captain McGregor did not make a claim for this aeroplane.'

Mac's account of this flight is as follows: In the early part of the afternoon he and Lieutenant D. C. ('Kiwi') Inglis went out on an independent patrol to try and secure an enemy aeroplane for the latter. Near

Armentieres they saw a large two-seater engaged in artillery observation. These machines were dangerous to attack, as not only were they provided with guns operated fore and aft by pilot and observer, but the later models possessed no blind spots. Then too, they were frequently used as bait by enemy *Jagdstaffeln*, members of which would lie in the eye of the sun and wait for some unsuspecting pilot to attack, when they in turn would dive on the unwary pilot.

Tactics of this nature were well known to all pilots. However, after careful examination of the sky in the region of the two-seater, Mac decided that no enemy scouts were in the vicinity. With 'Kiwi' close behind him, he dived on the unsuspecting German and fired a short burst with the intention of disabling the observer so that Inglis could get on the tail of the enemy and shoot it down. Mac got his burst in too far forward and killed the pilot, who disappeared into the cockpit, and so jammed the controls that the machine commenced to fly to the left in a wide circle without appreciably losing height. Mac and 'Kiwi' dived repeatedly, but were unable to bring the aeroplane down, as they could not get close enough for a decisive burst owing to the accurate shooting of the observer. Finally they were forced by heavy anti-aircraft fire to relinquish the attempt. When Mac last saw the machine, after he had turned for home, it was still flying in a wide circle.

At afternoon tea Mac mentioned the incident, whereupon another pilot set out to try his luck. On his return he claimed that he had shot down this machine, and credit for its destruction was granted him when one of out anti-aircraft batteries confirmed that it had crashed. Several days later Mac learned from the battery commander that the two-seater had fallen out of control a few minutes after he had turned back for the aerodrome!

July 24th saw Mac secure his 11th and 12th victories. When on morning patrol near Kemmel, he shot down a Fokker biplane which was seen to crash by another pilot. In the same flight he drove down a second Fokker, which was last seen descending in a spiral with the propeller stopped.

On July 26th the squadron suffered its greatest loss, when Major Mannock was shot down. Mac, usually so slow to kindle and display enthusiasm, was deeply moved:

'We have lost our Squadron Commander. Went down in flames

after getting over 70 Huns, and so the Royal Air Force has lost the best leader of patrols, and the best Hun getter it has had. In another month he would have had over 100. However, unlike other stars, he left behind all the knowledge he had, so it is up to the fellows he taught, to carry on.

'I was with him a few days ago when he shot down what turned out to be one of the crack Hun airmen. It all happened 25 miles across the lines, over a Hun aerodrome; and it was a pretty sight to see the way the Bosche handled his triplane, pretty little thing, all black with a white tail. Although he put up a good fight he had somebody better after him, and it was not long before the Major got his position for a few seconds and, with both guns going, shot off his tail.'

The actual cause of Mannock's death is uncertain; but from available evidence it would appear that he was shot down from the ground, probably by anti-aircraft fire.

At 5.30 a.m. on the morning of the 26th Mannock set out with Lieutenant Inglis, the New Zealand pilot, to assist him to secure his first enemy machine. In the vicinity of Lestrem-Calonne they observed a two-seater, upon which Mannock dived, getting in a short burst which killed the observer. Inglis then attacked at close range, sending the machine down in flames. Mannock followed the falling enemy down, as was his frequent practice. When within 200 feet from the ground, his machine was hit and crashed heavily out of control, bursting into flames as it struck. Inglis circled above the wreck at a low altitude, but could see no sign of his Commanding Officer. With heavy heart he turned away to return to his aerodrome with the tragic news. When nearing the lines and at a low altitude his machine was shot through the petrol tank. He was therefore compelled to make a forced landing, which he achieved successfully, without injury to himself, some five yards behind the British lines.

Daytime fighter pilots had to deal with intense cold (they sometimes landed with frostbitten faces) and pressure changes causing earache, headaches and nose bleeds. Pilots flying bombers on night raids faced the same hardships –

in darkness. A.R. Kingsford was one 'who raided Hunland on many dark nights', initially flying two-seater, pusher-engined F. E. 2Bs and Cs. (In the 2B the observer sat in front, in the 2C it was the pilot.) But, wherever you sat, things could go wrong, as A.R. and his observer discovered during a raid on the Metz railway station.

Night Raiders of the Air
A. R. Kingsford

The night was good, and as A Flight was leading, with ourselves flying second machines, I thought we would try out a new stunt, and instead of approaching the town from the S.E. we would go on past for a while, then turn and come in over from the north.

We went on undisturbed until just to the east of Metz, between the town and the Forêt de Remilly. We knew this wood well — shaped like a letter T, it was very familiar and showed plainly on the darkest night. It was also the hiding-place of several anti-aircraft batteries and machine guns, and we always gave it a wide berth. The enemy had apparently become wise to the fact that we always passed between this forest and Metz, for on this particular night we were suddenly confronted with three powerful searchlights, placed in the form of a triangle and working with their beams together, forming an apex. We wondered what had struck us when they first appeared, and Bourney, my observer on this night, who was new to the game, was tickled to death. He signalled to me that he wanted to have a pop at them. We were at five thousand five hundred feet; I pulled the joy stick over, and kicking the rudder, swung round on the one nearest Metz. The other two shut down and we slipped out of the beam, taking up a position to open fire. They picked us up again, and just as Bourney was about to open up, the familiar 'woof' of a bursting 'Archie' made the plane quiver. It had burst under the left wing and was unpleasantly near. Then two more shells burst within a few yards of us and the searchlights still held us. It was getting too hot, they'd hit us for sure if we didn't get out of that beam. Shells by the dozen seemed bursting all over the place and I yelled out to Bourney to hold tight, pushed the stick forward and the heavily-laden plane literally trembled with the strain of the high speed put upon her. The searchlight was a sticker and followed us, there seemed no getting out of that beam. Pulling the stick back, the machine zoomed up a couple of hundred feet, finishing in a climbing turn and eventually dodging the light. Bourney leaned over and shouted:

'Cheeky beggar that, like to give him one.'

'Right O!' I yelled. 'Be slick about it,' but immediately we turned, he swung the beam towards us. He was not quite so lucky this time, however, for we were dead over him and he didn't pick us up. Bourney pulled the release and let go a twenty pound Cooper bomb from the right under wing. We watched, no burst, and I switched my torch on to the bomb rack, to find that no bomb had dropped. The rack and the left carrier were both full.

'Try number two,' I shouted to Bourney, and we came over the search-light again and let go. There was a tremendous flash and then all was dark; must have been a good shot and blown the thing to smithereens we thought. Our patience had been rewarded.

We made on northwards to Metz, and to our left there was nothing but darkness, not a light to be seen anywhere. At Thionville, we saw that things were well lit and we were fascinated. What a target! We continued towards it, and then, as though one hand controlled all the lights in the town, they went out, and darkness reigned on both sides.

About ten miles north of Metz, we turned due south and came straight in over the town at four thousand feet. We knew the position of the station, and throttling the engine down, we lowered to two thousand eight hundred and were over our target. Away went four bombs in quick succession, bursting one after the other, and away to the left of the main rails, we took up a good position for our big bomb. Searchlights, darting here and there, were very busy, but somehow missed us. Bourney waved to me. I levelled the plane up and away went two hundred and thirty pounds of explosives in a good burst. Some flaming onions came near, Archie was busy, and other machines were around apparently, for we could see their bombs bursting. Turning south once more, we let go the remaining three, although we could only see two of them burst. Pulling away from our target, we found that the near bomb on the left rack had failed to drop. Bourney tried it again and again, but without avail. He realised the danger of landing with a bomb that had failed to leave the rack and that the jerk of the machine when touching earth would be just sufficient to send it off. Poor old Scuds had met his death through the same thing.

Telling me to keep the machine steady, Bourney climbed out of the nacelle in the darkness on to the lower plane and laying flat, he released the bomb from the rack and calmly climbed back. A false step would have meant Bourney's exit from this world, but it was all in a night's work to him. Back in the mess, I shouted him a spot for a plucky effort, and he made me promise not to tell the boys — he was no glory seeker....

Some time later, having completed training on larger Handley Page 0400 bombers, the night attacker was himself attacked.

On our way back, just before dusk, we had arrived at an aerodrome north of Amiens, where two Squadrons were operating, one flying Bristols and the other Camels. We intended staying there for the night, continuing our journey the following morning. The hospitality of the Air Force, so well known during those trying days, was extended to us, and after dinner we found ourselves enjoying a game of poker with three Americans, members of the Camel Squadron. The game carried on until about half-past eleven, things were exciting and an ace pot was in progress, when an orderly came in and informed the tall guy whom the others called 'Swift' that he was wanted in the orderly room. We dropped cards and had a spot, waiting for his return, when he blundered back into the mess, flying coat on, cap and gloves under his arm and announced:

'Some Gothas coming over — got to go up after them,' and disappeared.

'That sounds exciting,' I said.

'It may be, Swift put up a great show the other night,' one of the others remarked.

'Let's go out and see what's doing,' I said, and out we all went.

At the hangars there was great activity, but an absence of lights of any description. How these fellows managed was astounding. Three Camels were run out in no time, and our friend 'Swift' climbed into the cockpit of the first one. The engine was started, tested full out, he signalled, the chocks were pulled away, someone yelled 'Good Luck', once again the engine roared full out, and in a second he was gone. He didn't seem to study the wind or anything else, he just went, disappearing into the darkness of the night.

Away to the east, the booming of guns could be heard, and the halo of light indicated the direction of the line. Ten minutes or so passed and nothing happened. Groups of men stood about the aerodrome chatting, casting occasional glances skyward, listening intently for the drone of any engine.

The first indication of anything approaching was the flashing of a searchlight some miles northward and we all gazed intently in that direction. Searchlights flashed every now and then, scanning the heavens, and a beam not two miles away from us warned us that the plane was approaching in our direction. Then the familiar drone of the Gotha twin engines could be heard.

'Wonder where Swift is?' a voice near to me muttered. The drone was now much louder, coming right towards us. Two searchlights only a short distance off lit up, scanning to and fro the sky. The drone was now right overhead, there was no mistaking these twin engines. Suddenly the beam to our left caught a glimpse of the plane and instantly the second beam flashed round, forming an apex, and there they held him at four thousand feet. Immediately a whole army of guns boomed forth, but the big silvery-looking plane flew straight on, with shells bursting all round her. The beam still held him and the firing continued, when slightly higher and behind, a tiny green light appeared. The guns instantly ceased their firing and 'There's Swift', voices exclaimed. Then above the drone of the engines, the rat-tat-tat of a machine gun could be heard. We caught a fleeting glimpse of the Camel as it flashed through the beams, another burst from the gun was heard, then a reply from the enemy plane.

'He'll get him', someone near me said; speculation was running high, everyone was excited and bets were made. The searchlights never let that unfortunate plane go for an instant, they did their job well. Another rat-tat-tat followed by another in quick succession — the big machine seemed to rock.

'He's got him', someone yelled. Something was wrong, it seemed to almost stop, then a burst of flame shot out from the body of the machine, one wing dipped, the flame grew bigger, the plane started to fall, the nose of the machine pointing to earth. It was diving at a terrific rate, the whole thing enveloped in a mass of flames, down it fell, the beams following. As the blazing machine struck earth, there was a tremendous crash, flames shooting high as the benzine tanks burst, illuminating the country for some distance around. No one spoke, there was a dead silence save for the drone of the Camel's engine.

I stood as one hypnotised, gazing at the distant flare from the burning machine. What a glorious, yet dreadful sight it had been — four men — enemies, true, yet gallant — had been sent to Eternity in as many seconds. I shuddered at the thought, for had not the enemy been after us, trying and hoping to despatch us in the same way. I turned away just as the spotlight was switched on for the Camel to land. He made no circles of the aerodrome, just came straight in and landed, all done in a matter of fact sort of way, Swift hopped out, while the boys all gathered round to hear his story. The early hours of the morning still saw us sitting around in the mess yarning. Swift got the D.F.C. for this effort, and it was well earned. Camels were tricky to handle at any time, let alone at night. Flying Camels at night was little short of suicide, but it was done.

Finally, a tale from the Royal Naval Air Service about an air raid on a target in the south of Turkey. It was first published during World War Two in the RNZAF's magazine *Contact* and later reprinted in a book of stories, all by the same author.

Flights and Fancies
A. R. Grimwood

The other day I asked the Old Timer when he had suffered his worst scare in a quarter of a century of flying.

'Well, that's pretty hard to say,' he replied, 'but I don't think I ever had the wind-up more in my life than I did when I was bombed in the air.'

'Bombed in the air! How did that occur?'

'Sounds a bit strange, doesn't it?' he commented. 'But it all happened in the simplest of circumstances. To begin, then, I was stationed early in 1917 on the island of Mitylene in the Aegean Sea. We were flying

Sopwith One-and-a-half Strutters carrying four one-hundredweight bombs at a cruising speed of something like 75 knots. Our main task was keep the railway line to Smyrna unserviceable; and, as the line crossed quite a few rivers, bridges were the most attractive targets.

'Sometimes we bombed the mouths of tunnels. In fact, there is a certain high-ranking officer in the R.N.Z.A.F. to-day who made quite a sport of bottling trains up in these tunnels. His method was as artistic as it was effective. By spraying the tender of the onrushing train with machine-gun fire, he encouraged the enginedriver to smarten his pace towards the shelter of the little hole in the hill. Having seen the train safely in its burrow this officer would nip off to the other end of the tunnel and with a well-placed bomb effectively close the outlet. Then he would hasten back to the entrance and repeat the sealing process, no doubt to the considerable discomfort of the train crew and any others who were unfortunate enough to be passengers.

'However, that is just by the way. My story concerns a raid on the Manissa railway bridge. I might say that on this occasion I was the leader of a flight of my own and, as I rather fancied my skill at bombing, I planned the attack of my flight in such a way that, at a given signal, the other bombers were to form in line astern behind me, and not use their bomb sights, but release their bombs at certain intervals, after my load was away. Approaching the target I gave the signal for the flight to form line astern. There was a fair amount of anti-aircraft fire; but I was too engrossed with my bomb sight to notice what was happening around me. The other pilots were, with one exception, experienced men; and I knew that they would carry out their instructions meticulously. I had allocated the second position in the line to a new pilot, because I thought it would be easier for him to follow his instructions if he had not to consider more than one machine ahead of him.

'The bombing orders required No. 2 to release his first bomb three seconds after my first one went away, and No. 3 to drop his first one three seconds after the first from No. 2, and so on throughout the flight of seven bombers.

'Well, I released my four bombs at one-second intervals, but immediately after my fourth one went my aircraft was violently hit by something. A large gaping hole appeared in the port upper wing and the leading edge of the lower wing was severely crumpled, whilst several bracing wires were completely severed. The aircraft, of course, went into a violent left-hand spin and was almost uncontrollable. After spinning from eight thousand feet to eight hundred feet, I managed to get some control over the machine; but even though I gave it full right rudder it continued to turn left. There I was, eighty miles from my base with a badly-damaged aircraft that could only turn left, and which I could barely fly at the minimum flying speed.'

'Good heavens! How did you get out of that mess?' I could not help exclaiming.

'It wasn't too easy, but, by flying on a course that looked like a flattened coil spring, I arrived back at the aerodrome two hours overdue and with barely enough gas to taxi the old bus to the hangar. I must have done hundreds of left-hand turns on the way home. Was I thankful there were no enemy aircraft about? Yes, please!'

'You must have been scared all right. What were your feelings when your aircraft was disabled?' I asked the Old Timer.

'It all happened so quickly that I was taken by surprise, and I was so busy trying to get the aircraft out of the spin that I don't think I had time to feel frightened. Actually I had the most peculiar sensation of being an eye-witness of the whole incident and marvelling at my own efforts to save myself. I seemed to be speculating rather casually as to whether the 'plane would crash into the river or upon on of its precipitous banks. Until I finally got control of the ship I was automatically doing the right things. But when I had saved myself from immediate disaster, the fear element crept in, and I can honestly say I have never been so scared in my life as I was on that crazy trip home. It has always been a mystery to me how the broken wing lasted the distance.'

'What caused the damage? Were you bombed?'

'Yes, I was coming to that,' continued the Old Timer. 'The inexperienced pilot of the second aircraft in the line was so upset by the anti-aircraft fire that he overtook me and actually released his first bomb on top of me. Fortunately, it did not fall far enough to become "live" otherwise I should have been a little more detached from the scene. The last the flight saw of me was my downward spin, and when I failed to turn up with the others I was reported lost on operations.'

The Old Timer smiled broadly. I could tell by the twinkle in his eye that this was not the end of the story.

'The sequel had its lighter side, too,' he added. 'About three weeks afterwards the mail from England ceased to bring anything for me. Apparently my lady friends were neglecting me, so I wrote to the favourite one and much to my surprise she informed me by the next mail that I had been reported as "missing, probably killed." This was bad enough, but when my pay ceased I thought it time to make investigations. I discovered that the signal sent to the Admiralty reporting my disappearance had not been cancelled upon my return to life. Needless to say I hastened to have myself officially resurrected!'

PART THREE

NEW
HORIZONS

When the War to End All Wars ended, some pilots found a new enemy — distance. The Atlantic was crossed, first in a series of 'hops' flown by US Navy Curtiss NC-4 flying boats and then non-stop by Alcock and Whitten-Brown in a converted Vickers Vimy bomber. They left Newfoundland on 14 June 1914 and crash-landed in an Irish bog 16 hours and 27 minutes later.

In 1919 commercial air travel began, with both German and French airlines taking off in February. On 25 August 1919 the world's first daily scheduled air service (between London and Paris) was launched, once again using a modified wartime aircraft. A one-way ticket cost £21 and the flight time was two and a half hours.

No such excitements were available for the travelling public in this part of the world. The New Zealand Government didn't know what to do with aviation, civil or military. In March 1919, RAF Lieutenant Colonel 'Zulu' Bettington was invited here to advise the government 'on the defence aspect of aviation'. When Bettington's detailed report recommended an air force with a minimum of seven squadrons and at least one servicing depot, the politicians promptly told him to give them something cheaper.

Later that year, the British government offered New Zealand 100 war-surplus aircraft to start its own air force. After six months of dithering, Wellington advised London it wanted only 35 aircraft and eventually received just 33 — most of our quota having been gratefully snapped up by other countries while our chaps were still scratching their heads.

They didn't stop when the first of those 33 planes arrived in 1920, either. Still unsure how to use them, the Government kicked for touch and loaned them to the aviation-minded.

Slowly things started to happen. In Auckland the Walsh Brothers' New Zealand Flying School began airmail flights to Dargaville and other Northland settlements. On 25 August 1920, an Avro 504K from Sir Henry Wigram's Canterbury Aviation Company trundled across Cook Strait, also carrying airmails.

Further south, in Timaru, Rodolph Wigley was starting the New Zealand Transport Company (forerunner of the Mount Cook group) and in 1921 an NZAT DH 9 attempted the first one-day flight between Invercargill and Auckland. Sponsored by Creamoata, the plane had the name proudly displayed on its fuselage and wings. It left Invercargill on 24 October and reached Timaru where, sadly, bad weather left it grounded for the rest of the day.

Next morning pilot Bert Mercer took off with two passengers, Herbert Fleming from Creamoata and Rodolph Wigley himself. The trio reached Wellington by lunchtime and Auckland at 6 p.m. As they flew up the North Island, Herbert Fleming dropped silver spoons attached to tiny parachutes to promote his fine porridge.

In 1922 Rodolph Wigley announced plans for a regular inter-island passenger service. Noting 'the speediness and the ever-increasing reliability of aerial transport', he declared 'the time is ripe to get New Zealand ready for aeroplane development ... compared with motor cars, he had marked the absence of expenditure on tyres, the absence of dust and vibration, which also reduce running costs, and the absence of expenditure on making and maintaining roads.' Alas, local investors weren't ready for this radical vision and the idea languished.

But one New Zealander did found a large and thriving airline, although he didn't launch it here. Lowell Yerex started his airline in Central America. This is an amazing story and what is equally amazing is that Lowell Yerex is another who is virtually unknown in his own country, when he should be a household name. Judge for yourself.

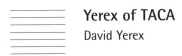

Yerex of TACA
David Yerex

Lowell Yerex was born in New Zealand on July 24, 1895.

He left the country of his birth with an elder brother at the age of 16, to attend university at Valparaiso in Indiana, USA, went on from there to teach in North Dakota, and in the First World War became a fighter pilot, training in Canada and finally seeing action on the front in France.

He was shot down, captured, and held in prisoner-of-war camp, and after the war returned to Valparaiso where the remainder of his family were then living. He took them to San Francisco to seek a passage back to New Zealand, but was told they might have to wait a year for a ship.

Lowell Yerex then joined forces with an American promoter and started an aerial circus which, for five years, toured California and the neighbouring states, giving hair-raising shows from small local airfields.

He married and, tiring of the flying circus life, took out a vehicle franchise in New Mexico. That business failed and he went back to flying, this time as a commercial pilot with a new airline which had opened up in Mexico.

When that venture folded, he became pilot for two American adventurers who forgot to pay him, but who eventually agreed to his taking a half share in the plane in recompense.

With that plane Lowell Yerex set himself up in Honduras as a freelance operator, a 'bush pilot'.

Within ten years he built his company into the largest airfreighting operation in the world, and the largest privately-owned airline.

With it he eventually linked the United States with the whole of Central America and the South American continent, rivalling for a short time the might of the US international flag carrier, Pan American Airways.

He was involved in a Central American revolution in which he saved an elected government from a military take-over, and lost the sight of an eye to a sniper.

He became a multi-millionaire living in plush splendour in New York, and he lost it all because he was British — Washington determined to get rid of him and Whitehall sacrificed him.

The manner of his rise, and of his fall, was the practical evidence of his faith in the principle of free enterprise ... and the death of that same principle at the hands of political expediency.

Lowell Yerex was a man with a simple faith; one that he had emblazoned boldly on his personal aircraft ... the insignia of a knight in armour, mounted, lance to the fore, and the legend 'Al Audaz Fortuna Favorece' ... Fortune Favours the Bold.

Lowell Yerex launched his own airline in 1931. He called it TACA — Transportes Aereos Centro Americanos. A year later, coming to the aid of the newly elected Honduran president, he was shot in the face while using one of his transport aircraft to attack rebel troops. Yerex described the incident in a letter to his mother: '... several bullets hit the ship. One of these ... smacked me in the face just at the moment we were about to drop a bomb. Guy Moloney does not fly pilot a ship, so it was very embarrassing. My right eye was gone and my left was filled with blood from the two-inch wound caused by the bullet, which fractured my skull and hit the gasoline tank in the wing.

There was no place to land, as the country was very rough, but with Guy holding a handkerchief tight around my head, we were able to make Tegucigalpa, 25 miles beyond the mountains over which we had been fighting ... We made a fair landing, and I still do not remember just how I did it.'

Mind you, Yerex was no stranger to risk. In 1920, stranded in San Francisco and waiting for a boat to New Zealand, he'd met Ivan Gates, a one-time used car dealer now offering the adventurous $1 aerial joy rides. With the war over, Gates insisted that America wanted entertainment and excitement. He decided to launch a flying circus and Lowell signed on.

Yerex put all his savings into the purchase of a plane — a Jenny — and the venture got under way. Besides being the first-string pilot, along with Clyde Pangborn, Yerex was also the business manager, handling the partnership's finances.

Gates was responsible for what he was a master at ... talking people into letting them use their airfields, organising publicity, turning every aerial manoeuvre into a 'death-defying stunt' by dressing it up. He was a clever publicist who could convey his own enthusiasm to others.

The aeroplane itself was still something of a novelty, but with the addition of Gates' promotion of the war aces — Yerex became a Captain overnight — and their death-defying acts, the venture flourished.

The ridiculously flimsy, spluttering planes became the symbol of excitement, and their swashbuckling pilots in their cavalry boots, their leather jackets, their helmets and goggles, became a new breed of heroes.

Gates hussled unceasingly, organising, arranging, bullying, pleading, while Yerex quietly went about developing the acts, arranging for the machines and their crews to move on to the next one-day stand.

The crowd wanted thrills, and if a stuntman fell to his death from time to time it only heightened their appreciation, and their willingness to pay to be shocked.

But like stuntmen of the movie world of later years, the aerial circus stuntmen and their pilots contrived to give all the appearance of standing on the edge of the abyss without ever actually stepping over.

New pilots joined the team, and among them came the New Zealander,

Robert Hector Gray, whom Yerex had last seen when together they left a prison cell in Germany.

There were also stuntmen and ground crews to be hired, as well as the 'advance men' who arranged bookings and promotions in advance of the circus as it barnstormed around the country.

The crowds came, and the money rolled in, but it poured out nearly as fast. Most of the takings went into keeping the planes in good trim, for fuel and publicity and transport from one airfield to the next, so that there was little left for high living.

But for the moment it was enough for these young men to be able to fly, or be part of it, living in a world apart from the common herd, sharing a sense of superiority when they looked down at the upturned anxious faces, like a floral tribute to their daring.

They paid a price of course.

Yerex was flying a Hispano-Squiza Standard at a show in San Jose, California, where one of the stuntmen, Jinx Jenkins, had decided to produce as a finale a 3000-foot parachute drop. Before a crowd of 20,000 Yerex climbed to the pre-arranged height, carefully judged so that Jenkins would drop and, at the last moment, open his chute over the centre of the field.

But the chute did not function, and Gates went looking for another stuntman.

Another stunt they developed was for Yerex, with a ladder suspended from his lower wing, to fly in above a second machine on which the stuntman stood on the top wing. The stuntman would grab the ladder as Yerex flew into position, and climb from one plane to the other in mid-air.

That gave rise to what was almost certainly the first mid-air refuelling operation, when a stuntman would climb the ladder with a can of gasoline strapped to his back and, when the pilot of the top machine turned off his engine, climb out onto the engine cowling and pour in the fuel.

One of the most successful stunts with the thrill-seeking crowds, was to have the wing-walker suddenly slip and fall from the lower wing and apparently, to those below, plunge towards his inevitable death.

The large number of women who fainted, was considered evidence of the success of this stunt, but they failed to see the remarkable conclusion of this feat when the stuntman came to an abrupt halt in mid-air, and then climbed back up the rope attached to one ankle to the plane from which he had fallen.

Yerex was also piloting the plane when one of their star daredevils actually rode a bicycle from one end of his top wing to the other, while a cameraman in a second plane filmed the event for showing in movie theatres around the country.

When Yerex suggested that a wire should be run along the top wing, to which the bicycle could be attached, the stuntman dismissed such 'sissy' safe-guards, only to be coldly informed by his pilot that the wire was to stop any aerobatic bicycle from ploughing into the tailplane of a valuable aircraft.

Others had similar ideas. In New Zealand, 'Mad Mac' McGregor was stunning crowds at air shows, and offering the odd joy ride.

Mac's Memoirs: The Flying Life of Squadron-Leader McGregor
G. H. Cunningham

When airmen foregather, many tales are told concerning various exploits of Mac's during his early days of commercial flying. Many are obviously apocryphal; others genuine enough, but difficult now to fix as to time and place. The following have been selected from dozens of authentic incidents known to his intimates. The best, for obvious reasons, cannot be reproduced.

When on a tour of Taranaki with a Hamilton Airways machine, Mac was due to conduct passenger flights at Waitara on the afternoon of October 1st, 1929. Anticipating at most but two or three passengers at this centre, he had made arrangements to stay the night at New Plymouth. Arriving with his wife as passenger at the landing ground, Mac found to his surprise about 20 Maoris anxious to sample flying. Prominent among these was a huge man — weighing at leave 20 stone, and with an 'equator' to correspond — accompanied by his bride of a few weeks, a sophisticated wench of about 18 summers. Most booked for ten-minute flights, among the first to do so being the large Maori and his wife. The first flight was made with the bride and, as was his custom when in this locality, Mac flew with her over the town of Waitara, then back to the landing field. The husband then hastened forward, but to his chagrin was quite unable to enter the cockpit of ZK-AAS. His wife therefore made a second trip half an hour later, so that the ticket should not be wasted.

Finally, when all had been accommodated, before returning to New Plymouth Mac offered the bride a third flight as a wedding present. On this occasion he flew on a different route, towards Mt Egmont, and remained away for about 20 minutes. After the machine had disappeared from their view, one Maori turned to the husband (the Maori is adept at 'barracking') and said: 'I think Mac clear out with your wahine, Ehoa.' Whereupon our oversized friend, with a sly, sidelong glance at dainty, well-dressed Mrs McGregor standing nearby, replied: 'Never mind, other wahine will do!' Dora was quite relieved to see the plane returning a few minutes later.

On March 26th, 1930, Mac flew from Murchison to Christchurch over the Lewis Pass. This, incidentally, was the first occasion on which an aeroplane had negotiated the pass, which lies in the Alps at the head of the Cannibal Gorge, between Reefton and Hanmer. It is not easy to traverse by air, as it is narrow, flanked by high peaks, and contains in its narrowest portion an awkward right-angled bend.

When Mac passed into the head of the pass he found that the wind was blowing at gale force and in consequence his machine, ZK-AAS, was unable to make headway. He could neither turn, because of the narrowness of the pass, nor climb, on account of the peaks with which it is beset; so was forced to sit suspended in the air like Mahomet's coffin, for what seemed to him an interminable period. Finally, by careful manipulation of the controls, he managed to creep slightly to one side and thus escape the full force of the gale. He was then able to traverse the neck of the pass, pass the bend, and dive into the comparatively open country of the Hope Valley, where the wind was less strong.

Mac felt certain that he would be forced to land through exhaustion of his petrol supply: and as the floor of the pass is covered with scrub and bogs, and the sides with heavy beech forest, the prospect was distinctly unpleasant. His passenger's thoughts, unfortunately, are unrecorded!

In April, 1931, Mac spent a couple of days making short passenger flights at Ikamatua, in Westland. On the afternoon of the first day business was brisk, for a large crowd had collected to see the machine, aeroplanes then being rare in that locality.

Mac noticed some restraint among the crowd the following morning, few coming forward for flights. He requested his assistant Eric to ascertain the trouble, and learned that behind it was a local 'bush lawyer', who claimed that the fee charged for each flight was too high. Cigarette in hand, Mac strolled over to this individual, and after listening for a while to his harangue, offered him a flight in the Avro. With the prospect of a free flight the 'bush lawyer' climbed into the cockpit, where he was securely fastened into the shoulder straps by Eric.

When well above the field Mac asked the unsuspecting victim if he would care for a little stunting. The latter acquiescing, Mac commenced a slow roll, a most disconcerting manoeuvre to anyone experiencing it for the first time. The aeroplane turns so slowly that the passenger can feel it actually capsizing, and usually scrabbles frantically at the side of the fuselage for handhold. Finally he lands with a chug into the shoulder straps, which take his entire weight during the time the aeroplane is inverted.

Mac held his victim inverted for a few minutes, then proceeded to put the machine through several aerobatics, finishing with a spin from a considerable height. He then landed and, with the aid of Eric, assisted his victim, now violently sick, out of the machine and parked him in a handy patch of scrub bordering the paddock.

The discomfiture of this individual so aroused the sporting spirit of the crowd that they literally rushed the machine, no less than 56 passengers being carried that day.

We should not assume, however, that the thrillseekers were the only passengers in those early days. More pious persons also took to the air, one of them being the Roman Catholic Bishop of Auckland, Bishop Cleary. Usually his pilot was one of New Zealand's great pioneer aviators, George Bolt. Growing up in Dunedin, Bolt was fascinated by the balloonists of the day. Later, when the family moved to Christchurch, he experimented with balloons and then, in 1911, flew a glider he had designed and built himself.

In 1916 he joined the Walsh Brothers' Flying School as an apprentice aero-mechanic, learned to fly and made the first official airmail flight from Auckland to Dargaville on 16 December 1919. Later in life he was appointed Chief Engineeer of TEAL (Tasman Empire Airways Limited) and initiated research into the achievements of Richard Pearse.

But let's get back to Bishop Cleary, who approached George Bolt with an inspired plan ...

George Bolt — Pioneer Aviator
E. F. Harvie

He had a specific purpose. 'He thought,' said George Bolt, 'that he might save a lot of time by using an aeroplane to get around parts of his diocese and wanted to find out first whether flying would agree with him. So they asked me to give him a short run in the Boeing.' His next letter home told what happened.

'I had another good old engine smash this week,' he wrote. 'I was out in "F" with a bishop as passenger and hadn't been in the air more than a couple of minutes when the crankshaft suddenly broke. There was a bang and a puff of smoke and the big bus just about shook herself to pieces. The whole front half of the engine burst through the radiator and, with the propeller still on it, flew into the sea. Things were only middling for a few seconds, I can tell you, but I managed to make a good landing and we were towed back to shore.

'The bishop got down on one float, had a look at the damage and then said, simply: "Well, well! That *is* a pity." But he didn't seem at all frightened and told Leo afterwards that he wanted to fly to Mercury Bay next week.'

Writing to his sister on 9 March, Bolt enclosed a photograph: 'This is how the engine looked before we lifted it out of the Boeing. Only about £900 worth of damage — cheap smash, eh? But this was my height-record engine and I'll have to make some souvenirs — little aluminium propellers, or something — out of the broken bits. We've just fitted another engine and "F" will be flying again tomorrow.' ...

Bishop Cleary's startling introduction to aviation had not upset his intention of using aircraft, whenever practicable, to take him to various parts of his diocese. Some districts were accessible from Auckland only over unformed roads often made impassable by heavy rain, or by travel on horseback over rough bush tracks. The Coromandel settlements, however, were within easy reach of Auckland by aeroplane and at 7.33 on the morning of 11 March 1919 His Lordship, with Vivian Walsh as pilot, left Kohimarama in the Boeing floatplane 'F' to visit some of them.

In ideal weather they rounded Cape Colville at 8.13 but a few minutes later were obliged to alight to recover some article, never publicly identified, which had either fallen, been blown or accidentally been dropped overboard. They did not therefore reach their first planned stop, Kuaotunu, until 8.37. After receiving a great welcome from the entire community gathered on the foreshore, the fliers left on a 20-minute run down the coast to Mercury Bay.

To the large crowd waiting at Whitianga, the Boeing, moored close inshore at Buffalo Beach, instantly became as big a centre of attraction as the bishop or his pilot, especially among the schoolchildren. Reported the *Auckland Star*: 'The educational value of the seaplane exhibit was much appreciated by the teachers, some of whom made it the subject of a composition, others of direct drawing or other lessons.' Nobody in those parts had seen an aeroplane before.

At 2.30 pm the fliers re-embarked and made a 19-minute run to Tairua, a settlement normally accessible, according to the paper, 'only by horse track and sea, with a bad bar.' Describing the arrival, its correspondent reported that 'Strong air eddies and cross-currents from the neighbouring hills were successfully negotiated and the whole populace turned out to see the visitors.'

After conducting a confirmation service the next morning the bishop was ready to return to Kuaotunu for a similar purpose, but a strong blustery wind now delayed his departure for two days. On Friday morning, 14 March, after 'a rapid flight of 27 minutes', the fliers alighted at Kuaotunu, stayed a couple of hours and then went on to Coromandel. Late the same afternoon they set out for Auckland. His mission, mused the bishop, had been a great success. Yes, but the voyage had yet to end.

'Just as they got within sight of home,' wrote Bolt a day or two later, 'they ran out of juice and a backfire in the carburetter set the engine alight. Walsh got down quickly and put the fire out with an extinguisher and little damage was done.' Despite this experience, Bishop Cleary had nothing but praise for travel by aeroplane and declared his intention of doing a lot more. He was not to know, however, that serious illness was soon twice to confine him to hospital for lengthy periods and that it would be a full year before he could make another aerial voyage....

Bay Of Plenty Venture. In March 1920 Bishop Cleary, not long recovered from serious illness, was ready to make another pastoral visit to distant parts of his see. But he now felt unable to face travel over almost impassable roads in some parts of the country and only 'a scarcity of benzine and the fact that the only machine in commission was in requisition by the Government for its aerial mail services' had prevented his using an aeroplane sooner. This time he proposed to visit settlements in the Bay of Plenty.

Cleary had again offered to carry mails free of charge, and at 7 o'clock on the morning of 12 March a launch took several bags out to the Boeing 'F', moored near Queen's Wharf.

With Bolt as pilot, the aircraft departed at 7.16 and in ideal weather rounded Cape Colville 34 minutes later. Circling 800ft above tiny Port Charles at 8 o'clock, he dropped a bag of mail and then climbed to 3,300ft to follow the coastline as far as the Bowentown entrance to Tauranga Harbour, where a detour was made to circle Katikati at 1,800ft. Two hours and 19 minutes after leaving Auckland, he alighted at Tauranga, where no aeroplane had been seen before.

From an alighting point opposite Victoria Wharf Bolt taxied in to a beach near the steps in front of the Commercial Hotel, and while the bishop handed over a 240-letter mail and got busy with his camera, children granted leave from the district high school swarmed round the Boeing and plied Bolt with questions. To speeches of welcome, the bishop and pilot replied and were given lusty cheers. Since they intended to return later, their stay was short. 'The machine,' reported the *Bay of Plenty Times*, 'rose beautifully, crossing the isthmus over Mt. Drury [*The Mount, today*], and then followed the coast.' Thirty-three minutes later, at 11.48, it passed over Matata and at two minutes past noon alighted on the river at Whakatane.

While taxiing, Bolt noticed a rock marked with a buoy near the wharves and after stopping the engine threw a rope to a Maori in a nearby

dinghy, who made the Boeing fast. 'This,' wrote Bolt, 'was the sacred rock
to which the legendary *Mataatua* canoe had been tied at the end of its
migratory voyage down the Pacific from Hawaiiki [sic]. The Maori in the
dinghy, Rewi Hetariki, told me he was a descendant of a chief who had
come to New Zealand in the *Mataatua* and was very pleased I had chosen
that rock for an anchorage. He passed the word round, and in no time all
the Maoris were calling me Toroa the Second, after the early chief. They
named the Boeing *Whakarererangi*, floating sky canoe, or *Manureremoana*,
bird floating on the sea.'

After delivering a 230-letter mail the visitors lunched at the Commercial Hotel and at 2 o'clock left for Opotiki. 'A little after 2,' reported
the *East Coast Guardian*, 'the ringing of the fire-bell announced the fact
that the seaplane had left Whakatane. Business was promptly suspended
and a large crowd made haste to the waterfront to witness the landing' at
2.25. Carrying a 225-letter mail, the fliers went ashore by launch.

During their journey from Auckland, Bolt, Bishop and Boeing had
covered 240 miles, the greatest distance then flown in New Zealand in
the course of a single day.

After a welcome from an enthusiastic crowd, including 'local Catholic
school children in gala attire', the bishop began his pastoral rounds
while Bolt went for a car ride. At an official function that evening, the
bishop referred to 'the current benzine famine' and suggested alternative
sources of fuel: 'A good product being extracted from the gumfields; vast
quantities of coal slack, from which benzole could be extracted; and
heaps of sawdust from which a product containing alcohol suitable for
locomotives could be obtained.' But, said the *East Coast Guardian*, 'He
added a warning that as soon as production from these sources became
a menace, the oil trusts would attempt to burst them up. He hoped that
the Government in power would be strong enough to prevent this. He
was getting old but, before departing hence, hoped that younger men
would push this matter along.'

The health of the pilot was proposed. 'Another speech here,' wrote
Bolt. 'No good to me.' But he had managed to say that this was 'a good
coast for an aerial service'.

At 9.28 on Saturday 13 March the fliers left Opotiki with 100 letters
for Auckland, and 19 minutes later alighted at Whakatane, where Bolt
took the Harbour Board's engineer for a flight to observe 'the disposition

of rocks and shoals at the harbour mouth'. The engineer saw obstructions that he had known nothing about. (In 1948, to Bolt's dismay, the Mataatua rock at Whakatane was demolished by the Navy because it formed an obstruction to shipping using the river.)

'If I had been looking for passengers at Whakatane,' he wrote later, 'I could easily have brought home £100 from £5 trips.' During that stay, Noah Jonassen, the former 'Aerial King', had come to see him and shown him round the place by car. They spoke endlessly of earlier gliding activities in Christchurch.

With 67 letters for Auckland, Bolt left for Tauranga at 12.10. The *Bay of Plenty Times* reported: 'At eight minutes past one, the *Ngapuhi* sounded its whistle and, dexterously handled by Pilot G. B. Bolt, the seaplane touched the water with the grace of a sea bird.' During the afternoon, Bolt took the bishop to Te Puna, 5 miles west of the town, where they stayed for an hour. Next day, Sunday 14 March, they flew to Matakana Island. Continued the newspaper: 'Pilot Bolt circled and dropped the machine, the Maoris being delighted.' But he didn't like that report. 'I haven't dropped the Boeing yet,' he told his family. 'At Tauranga, I took the earliest settler for a flight. He fought in the early Maori wars, is now 90 years old, and still pretty active. Those Maoris were great. They gave me a *haka* and a lively welcome as "Captain of the Big Bird"'.

At a function in Tauranga that Sunday evening, Dr Cleary was presented with a gold watch and George Bolt with 'a handsome pair of ebony hairbrushes inscribed with his initials on silver shields'. 'Another rotten speech, of course,' he wrote later. 'Still, those people had been very good to me.'

At 6.47 on Monday 15 March the fliers left Tauranga, passed Tairua at 7.37, Kuaotunu at 7.58, Cabbage Bay at 8.14, and reached Auckland at 8.43.

Said the bishop to the *Auckland Star* on ending his voyage: 'Travelling at 60 miles an hour or more through the clean and dustless upper air provides a tonic effect.' …

Taharoa Taniwha. As destinations for his third aerial voyage to minister to the faithful, Bishop Cleary had chosen settlements in the Raglan and Kawhia districts. At 7 o'clock on the morning of 12 April 1920 and with a couple of bags of mail which he had again invited the Post Office authorities to allow him to carry free of charge, he reached the man-of-war steps in downtown Auckland to join Bolt in the Boeing seaplane 'F', moored nearby.

Reported the Catholic newspaper, *The Tablet*: 'The airship left Auckland in beautiful weather, the morning being still and clear, but a dense fog in the Manukau Harbour frustrated an original intention of circling over Waiuku, as all landmarks were obliterated. The voyage continued down the harbour, through the Manukau Heads, and via the coast to Raglan where a mail was dropped at half past eight.'

'Due to fog on the water and over most of the land near Raglan,' explained Bolt, 'we couldn't alight to deliver the mail. But I could just make out part of the coastline near the harbour entrance and decided to take a chance by dropping the mail bag, with 32 letters in it, over a place where I fancied the post office stood. But that was a bit unfortunate. After we had reached Raglan next day, we learned that it had hit the ground less than ten feet from an old draught horse which had immediately dropped dead from fright.'

'The seaplane,' continued *The Tablet*, 'then rose above the intervening land, crossed the Aotea Harbour and the narrow isthmus, finally taking the water easily and gracefully opposite the Kawhia wharf at 8.55. A large crowd had assembled there and the arrival of the bishop was greeted with hearty cheers.' Because he was due to return later, the bishop made only a short stop at Kawhia. 'After partaking of breakfast, he left for Taharoa.'

Lake Taharoa, about a mile and a half long and one-third wide, lies a short distance from the coastline four miles south of Albatross Point, and at the tiny Maori settlement nearby Dr Cleary had arranged to conduct a confirmation service.

As the Boeing alighted on the muddy brown waters of the lake, sending rooster-tails of spray flying in its wake, about fifty settlers, mainly Maori people who had come from all parts of the surrounding district, gathered by the lakeside reeds to watch while several men went out in the only craft they had to bring their distinguished visitors ashore. Some had come on foot, others, singly or in tandem, riding bareback on horses of strangest pedigree, others still, crowded into drays drawn by oxen or standing closely packed on crude horse-drawn sledges.

Clinging to the folds of their mothers' dresses, wide-eyed Maori children, barefooted and in tatters or regaled in oldfashioned Sunday best, kept strangely silent. Others, terror-stricken on seeing what could only be that dreaded water-*taniwha* their elders had spoken of in hushed voices and which would surely gobble them up unless they were to behave, ran away to hide in nearby bushes and scrub. Yet, from their places of refuge, it seemed that the old man whom they now watched descending from the monster's back was not the least afraid of it, and the young man who had come with him was even patting the *taniwha's* side. No longer did it rush through the water, charging shoreward with a roar like a maddened bull; it seemed to be fast asleep. One by one, they rejoined their elders.

In a great Maori canoe, laboriously fashioned in days that nobody could remember by men using crude implements to hollow out the massive trunk of a giant tree felled in a nearby forest, the fliers were brought ashore. The craft had no oars or sails, and was simply poled along like a punt, in which every occupant stood. A shy little company then led their visitor uphill from the water's edge to the Taharoa schoolhouse for a church service, followed by a simple but dignified Maori welcome and lunch on the *marae*. One by one, each settler lost his shyness, addressed himself to the bishop who, he was later to confess, had been 'deeply touched by the kindness of such genuinely affectionate people'. He would have liked to remain with them longer but even now, back in Kawhia, others were waiting and it was time for the flying *taniwha* to bear him away.

On his return to Kawhia, the bishop received another warm welcome and spent the night there while the Boeing rode at anchor on the harbour. Next morning, at 9.25, the fliers made a 27-minute run north to Raglan, where the bishop was kept busy till noon. At 1.29, just after lunch, he was off again and after making a brief alighting at Waiuku, on the Manukau Harbour, reached Auckland at five minutes to four.

Back at Kohimarama, the Boeing, like the rest of the school aircraft, was now made spick-and-span for an important event. On 24 April 1920 the great battle-cruiser HMS *Renown*, with Edward, Prince of Wales, aboard, entered the Rangitoto Channel to be escorted into Auckland by hundreds of gaily-decked surface craft while four aeroplanes, each trailing a coloured pennant, wheeled in lazy circles overhead....

In mid-winter 1920 Bishop Cleary completed arrangements to visit another part of his diocese: the distant north-western settlements of the Hokianga and Whangape, then accessible from Auckland only by slow, varied and often unreliable means of surface transportation. By this time he had come to respect young Bolt's proficiency as a pilot and reputation for acting coolly and resourcefully whenever anything might go wrong, and was prepared to accept the risks of unforeseen circumstances which might result in being forced own in inhospitable country far from the amenities and comforts of civilisation. A firm friendship now linked them both and the bishop's request that, if possible, Bolt be permitted to take him north, was acceded to by the Walshes who shared His Lordship's confidence.

With mails which the bishop had again offered to carry free of charge to the Postal Department, the pair set out in the Boeing 'F' at 8.5 [8.05] am on 21 July and 80 minutes later alighted at Pouto, near the Kaipara entrance, to deliver the first of the bags. Twenty-five minutes later, they resumed their journey, dropping another bag as they passed over Te Kopuru and heading for the west coast beaches leading northward to Whangape.

'At about 10.55,' reported Bolt, 'we were close to the Hokianga entrance when a valve rocker arm broke and I had to set the Boeing down just inside the heads. Some men in a boat seemed to realise that we needed help and came over to take us in tow to a small bay sheltered from the wind and sea. After going ashore I borrowed a broken-down horse from a local farmer and though I didn't know the first thing about riding managed to reach Opononi, a couple of miles away. I was after some borax, pieces of steel plate and some brass. The only brass I could find was part of an old tap, but that was good enough, and after rein-forcing the rocker cage with strips of mild steel plate, I brazed together the bits of the broken arm in a hole by the water's edge.

'By this time we had begun carrying pigeons with us on long flights, and from the Hokianga I released a young male we called Billy, with a message tied to his leg, asking our people at Kohimarama to send up a new rocker assembly as quickly as possible.

'We spent the night in the district, finished temporary repairs the next morning and at 2.14 left for Kohukohu, which we reached in 16 minutes. A bit later we left for Whangape, where the settlers had been keenly looking forward to our arrival. Between the tops of some hills on the approach to the alighting area was a telephone line which the Maoris were certain we would hit because we didn't know it was there. So determined were they not to let anything stop our coming in that they simply chopped it down. We landed at Whangape at 3 o'clock.

'Most of the people there had never seen a motorcar, let alone an aeroplane. Horse- or ox-drawn sledges were about all they knew, so our arrival was quite an event. A party of Maoris came out and took us ashore in a dinghy and, in their shy but kindly way, treated us as though we had come from another world. It was embarrassing, in a way, to see some of them kneel down and lower their heads as we walked along the road. They gave us the best of all they had, even providing us with a house where they came and did the cooking and looked after us with great consideration. The meals were cooked in Maori fashion and eating them was a wonderful experience. I'd never met more generous, kindly people.'

Judged by standards in the world outside, life in Whangape bordered on the primitive. Yet the little settlement boasted a picture house of sorts, an ancient barn in which on rare but joyful occasions a magic-lantern show was held. On a dusty, mildewed sheet an ancient projector shed its feeble pool of light while the audience enthralled, holding unfurled umbrellas above their persons, paid scant heed to a flock of raucous starlings which nightly lodged in the rafters and on show evenings seemed to delight in rudely excreting as they poked fun at the goings-on on the screen.

At Whangape, engine-starter trouble kept Bolt busy for most of the night of the 22nd, but he managed to rectify it, and at 9.8 [9.08] next day took the bishop back to Kohukohu....

Dr Cleary had now made plans for what was to be his last flight to diocesan outposts, and was expected to visit Whangarei, Russell, Whangaroa, Mangonui, Kaimaumau, Lake Ohia and Ahipara before going on to Houhora and Parengarenga, tiny settlements on the Aupori Peninsula in the most northerly part of New Zealand.

Usually distinguishable by the frock coat, breeches and gaiters worn as symbols of his high ecclesiastical office, the bishop was by nature the very antithesis of his rival but less experienced air traveller, Colonel Allen Bell, who also favoured breeches and gaiters but of a kind appropriate to the normally more ungodly cavalry. Bishop Cleary had not forgotten the competition proposed by Bell a year before — to see which of the two could fly further north than the other — but had never intended to use it as an excuse for making a particular journey.

Nonetheless he derived some satisfaction from considering the prospect of a trip which, his Maker, pilot and aircraft willing, might take him further north than Bell had flown with Bolt on 9 April 1920 to reach Unahi, near Awanui. From His Lordship's viewpoint, Parengarenga, some forty miles beyond Awanui, was about as far north as any Christian of the day should be found wrestling with problems of faith and soul. That aside, advancing age and two serious illnesses had begun to extract their toll and he was anxious to fulfil an earlier promise to visit his people in the northernmost part of the country before it became too late.

Originally, it had been expected that he would go north in the Supermarine with Hoare as pilot-in-command and Bolt as reserve, but there was a last-minute change of plans and, after a short run with Hoare, His Lordship made the principal journey with Bolt. The flight, thought Bolt, would prove a useful shakedown cruise for the newly-acquired flying boat which, according to the *Auckland Star*, had 'recently developed a speed of about 70 knots in a trial flight' and was regarded as being the ultimate in comfort for passengers and as mechanically reliable as any aircraft in use in New Zealand.

The flight to the far north began at 3.40 on the afternoon of 15 April. Sailing effortlessly off the Waitemata Harbour, the graceful Supermarine lazily climbed away past North Head, Cheltenham and Milford beaches. In a little notebook he kept in one pocket, Bolt dutifully recorded their progress: 'Kawau: 4 pm; Takatu: 4.5 [4.05]; Mangawai: 4.25.' From time to time he would scribble a note and pass it on the end of a stick to the bishop, up in the bow: 'Hen and Chickens on right; Moko Hinau on horizon beyond', for instance. Seventy minutes after departure from Auckland they alighted at Onerahi, secured the boat for the night and drove into Whangarei.

The weather was fine and clear when, at 9.40 next morning, they continued their journey and both were able to relax, Bolt nonchalantly devouring bars of his favourite chocolate while His Lordship, blinking contentedly behind his spectacles and superimposed goggles, took in the beauties of land, sea and sky as he planned his ministrations in the north.

In the midst of their contentment and while rounding the precipitous face of Cape Brett at 10.45, they were startled when their Beardmore engine suddenly lapsed into silence. Bolt had no alternative but to alight on open water beyond the shelter of the Bay of Islands. After getting down without incident, he realise what had happened. 'Through some misunderstanding at Kohimarama,' he wrote later, 'the staff hadn't filled our tanks. We had simply run out of fuel.'

'Riding a big ocean swell in a brisk offshore wind,' he continued, 'we began drifting rapidly out to sea. Fairly shallow water appeared to cover a reef we were about to cross but our anchor line was too short to reach the bottom and I had to act quickly. Without a second thought, I cut the elevator and rudder cables adrift, joined them together and tied them to the anchor line to make sure we hooked the reef. We had no sooner succeeded in doing so than a launch with four men in it came out from the shore to find out what was wrong. They offered to tow us to Russell.'

During a monotonous, 6-hour, 20-mile haul, Bolt spent the entire time sitting astride the tail, whence he could watch the operation closely and give occasional directions to the boatmen. 'At high water,' he said afterward, 'we beached the Supermarine at Russell where on a soft pebbly bottom she was held by six improvised anchors with lines allowing her to rise and fall with the tide. I then had the job of replacing the rudder and elevator cables whose end fittings and turnbuckles I had purposely left attached to the control surfaces. I managed to find some lengths of fencing wire, and after making loops at their ends joined them to the fittings, tightening everything up afterwards with the turnbuckles. With those temporary repairs we were able to continue our journey and later get back to Auckland.'

After a 15-minute test flight on Sunday 17 April Bolt took off with the bishop at 2.22 pm and 33 minutes later set the Supermarine down on the Whangaroa Harbour. But stormy weather during the next four days obliged the bishop to fulfil public engagements by boat or car, so depriving him of the opportunity of flying further north than Bell. Still, that mattered little; another time, perhaps. Riding at anchor in a choppy sea, the Supermarine now sprang a leak in her hull, which kept Bolt perpetually busy bailing out the bilges.

On 22 April they began the long flight home. 'Left Whangaroa 7.33,' wrote Bolt in his notebook, 'arrived Russell 8.29. Left 10.3 [10.03] and landed on coast 10.45. Started again 11.5 [11.05], arrived Onerahi 11.40.' Why that landing on the coast?

'North of Whangarei Heads,' he explained afterward, 'we struck engine trouble and had to alight on the sea. One of the cam followers had broken but we were able to keep going on reduced power and get to Onerahi. I then drove into Whangarei where, in a small engineering shop, they lent me some gas bottles and a torch and I welded the broken parts together.'

At 11.6 [11.06] on the 24th they left Onerahi and 67 uneventful minutes later reached Auckland. So ended Bishop Cleary's last aerial voyage. Bolt had great admiration for the kindly old man: 'He was a wonderful passenger who didn't mind lending a hand when things went wrong. And though we sometimes went through some sticky situations together, nothing ever seemed to frighten him.'

If the bishop was calm, others — including Colonel Allen Bell — were positively lyrical. Bell was a staunch advocate of Northland (he coined the phrase 'the winterless north') who flew with Bolt in 1920, and later described his trip in the *Northland Age*: '... have you ever experienced the feeling when the woman you loved said yes? Have you ever drunk deep from a flagon of rare, forty-year-old wine? Have you ever experienced the feeling when you come out of action and rejoined your regiment unscathed? If you have not, then you cannot realise the joyous feeling of ecstasy when you rise a few thousand feet into the air and rush through it at the rate of 70 miles an hour on your first aerial flight.'

Here's an equally enthusiastic account from the *Otago Daily Times* of 1 February 1936 ...

The Winged Express
Adventure of Flying
Height and its Reactions
Safe, Swift and Absorbing Travel
Otago Daily Times, 1 February 1936

If anyone wishes to become a real unwavering cynic, cultivating the unamiable quality of a thorough contempt for his species;

if he wishes to realise and become a convert to the truth of the common-places of the preacher about the utter nothingness of the things of this world, let him go thousands of feet up in the air and look down to Mother Earth. If he wishes to enlarge his views on life, to spring out of his narrow circle of ignorance and prejudice; if he wishes to divest himself for a few short hours of the depressing feeling of adoration which he conceives for himself on the ground, let him just go flying in an aeroplane, far above the house-tops, higher than the misty mountain tops. Seen from such an elevation, the prince's chariot and the huckster's cart, the glossy citizen and the tattered tramp, the noble mansion and the tottering tenement, one's dearest friend and one's bitterest enemy, are all merged in one mass of indistinguishable equality. Not even the roar of the accumulated voice of a city can reach the ears. There is nothing but spectacle, and one finds one's self looking on at it all with the silent contempt of the gods on Mount Olympus. Those black dots that hurry and wriggle through streets that look no wider than the passages of a beehive, what are they? Men with immortal souls, centres of happy households, but seen from the trifling elevation of a couple of thousand feet they sink into the most insignificant beetles that ever crawled on the earth. Nothing can be expected to induce a firmer belief in one's immeasurable superiority over his diminutive fellow-worms than this business of flying.

An Exhilarating Adventure
It is all so simple, so safe, and so exciting. And yet there appears to be a supposition among many people that it is an ordeal. Experience proved otherwise when a Daily Times reporter made the flight from Dunedin to Palmerston North and back within 24 hours this week in one of the luxurious De Havilland express air liners of Union Airways of New Zealand, Ltd. Impression follows impression as swiftly as the imagination can cope with them, and one learns anew how multitudinous are the thoughts that rush through one's small head. The experience of flight is immensely exhilarating, giving a sense of glamorous adventure, even of a vague heroism. One feels himself in a great void that can never be translated into the grim reality of the earth, but merely leaves a feeling of utter remoteness as if the air were a separate world. Sunshine or fog, wind or rain, the air has much to offer.

The Glow of Nature

Fog or cloud that great wall into which the huge winged machine is now heading may be; but what a noble barrier it is! Rising high into the purple heavens in which the imagination may see more golden palaces and thrones and floating forms than ever artist dreamed of in his sleep, and which, when his feeble pencil endeavoured to put them upon canvas with all their beauty, height and breadth and depth degenerated into an earthy imitation of natural grandeur and immensity. Keep all the masterpieces of Turner — or any of the great colourists — down between the close walls and narrow streets of the city. Do not take them with you into the air to be shamed into insignificance by the glow of Nature.

And now watch the veil of fog and mist which covers the earth below, and under which you feel, if you cannot hear, the murmur and throb of teeming life — see it float away like the flowing skirts of an archangel's robe, revealing churches, bridges, houses, shipping, streets, rivers and men, and then decide whether you would give up even that fog, with all its ever-changing, glowing, Rembrandt-like effects for all the artificial beauty and liveliness which man gathers round him in the valleys of coughs and respirators which he calls his cities. For the love of art and Nature say 'Never.' But to those who do not know the thrill of flying these things cannot be known.

Vagrant Impressions

And as you look downwards on the sea of roofs below there are infinite directions for the mind to take. What a fantastic variety of shapes and sizes the chimney pots have; what differing colours of red, green, slate and plain rust the rooftops display. Imagination steps in and represents to you all the fretting and fuming and worry and care which were endured in the building of these houses and the raising of these chimney pots. And the puppet men hurry to and from among the puppet houses and shops. You feel like one who sees a great battle from afar off — sees a puff of smoke and the closing together of two lines of men and watches with the calm unruffled serenity of an Egyptian sphinx because just tiny little things mean nothing to you. If the black dots in the deep distant streets were to hustle and fight and destroy each other, like the animaliculae [sic] in a drop of water, you would probably laugh at them, as you would laugh at the insect battle when revealed to you by the powers of the microscope. Go flying and see for yourself, and learn how completely and utterly man lies under the curse of excessive smallness compared with Nature.

The Speed of Wings

Still the reactions to flying are by no means everything. There are many other aspects of the most modern means of travel and all are emphasised by the service which Union Airways, Ltd., has just established between the North and South Islands. The speed at which everything is done impresses the passenger first of all. The north-bound liner leaves Dunedin each day at about the same time as the Christchurch express. While the biggest locomotive in the railway service is still puffing laboriously along somewhere south of Timaru, the air express is disembarking its passengers at Palmerston North, hundreds of miles away. Land and sea, mountains and rivers, are all the same to the aeroplane, which takes them all in its stride. There are no corners to negotiate, no gradients to climb, no bridges to cross. The great machine is moving at 80 miles an hour before it leaves the ground, but once in the air it shows that such a rate of progress is actually very slow. For league after league the great red and white and silver birds of the Union fleet cruise along at a steady 140 miles an hour, and if an extra strong head wind or any other impediment be encountered there is a further 30 miles an hour in reserve to be called upon whenever minutes have to be made up. No waiting, no delay, starting to the minute and arriving on the tick of time, the aeroplane represents the acme of swift travel.

Variety of Route

Then there is another great advantage which flying has over all other modes of transport. The motor car must follow the same road always, the train must keep to its rails, and the steamer has its regular course. Not so the aeroplane, which may, as was the case in the flight which occasioned these reflections, change its actual course every time it goes from one point to another. Northward bound Oamaru, Timaru and Ashburton may lay right beneath the wings; southward bound they may be just glimpsed and no more from a distance of 25 miles out to sea. And always there is something below to catch the eye, even if it is only thickly clustering clouds or driving mist and fog which whisk past the windows of the cabin like wraiths, breaking here to show the dull green of water beneath or the more solid hues of farmland and hillside. Even the steamer, miles and miles out of sight of land, cannot convey the idea of the immensity of the ocean which is had by flying over it, and the railway train has never been built yet that can emphasise the enormous grandeur of the inanimate — scarred mountain, reeded marsh, wide plain, or shadowy forest — in the way that an aeroplane can.

Safety and Comfort

Many people still think that to travel through the air at several thousand feet demands special attire, careful wrapping up, and a host of safeguards against cold. The idea is entirely erroneous. The modern air-liner asks no favours. The same wardrobe that suits motoring or travelling by rail may be used in the air. The enclosed cabin of the De Havilland express liner is warmed and automatically ventilated so effectively that the keen, rare atmosphere 10,000 feet up in the blue vault of heaven, above the Kaikoura mountains, is powerless to cause one shiver. Nor does this virtual sealing up of the passenger against the elements have the effect of cutting him off from the wide world outside. One gets the feeling of being of the outside scene although actually cut off from it. The scent of the sea beneath, the forest's cool secrets, the smell of sweet pastures and the odour of lowing herds seem almost as if they were inside the machine.

Of the safety and comfort of travel in the De Havilland air-liners of the Union Airways Company there can be no doubt. Height in an aeroplane gives no sensation of dizziness, as there is no connection with the ground as there is in the case of looking downwards from the top of a high building. The liner has four 200 horse-power motors on any two of which it can continue to fly if any of them break down or cease to function. The pilot is in constant two-way radio communication with the ground, and is therefore able to keep himself posted concerning weather conditions and possible changes en route. One has only to go into the cockpit of the big machine and watch the pilot at work to see how completely under control the great engines are. Every modern device for the efficient running and control of tremendous power that is generated is installed on the elaborate switchboard. Compasses, meters, gyroscopes, gauges, and speedometers all combine to make safe and comfortable flying. Lifebelts are carried under every one of the luxurious chairs in which the passengers recline, and these can be donned at a second's notice.

A Bird's Eye View

No better survey of the primary industries which the fertile soil of the Dominion supports could be made than that which is presented by a flight through four provinces. The alluvial flats of the Taieri Plain, basking in sunshine or seen through mist, seem to breathe prosperity and productiveness. The poorer and lighter hilly country immediately to the north of Dunedin is less sightly, but the rolling downs of Waihemo and the southern part of North Otago fill the eye completely with their mixture of arable and pastoral farming. There is something inspiring about a 50-acre wheat crop in stook, as so many scores of them are at the present time in both Otago and Canterbury, and it is more than interesting to watch the dust and the flutter all round the threshing plants that can be seen a dozen at a time on the Canterbury Plains. The harvest is at the full in the north and as far as the eye can see there is a patchwork of green and gold from the sea to the distant mountains. The russet gold of stubble and stook contrasts picturesquely with the green of pasture and crop. Then the scene changes again, and instead of farmland and down there is nothing but steep mountain sides, through which rivers wind in and out their waters glinting in the sunshine and giving life to trees and sheltered glades and overhanging shades, but mostly there are only rough rock faces, innocent of vegetation, scored by landslides and water courses, a veritable wilderness. Then come the lowlands of Marlborough, with Blenheim looking very forlorn and not very big in the middle of them. After Blenheim, Cook Strait and the Sounds, seen from a distance it is true, but a sight to remember and looked for again. Neither island is ever out of sight, and just as the headlands of the South begin to get a little blurred, it is possible to determine the topography of the western coast of the North Island as the plane, still flying over the water, passes Cape Terawhiti and moves up past Titahi Bay, Plimmerton, Paekakariki, Otaki, and Levin, passing directly over Kapiti Island to reach Foxton, where it turns eastward to Palmerston North, over the swamplands of the Manawatu River country, where acres of drying flax can be seen spread out like raiment in the sun.

And as the aeroplane draws up in front of the hangar at Milson

Aerodrome with a triumphal sigh one realises that half the country has been traversed in a few hours.

A Great Organisation

Simple as it all seems, there is something admirable about the organisation that brings it all about. The passenger is inclined to take everything for granted and to forget that behind this speedy transport there is a vast deal of work which keeps the service alive. The pilots have their job, and that they do it well is shown by the fact that they fly to the minute everyday. Their chief concern is the ship and its timely arrival at its destination, but that does not mean that they have no time to worry about the comfort of their passengers. Everything possible is done for the welfare of those who use the service, and no courtesy is withheld. But the passengers have to be collected and taken to the airport and at the other end transported to their destination. All this is done by a tender taxi service in each centre on the route, so that nobody has any excuse for missing his plane or being late. The ground staff of the company is always on hand on the arrival of each machine, and until it takes off again there is a scene of orderly bustle everywhere until refuelling and the dozen and one other necessities have been attended. Passengers and their baggage have to be weighed, the load properly distributed, weather reports received and studied, and the machine looked over. And yet it is all done in 10 minutes.

In 1930 New Zealand surgeon R. Campbell Begg was also lured aloft during a visit to England.

The Secret of the Knife
R. Campbell Begg

On the way up to London a few days later, I was browsing through the morning papers when an exciting advertisement caught my eye. It announced: 'Zeppelin Cruises'. The *Graf Zeppelin*, which had just come across, was leaving Cardington for Friedrichshafen on 26 April taking passengers. This was tempting — I had never been in the air before and here was a good chance to begin. When she flew out of England I was on board.

I glanced away for an instant, and when I looked out the window again, the men below had become midgets. The great airship hung motionless, 600 feet above the ground. The engines in the gondolas started — one, two, three. There was trouble with the fourth, but after some delay it, too, roared into life.

We scudded along over the English countryside, gay and beautiful with all the green of early spring. In the cabin, slung under the nose, it was very quiet. Behind, the din must have been deafening; I judged this by the behaviour of the animals below. Poultry scurried into their sheds, horses reared and cattle ran aimlessly about in terror.

Presently the winding Thames appeared; beside it, the Tower showed up like an architect's blue-print. Traffic had almost stopped, hundreds of white patches indicated faces staring up. Londoners had seen German airships before, but never one that had come in peace. History was being made.

It hadn't been easy to get on. The agent of the Hamburg–American Line needed a lot of persuasion. The advertisement was just for publicity. It was really propaganda. Only twenty passengers were carried and all the places had been already taken, for the most part, by the official guests of Dr. Eckener. I spoke airily of the New Zealand Prime Minister, and his interest in airships as a possible solution for the isolation of his country. There might be orders. No one would have been more astonished than the Hon. W. Massey to know of these views attributed to him. But it did the trick with Neuerburg who was handling the bookings. He promised to get in touch with Cardington and see what could be done. The same evening he rang to say that one of the crew was being left behind and my passage was confirmed.

The next day, I boarded a bus, gaily bedizened with bunting, and carrying a huge pennant with the name *Graf Zeppelin*. As we neared Cardington, the crowds became denser and denser and there was difficulty in getting through.

At five o'clock a tiny speck appeared over the wooded ridges to the south. It increased rapidly in size, and soon the great gas-bag loomed overhead — the whole 250 yards length of her — aluminium duco shining and glittering in the rays of the westering sun.

She put her nose into wind and dived at a steep angle towards the field, discharging, as she did so, half a ton of water-ballast, which drenched the soldiers waiting below to seize the hawsers as they came down. She always did this, like a badly trained cow entering her byre. Her German landing crews had become wary, staying out of range till she had relieved herself!

It was Ernst Lehmann who had brought her across, but Eckener himself took over for the homeward journey. The passengers were stowed on board, one at a time, and put in their places as carefully as eggs into a basket. Then the door was closed and we were off.

Three times we circled London, then glided down the Thames over Sheerness and the Medway. No one seemed to worry about this foreign review of our naval defences! Dover Castle on its headland was the last English landmark as we went out to sea. The setting sun, large and red, glinted on the miniature steamers cruising up and down the Channel, each dragging the white plume of its wake.

The dinner-gong sounded. In the spacious saloon were two tables — Dr. Eckener presided over one; Captain Flemming, second-in-command, over the other — just like in a liner at sea. There was a modest three-course menu but the two-page wine-list was impressive. The passengers were too excited to eat much — keen to get to the sloped windows and see what was passing below.

The sun had set behind the English coast; there was just enough light to see the historic Zeebrugge mole and the Bruges Canal, scenes of the heroic British action of the First World War. After leaving Brussels we dropped mail, specially stamped, by parachute. I turned in late and slept soundly under the feather-light Jaeger blankets.

At four o'clock I looked out — the first green of dawn was in the east. Banks of mist filled the valleys. At five, we passed over the airfield at Friedrichshafen. We were two hours ahead of schedule; there was no one about. This was lucky — we got an extra jaunt. The *Graf* put up her nose, soared to 8000 feet and glided along the glaciers and crags of the Alps around Lake Constance.

When we returned in a couple of hours, there was still some fog, and we felt our way in with the echo-gun banging away at intervals. A rift appeared and we dived through. Ropes went spinning down. The landing crew — 200 of them — pulled us in till we swivelled at rest on the bustle-like keel under the bow.

Many of the passengers were leading figures in the British world of aviation — Sefton Brancker, Lord Sempill, and V. C. Richmond, the designer of R101, among them. I was rather an anomaly in that distinguished company; but Dr. Eckener invited me to the dinner which he was giving that night to his official guests.

A motor launch took the party out on to the lake to see the new giant Dornier flying boat. She had lifted 120 passengers into the air the day before. The craft, as she floated at anchor, looked gigantic: the wing-spread was 156 feet; length, 130 feet, and height 33 feet. Perched in a transverse row across the top of the fuselage were six pairs of engines — six propellers in front and six behind. She was a remarkable prototype for that period, but the air resistance must have been colossal. Her original engines were afterwards replaced by Rolls-Royce ones in an endeavour to give her more power. She was scrapped, however, after one precarious journey to South America.

The dinner was interesting. I was roped in as interpreter. Hugo Eckener, a striking figure with pouched eyes and small goatee beard, was supported by Lehmann, stocky, alert and an intellectual. Other famous zeppelin men there were Flemming — burly, sailor-like and hearty — and von Schiller with a record of thirteen raids over England.

The ladies were also present: my place was between Frau Flemming and Lehmann. The former spoke chiefly of the iniquity of the Weimar Republic and the lice her daughter got in her hair at the common school. The latter, who spoke idiomatic English, told of his many adventures in wartime zeppelins, including the story of a man in an observation basket who couldn't be pulled up again and had to be jettisoned.

Eckener spoke enthusiastically of the future of airships for long-distance travel over oceans. I asked him about the possibility of bringing the *Graf* to New Zealand. It was late in the evening. The wine had been good and plentiful, we were all happy and optimistic. As I said good-bye and thanked him for the evening, he responded cordially:

'Aufwiedersehen, Herr Begg, in Neu Seeland!'

A pleasant fancy, which was never realised. For six years, the *Graf Zeppelin*, and later the *Hindenburg*, plied the Atlantic. In the end, at Lakehurst, New Jersey, on 6 May 1937, the *Hindenburg*, struck by lightning, burst into flames and perished together with twenty-eight of her passengers and crew. Our ship, the *Graf*, was grounded and dismantled and that was the end of the German dirigibles.

Lehmann died of his burns in the holocaust at New Jersey, Flemming and von Schiller passed out of my ken. I page down through the years to follow Eckener — he survived the Second World War and Hitler's displeasure. He had refused to demean his ship by flaunting Nazi slogans from her. He died as recently as 1954, a sad and disillusioned man. With the triumph of the long-distance planes, he saw his dearly held views on the future of lighter-than-air refuted, and his life's work turned to dust and ashes. He was glad to go.

The Zeppelin spoilt me for surface travel; I had got the 'bug' in a big way. I couldn't fly from Friedrichshafen to Berlin — there was no service. So I crossed the Bodensee (Lake Constance) to Romanshorn on the Swiss side, and took train to Zürich.

The next morning, I drove out to Döbingen airport to catch my first aeroplane. She was a single-engined job with a cruising speed of ninety miles an hour and an open cockpit in front of the cabin. Her engine was water-cooled. I climbed aboard; the pilot went through his 'vital actions', thundered her up to full revs as we paused at the end of the taxi-ing run, and lined up for the take-off. With a deafening clatter she sped down the grassy runway. Slowly the tail came up and we roared and bumped into the air. What a contrast to the quiet lift of the *Graf*!

I looked round the cabin — there were only two other passengers, a farmer and his wife. It was their first flight. There were no crowds waiting at the booking offices in 1930. At first, all went well: we jerked up and down in the turbulence, then cleared the ground radiation and smoothed out into straight and level flight.

I was on the port side, the Swiss couple on the starboard. It was too noisy for talk, but we made little gestures and pointed down excitedly to the scenes far below. Presently we were over Lake Constance at an altitude, I judged, of some 4000 feet. Suddenly the nose turned through an angle of 180 degrees and the plane was clearly on the way back to Zürich.

The door of the cockpit opened and the face of the co-pilot appeared, gloomy and worried, or so I imagined it. He pointed to a red sign that had just flashed on; 'Landung. Bitte anschallen.' (Landing. Fasten seat belts.) We were rapidly losing height. Below, I could see houses, fences, pigsties and cowsheds but no open space. The roar subsided and a disquieting silence filled me with foreboding: I was too ignorant to know that the pilot had only throttled back his engine.

The Swiss couple appealed to me. What was the matter? Why were we going back? Was everything all right? 'Yes. Certainly. *Jawohl*.' I tried to keep the tremor out of my voice; I was scared stiff. I felt sure that my first aeroplane flight would be my last. We were getting quite close to the ground now. Still nothing but buildings and walls — no hope here. I craned my neck to look ahead for a clear patch, but could see none. We had had it!

The wheels were almost touching the tree-tops. The *frau* screamed and squirmed within her belt. I braced myself for the end. Bathos! The plane flattened out, touched down and taxied to a standstill; we had sneaked back into the airfield after all. The two pilots hadn't bothered to tell us what they were at. Whew! what a relief! But it had been a bad quarter of an hour.

The captain was still quite casual. There had been a leak in the water-cooling system, somebody hadn't tightened up something enough. I couldn't follow his German technicalities. I gathered that he was afraid that the over-heated engine would seize, and had decided to come back.

There was nothing to worry about! Everything would be fixed up in half an hour. All very well for him — but I *was* worried. I didn't feel a bit like entrusting myself to that contraption again. But the canny Scottish blood ran strongly in my veins. I had paid my money; I was expected in Berlin that night — so I climbed in. Now, I was the only passenger; the farmer and his wife had had enough for the day.

Not a hitch! At seven in the evening we swept past the high buildings to a perfect landing on Templehof airfield. I glowed with euphoria: I had sampled both airship and aeroplane and here I was. Fit as ever and all in one piece!

In the years after World War One, officialdom may not have been keen to promote aviation but, as time went on, it did what officialdom always does — introduced rules, regulations and procedures. A Southland pilot, Arthur Bradshaw, was involved in one such procedure, the first of its kind in New Zealand.

Flying by Bradshaw: Memoirs of a Pioneer Pilot 1933–1975

Arthur Bradshaw

It was about 2.30 pm on 30 December 1936 when I took off with four passengers from Invercargill in Fox Moth ZK-ADC, which I had hired from the Southland Aero Club. Our destination was the Franz Josef Glacier, making a brief stop en route at Big Bay, which is about 30 miles north of Milford Sound. One passenger had arranged to disembark at Big Bay and join Davy Gunn's tramping party. Another intended to take her place and continue with us to the Franz, where we were booked to spend the night, intending to return to Invercargill the following morning.

The flight was uneventful as far as Big Bay, where we had to make a circuit over the sea prior to landing on the beach. As we were making our final turn, the aircraft became slightly nose-heavy, and suspecting that a passenger had leaned forward, I eased the control column back slightly. Without any further warning the aircraft stalled and commenced a spin to the left. As we had little altitude, I realised that a crash was inevitable. I corrected the spin, applied full power, dived steeply, and at the last moment, pulled the control column hard back. This flattened the attitude of the aircraft so that it squashed on to the sea rather than hit nose first.

We landed in about four feet of heavy surf. Apparently I was thrown out on impact, and found Sister Buckingham to be in the water alongside me. Although she weighed more than 12 stone, I somehow managed to get her ashore. Later it was found that her injuries were fractures of the left forearm, femur and pelvis, a severe scalp wound, and a fractured skull. Davy Gunn had arrived on the scene by this time. He took Sister Buckingham from me, and I returned to the wreck, and in turn brought Bill Hunt and George Ross ashore. Bill's injuries were later found to be a fractured vertebra and rib, and some chest trouble. George had a fracture of the left thigh, and a fractured sternum.

Once more I returned to the wreck, and after a good deal of trouble I managed to locate the last passenger, Walter Sutton Jones, and success-fully got him ashore. His breathing had ceased, so I applied artificial respiration which I continued for over half an hour without any success.

It was later determined that he had died instantly from a blow on the left temple when we crashed. At that time my injuries were described as a severely cut head and hands.

With the help of Davy Gunn and two nurses, Miss Mehaffey and Miss Robbie, who were members of Davy's tramping party, the injured were transferred about half a mile to Davy's hut. There, these nurses did a magnificent first aid job on the injured with very meagre equipment.

At about 7 p.m. Davy Gunn, who had already done a day's work, set out on a marathon journey to get help. He walked about four miles to Lake McKerrow at Martins Bay, rowed a boat about thirteen miles up the lake, and then walked the steep, 30 mile track over the Main Divide to Marion Camp which is near the head of the Hollyford Valley. Davy covered this distance, which would normally take about three days, in a little more than 20 hours, some of it in darkness. This amazing, and truly commendable feat was later recognised and commemorated by the erection of an inscribed stone at Marion Camp. At a later date, Davy received a 1937 George VI Coronation Medal.

From Marion Camp the accident was reported to Invercargill. Ernie Clarke had been worrying about our failure to return and was about to commence a search in our Puss Moth ZK-ACX. The hotel at Franz Josef should have reported our failure to arrive as we had booked in to stay the night there, and they knew that we were flying from Invercargill.

The next day, Ernie in ZK-ACX, and an Aero Club Moth carrying a doctor, landed at Big Bay at about 6 pm, and the injured passengers were treated before being flown to Invercargill. On the following day, Ernie made two return flights in ACX, taking out the Doctor, Nurse Robbie, Walter Jones' body, and me.

I was taken to Invercargill Hospital and X-rayed, and it was found that I had a fractured spine. The Doctors said that it was a miracle that I was not paralysed through doing so much lifting after the crash. I remained in hospital for a few days during which time I was encased in a plaster cast from my shoulders to my thighs. The first attempt at a cast proved to be unsuccessful, but the second was satisfactory. I had a suit altered to fit my changed dimensions, and no-one, unless they had known me previously, ever guessed that I was encased in plaster.

There had never been a Board of Inquiry into an aircraft accident in

New Zealand before, but the Minister of Defence (Fred Jones) ruled that there would be one in this instance....

It was 23 June 1937 before it sat. Magistrate H. A. Young of Christchurch was the chairman, and with him were Flight Lieutenants J. M. Buckeridge, senior assistant to the Controller of Civil Aviation, and H. B. Burrell from the Air Force at Wigram. The Minister of Defence directed that the inquiry be held in camera, but the chairman, my counsel, and the counsel for the Southland Aero Club, all held that as the evidence given at the inquest into the death of Walter Jones had been made public, then the balance of the evidence should also be made public. They deliberated for hours, but it was finally decided that they could not oppose the directions of the Minister. The Member of Parliament for Awarua (J. Hargest) even brought this matter up in Parliament, but the Minister would not change his decision, although at a later date, he did admit publicly that he had been wrong. Mr Ron Bannerman of Gore, an air ace of World War One, and Mr Brian Hewat of Invercargill defended me while Mr H. J. Macalister appeared for the Crown.

The main theme of the prosecution was that I had flown the aircraft too slowly, thus causing it to stall. This was correct, but I held that, according to the airspeed indicator, I was indicating 75 mph when the stall occurred. I claimed that the airspeed indicator had been faulty, and had given me a falsely high reading. The prosecution claimed that a stall would occur at 70 mph, but I produced evidence that a plate screwed in the cockpit of the aircraft stated that a stall would occur at 50 mph when fully loaded. The inquiry would not accept this as evidence, so I tele-graphed Captain Mercer of Air Travel in Hokitika, and asked him to carry out a stalling test in one of his Fox Moths, but my request was ignored. I then cabled the manufacturers, De Havillands, in England, and they replied that a stall would occur at 50 mph when fully loaded. The Board would not accept this as evidence either, and then it commenced to dawn on everyone why this Board of Inquiry had been held in camera. There was a lengthy technical discussion on the design of airspeed indicators, and R. C. Kean, who in addition to being an aircraft inspector was the inspector of accidents, was called to give technical evidence. His expertise became very questionable when cross-examined by Mr Bannerman for his knowledge of airspeed indicators appeared to be sadly lacking.

When the inquiry was over, to counter the holding of it in camera, I gave a statement to the Press, telling of the plate in the cockpit of the Fox Moth, and of the cable to, and the reply from De Havillands. I also related other matters which came up in the course of the inquiry, and this publicity possibly eased public opinion.

After the hearing had been completed, I thought that then, surely, there would be no further delays, but it was 3 August before the Minister released the findings of the Board. I learned later that my dear mother rang Magistrate Young, and requested him to hurry the finding as I wished to get back to work. Apparently she received a very curt reply telling her that he had plenty of more important matters to attend to.

The Board found me guilty of negligence, but my licence was to be returned. I was not fined, and costs were not awarded against me. A rider was added 'Although seriously injured, and suffering much pain, he by almost superhuman effort, brought three passengers ashore, and so probably saved their lives.'

But my troubles were not over by any means. Mr Kean, who had already tried to denigrate me in every possible way, then moved to prosecute me for various breaches of regulations resulting from the accident. The main charge was that we were a few pounds overweight on takeoff from Invercargill, but this was irrelevant as we would have consumed approximately 100 lbs of fuel on the flight to Big Bay. Another charge was for failing to notify the Department of the accident within 24 hours of the happening, but as the Southland Aero Club were the owners of the aircraft, this charge was altered, and was made against them. The outcome of these hearings was that again I was not fined, and costs were not awarded against me. The Magistrate, during this case, made the commendation: 'I gather that the courage shown by the defendant was remarkable, and was characteristic of British pioneering. It was a degree of courage very few would have shown.'

My injuries had mended, and I became very impatient that my licence had not been reissued, and so Mr Bannerman arranged a meeting in Dunedin with the Prime Minister (Michael Savage). He was very sympathetic, and promised prompt finality regarding the reissue of my licence.

About this time our Puss Moth ZK-ACX had become due for the renewal of its Certificate of Airworthiness. The Department had inspected the aircraft, and had issued a defect report. We had completed the work required to be done, and had notified the Department accordingly, but they would not send an inspector to Invercargill to make the final inspection. Mr Kean was in Invercargill about this time, and he was asked to check ACX, but he refused, saying that he was now the Chief Inspector, and no longer did such work.

One month had passed since our meeting with the Prime Minister, but no word had been received from the Air Department. I became frantic and sent the Prime Minister a telegram: 'You promised me a decision re my licence STOP what are you going to do about it STOP I am desperate Bradshaw.'

The following day I received three telegrams. From the Prime Minister: 'The Honourable Minister of Defence will communicate with you today Savage'. From the Minister of Defence: 'Your licence has been re-issued, and will be posted to you today Jones Minister of Defence', and from the Air Department: 'Inspector will depart Wellington by air today to inspect your aircraft ZK-ACX Civil Aviation'. So, late in 1937, thanks to the Prime Minister, I was back in business with my reputation to regain, and my debts to pay.

With aircraft reliability (and range) improving, the 1920s and 30s became the golden age of record attempts. Few people have heard of Aline Barton (later Aline Quin) but she was New Zealand's first female trans-continental aviator. Having trained as a pilot here, she went to England in 1930 and bought her own Gypsy Moth biplane.

In 1931, she decided to fly to Kenya, taking with her a passenger who'd worked there and wanted to get back. They had their share of adventures en route, including an overly enthusiastic welcome near Sofia in Bulgaria. 'We landed in a valley about 1000 feet up among the mountains. It was a bit rough as herds of pigs had rooted all over the place. In a magical way locals came buzzing like flies and whilst my passenger went off in search of petrol I stayed to look after the plane. Hard work to keep the wings from being broken and torn to ribbons by the curious people, who even banged them with sticks to see what they were made of.'

Eventually Aline reached Nairobi and sold her aeroplane to a man who later crashed it. His wife was killed in the accident and he shot himself.

Equally unfortunate were Lieutenant John Moncrieff and Captain George

Hood. On 10 January 1928 they left Sydney bound for Trentham, near Wellington, in an American Ryan monoplane. They never arrived, and the crowds that had gathered to greet them, including their wives, slipped sadly away.

Aviators knew the Tasman was a cruel sea, but eight months after Moncrieff and Hood vanished, it was finally flown by the *Southern Cross*, a three-engined Fokker monoplane piloted by Charles Kingsford Smith ('Smithy'). Flying with him were two other Australians, co-pilot Charles Ulm and navigator Harry Litchfield. The fourth member of the crew was New Zealander Tom McWilliams, a radio operator selected by the Air Branch of the New Zealand Defence Department.

On 10 September 1928 the *Southern Cross* left Richmond air base near Sydney for Christchurch. To begin with, the flight went to plan, until the Tasman threw a freezing storm at them. To their dismay ice began to form on the cowling of the central engine, the controls became sluggish and the altimeter fell.

Smithy
John Stannage

Smithy could not see much of his wings because of the black immensity about him; but ahead, dimly, he detected the ice formation on the cowling of the centre motor. That meant that it would be forming everywhere. Soon he felt the inevitable sluggishness in the controls, and he watched with a sense of increasing dread the gradual fall of the altimeter.

In the cabin Litchfield and McWilliams were faring badly. The pilots were too busy to bother with messages and the two men, well aware of the increasing hazards, had no means of learning the real seriousness of their plight. It was freezing now. The swinging compass-needles gave Litch little chance of making anything but the sketchiest of dead-reckoning navigation, and the roar of static from lightning discharges blacked out any possibility of McWilliams using his radio equipment.

Smithy, concentrating on the faintly glowing dials of the blind-flying instruments before him, dropped to a lower altitude where he hoped he would avoid the ice. Suddenly the air-speed indicator plummeted from its steady 95 miles an hour to zero! Trained to place implicit faith in his instruments, he immediately pushed forward on the controls to counteract the implied stall. The machine hurtled downwards to gain flying speed. But the air-speed indicator remained at zero. In the few seconds it took him to realize that the pitot-head was blocked with ice the *Cross* had fallen from 8,000 to 2,500 feet.

Quickly Smithy brought her back to an even keel. Disregarding air speed, he kept her roughly at 2,500 feet by watching the engine revolution counters. If the revolutions crept up the machine was diving. If they decreased she was climbing. It was a crude method of flying blind; but it was effective until the ice melted and freed the pitot-head, which led the air to the air-speed indicator.

At the lower altitude the storm still raged. Lightning flashed and flared in a fiery confusion against the night. The wind roared in growing fury. But to Smithy's intense relief the machine had ceased to ice-up. The plane gradually freed herself from the burden of her unwanted freight, and soon the air-speed indicator read truly again; but the bumps were worse than ever. The pilots were strapped in their seats; but Litchfield and McWilliams in the cabin were tossed about like boots in a sack whenever circumstances forced them to release their grip on the cross-members of the fuselage.

It was the worst storm that Smithy had encountered in the whole of his flying career, and he began to wonder if there was any hope of relief. Something akin to fear laid a heavy hand upon him. They had escaped death by the narrowest of margins. What next? If the bank-and-turn indicator quit there would be no hope. They would dive precipitately into the cauldron beneath them. The whole machine was charged with static electricity: a flash and the three propellers were enshrouded in an eerie glow, blue, malevolent and full of hidden menace; another flash, and the glow was gone. A small spark, and that 500 gallons of petrol would send them sky-rocketing in a chaos of flying debris. The motors, too, were beginning to vibrate alarmingly. He had given no thought to the engines until now. But why that vibration?

It was found subsequently that the three micarta propellers had been so badly chipped by ice that they were unbalanced. Before the return flight could be made three new airscrews were fitted; but now in the deadly grip of the storm Smithy could only guess and wonder, in sickening alarm, how much longer the motors could carry on.

Charlie Ulm handed him a note which he read with difficulty. 'Lightning put both radio sets out of action.' A flare of lightning had struck the trailing aerial, burnt out both aerial ammeters and rendered both transmitters useless. McWilliams immediately dismantled the sets to remedy the defects. With the parts strewn wildly about him, he was thrown violently across the cabin by a bump which was by far the worst they had yet experienced. Some of his equipment went sliding down into the tail and was irretrievably lost. But he consoled himself with the thought that the radio had been little use so far anyway, so it was little loss!

In the cockpit Smithy and Ulm were fascinated by the static discharges which continued to light the leading edges of the wings, the propellers and the protruding points of the *Cross*. Smithy had read of such static haloes; but he had never imagined the phenomenon could be so awesome in its eerie brilliance. He speculated on the cause of it. Every metal part of the aircraft was carefully bonded together with earth wire. This was to prevent the possibility of a discharge spark between two unequally charged metal frames causing a fuel explosion. It was probable, therefore, that in a storm of this intensity the whole aircraft would suddenly absorb a very high electrical charge from the lightning-laced clouds, and the corona of light indicated that the charge was being loaded upon the surrounding air.

That, in fact, is what actually happens in such circumstances. It is possible for a machine to fly through atmosphere that is highly charged and accumulate thousands of volts, then fly into a layer of air carrying a charge of opposite polarity. Naturally there is an immediate equalization of the electrical disequilibrium by a discharge into the surrounding air. Discharge points are always the sharpest protruding parts of the craft. This static discharge caused considerable interference with radio communication until it was overcome by the trailing of a piece of wire from the extreme end of the tail.

At three o'clock the storm began to ease, and Smithy was able to climb again to smooth flying conditions at 6000 feet. The crew knew now that they were safe, and notes were passed to Smithy from the cabin to congratulate him on bringing the machine through an incredible ordeal. At 5 a.m. the skyline became visible in the first glimmer of dawn.

But even now ice still clung to some parts of the *Cross*, and the vibration of the motors continued. Smithy thankfully handed the controls to Charlie Ulm, and sat back with a sigh of relief as he reached for the thermos flask of cocoa.

A note came through from Litchfield. 'Watch clouds on starboard bow.' Four pairs of eyes swung as one to the far distant gleam of the snow-covered peaks lifting from a sea of rose-tinted cloud. The Southern Alps!

New Zealand appeared to be entirely covered with cloud masses; but presently, through a thinning bank, they glimpsed the sea and a long finger of land. Smithy lowered her slowly down and headed her towards Wellington, the capital city of the Dominion. The long, awful night was behind them. Their spirits rose in exultant gratitude, and Smithy promised himself that never again would he cross an ocean by night if there was a reasonable alternative.

Circling Wellington at this early hour, the *Southern Cross* was an amazing sight for the many thousands of people who were, by now, anxious for her safety on account of the lack of radio signals. She was the first multi-motored aircraft they had ever seen. In the morning sunlight she looked even larger than she really was, as she roared triumphantly on her way to Wigram aerodrome near the South Island city of Christchurch.

As most of us know, the other great Tasman flyer was Jean Batten, who now enjoys the posthumous honour of having a food court named after her at Auckland International Airport.

In the 1930s Jean Batten's solo flights stirred and thrilled human imagination and added glory to her sex — or so said New Zealand's Governor General, Lord Bledisloe, adding 'she has given a much-needed impetus to civil aviation and done honour alike to her country and to the British Empire'.

Contrary to popular belief, she wasn't the first New Zealander to fly the Tasman solo. That distinction belongs to Pat O'Hara, who flew from Sydney to Mangere on 18 October 1935. Alas, because his plane's certificate of airworthiness had been withdrawn he got into hot water after the flight and didn't try again.

Batten's flight across the Tasman was one of her last epic trips. She'd previously flown solo from England to Australia and back, and also taken her little Percival Gull across the South Atlantic to Brazil.

My Life

Jean Batten

A dreadful feeling of loneliness almost overwhelmed me as I left the African coast and steered the aeroplane out into the blackness of the Atlantic on a course for Brazil, nearly two thousand miles away. To the north I could see the blurred gleam of the lighthouse at Dakar sending its friendly beam out into the night. I switched off the navigation lights, for the lighted cabin seemed to make the darkness outside more intense as I peered vainly through the windows trying to distinguish the horizon. 'It must get light soon,' I thought, glancing at the clock, to realize that it was only twenty minutes since I had left Thies. It was no use looking at the clock, I decided: it only seemed to make the time pass more slowly. What was that? Did I imagine it, or were there really lights in that black void below? Yes, it was a ship. 'Looks like a steam-yacht,' I thought, glancing down at the small lighted vessel. I flashed the navigation lights on and off a few times, hoping for an answering signal, but there was none. They seemed very close to the coast, and must have been making for Dakar, and I envied the sleeping passengers on board.

The rain continued to fall, and thick clouds at a thousand feet forced me to fly low, so that all my concentration was focused on keeping the machine level and straight on its course while I patiently waited for the dawn to break.

Gradually and almost imperceptibly a grey light stole into the cabin, and I began to distinguish the white tops of the waves beneath. The rain had ceased, and as the light became stronger the scene which unfolded itself before my eyes in that grey dawn was one of majestic beauty. The sky was completely overcast, and in every direction there stretched the vast blue expanse of the mighty Atlantic. Not a ship in sight, nor any sign of life whatever. My only company was the roar of the engine as the aeroplane winged its way low over the ocean like a solitary bird. So completely isolated did I feel that to all intents and purposes I might have been the only person in the world.

The clouds began to break up, and soon the sun shone down from a clear blue sky with such fierceness that the heat in the cabin became quite suffocating. Visibility was good, for the horizon was a deep blue against the paler blue of the sky, and by the white-capped waves I saw that a north-east wind was blowing. Not that I should benefit very much from it because of the low altitude at which I was flying, and I knew it would drop completely as I neared the doldrums region.

The engine purred faithfully on, and with everything going so smoothly I was able to relax and eat some of the sandwiches that the Commandant had given me. The weather was too good to last, however, and a hundred and fifty miles out from Africa the sky became flecked with clouds and gradually completely overcast. Great black, ominous clouds were banking up ahead, and in the distance the leaden sky seemed to merge with the grey sea. The wind had by this time dropped altogether, for the sea was calm, looking almost as if oil had been poured on the angry waves, and all was still except for a long swell which gently rose and fell.

My position was 200 miles north-east of the equator, and I was entering the doldrums, or region of calms which had always been such a nightmare to mariners in the days of sailing-ships. As a child I had listened spellbound and thrilled to the tales of how my own grand-parents had set sail for New Zealand. Their ship, after being becalmed in the doldrums for weeks and encountering a fierce storm, had run aground, nearly being wrecked on a reef off the coast of Brazil. The ship had been floated off, however, and undaunted they had continued their voyage. During a terrific storm off Cape Horn, when giant seas had swept the decks of the little ship and the sails had been torn to shreds, the captain had died. After further adventures the ship had eventually arrived in New Zealand six months out from England.

The heavy rain-clouds ahead seemed to open and pour their contents down with such force that the rain resembled a great black curtain. The Pot au Noir.... Could I go round the storms? I wondered how far south they extended. No, that was out of the question. Altering course without radio to check one's position would only result in being lost in mid-Atlantic. Could I fly above that dark mass directly in my path? I remembered the weather forecast: '*Intérêt à voler bas.*' Either go back now while there is time or go through it.

Before I could think any more about it I had plunged into the pouring rain. Flying so low that at times I must have been less than fifty feet above the surface I tried vainly to keep the sea directly beneath in view, but suddenly lost sight of it altogether. For one terrible moment I thought the aeroplane would plunge into the water before I gave the engine full throttle and pulled the machine up into a climb. If I had to fly completely blind I should do so at a reasonably safe height.

The Gull roared up through the dark mass, until at 1000 feet I put the machine on an even keel and flew on. With both feet braced against the rudder-bar and my hand firmly gripping the control column I concentrated all my attention on the blind-flying instruments and the compass. Relaxing my grip on the control column every now and then so that I should not in my anxiety over-correct any slight error in steering, I flew on, unable to see a yard outside the windows, against which thundered the heavy rain, almost as if bent on destruction. Every minute seemed like an hour. Would I never penetrate that dark curtain of rain which seemed drawn round the machine?

Suddenly I saw the compass needle swinging slowly round the dial. 'It must be imagination,' I thought. Drawing my hand across my eyes I felt the tiny beads of perspiration on my forehead as the needle continued its ghastly movement. I was lost.... If I followed the compass now I should go round in a circle. 'It is all up now,' I thought frantically. The compass had swung round about 180 degrees. If only I could see the light instead of this terrible blackness enveloping the machine. I almost prayed to see the sky and sea again. No, I should not give in now: there were still the blind-flying instruments, and the machine was flying a straight course by the bank and turn indicator. 'I must not lose faith now,' I told myself. My eyes were staring at the turn indicator, but I realized that unless the compass righted itself it would not be possible to steer another thousand miles to Natal [a port in Brazil] on that alone. This was torture. The strain was terrific. The perspiration was trickling down into my eyes, and every muscle and nerve in my body was alert.... Were my eyes again deceiving me, for slowly but surely the compass needle was swinging back to its former position? Thank God I was saved, and within a few minutes the darkness outside the cabin gave place to light, and once more I saw the calm sea beneath.

All the muscles that had been taut for so long relaxed, and I sank back in the seat breathing a prayer of thanks. Taking out my handkerchief I mopped my forehead, and throwing open the windows let some air into the stifling cabin. I saw the compass needle steady, and once more thanked God for my preservation. I realized by my clock as I entered up the log that I had crossed the equator during the storm. The sky was still overcast, and my spirits sank as I saw more storms looming ahead. Very soon I plunged once more into a succession of heavy rain-storms, and although they were not so thick through, it was a strain blind-flying for so long. As soon as I would emerge into the light again from the nerve-racking experience of one storm it was to see another ahead. They looked something like huge black mushrooms, seeming to come up from the sea to join the clouds, resembling photographs I had seen of cloudbursts.

When at last I entered a fine zone I felt thoroughly worn out, but after some lunch and a drink of black coffee felt quite refreshed again. My altitude was 600 feet, and I calculated my position as about 1100 miles out from Thies. That meant approximately another 800 miles to the coast of Brazil.

The sun had penetrated the clouds, and was burning down on to the blue sea, which had lost its calm look and was now capped by myriads of white-topped waves. The sea became more turbulent, until at last huge waves left great trails of spray, which the wind caught and carried along like thousands of streaming white pennants. The strong south-easterly wind was now increasing in strength, and by aligning the nose of the aeroplane against the waves I could see that the machine was drifting northward. Even at the low altitude at which the Gull was flying I calculated that the present rate of drift would carry me well off my course. There was not another aerodrome north of Natal for hundreds of miles, and the petrol margin was not great enough to allow for any but the smallest error in navigation. Apart from this there was the record to consider, and any error meant loss of time, for as I was endeavouring to break the record of a multi-engined flying-boat equipped with radio and a crew of experienced men every mile I drifted northward of the course meant precious time wasted.

I spent the next few minutes trying to ascertain accurately the amount of drift, and calculated it at eight degrees to starboard. I decided to alter course eight degrees to port to compensate for the drift. This should take me to Cape San Roque, where I expected to make landfall. Leaning forward I unlocked the compass verge ring and set the machine on its new course. Vainly I searched the horizon for some sign of a ship, but there was no trace of any vessel. Time slipped by, and I felt very lonely, but comforted myself with the thought that after my terrible experience in the storms it was good to see the sun, the sky, and the sea again. 'Nine hours out from Thies,' I wrote in the log, and hopefully thought that if visibility were good I might see the coast of Brazil in under four hours.

Scanning the horizon for the hundredth time I caught sight of a small dark object in the distance. Were my eyes deceiving me or was it really a ship? Yes. As I drew nearer it was possible to distinguish the masts and funnel of a boat. It seemed too good to be true. For almost eleven hours I had been completely isolated from the rest of the world, with no one to talk to, no sign of life. The blue sea everywhere made me long for the sight of other human beings, a ship, or anything to relieve the monotony of the vast blue waste stretched before me. Jungle or desert stretches would be a pleasure to fly over compared with this.

The sea was becoming rough, and huge waves seemed to rise beneath the Gull, as if stretching up in an effort to grasp the machine which flew contemptuously out of reach on its lonely way. The ship was quite near now. It was a cargo vessel, evidently bound for Dakar, and my course lay right along the ship from tip to stern. I was almost breathless with joy, for the ship must have come from Natal, in which case I was absolutely on the right course. 'Unless it is from Pernambuco,' I thought, and a shade of doubt entered my mind, for perhaps the drift was not as strong as I had estimated and eight degrees' compensation was too much to allow. Glancing at my chart I saw that Pernambuco was 160 miles south. No, it was unthinkable that I should be that much off my course. The ship was definitely from Port Natal, I decided. As my altitude was still only 600 feet it was quite easy for me to see the name of the vessel, which I read with such joy and eagerness that it must be stamped on my heart for all time. The name painted on the bows read *Belgique*.

Figures on deck were waving wildly, so taking off my scarf I held it out the window and let it trail in the slipstream, and also dipped the aeroplane in salute over the ship. How I longed to circle, for although the crew must have been excited to see a small silver monoplane winging its way over their ship so far from land, their feelings were not to be compared with mine, so overjoyed did I feel at sighting the vessel. 'Wish I had radio and could ask them what port they are from,' I thought longingly as another doubt assailed me that they might be from Ceará or Maranhão, both hundreds of miles north of Natal. Thrusting the doubt from my mind, I decided not to let anything mar my joy at seeing the ship and at the realization that I was only about three hundred miles from land. Several times I looked back, until the ship was merely a speck in the distance.

Time seemed to drag terribly now, but perhaps soon I should sight Fernando Noronha island. This small volcanic island was shown on my chart as being about twelve miles long, with a cone rising to a height of over a thousand feet. In good weather it should be visible from a great distance, although, looking closely at the chart, I saw that it lay almost fifty miles south of my course and about a hundred and fifty miles from the Brazilian coast.

The sky was growing once more overcast, and I was not going to reach the land without another battle with the elements. For the next two hours I flew through one tropical deluge after another, until I felt terribly disappointed at missing a sight of Fernando Noronha island, and very tired at the continual blind flying, which after twelve hours in the air seemed even more difficult than ever.

Emerging once more into the light after a particularly heavy downpour I saw a faint yellow line on the horizon ahead. Was it really land, I asked myself. Glancing round the skyline I saw a similar line, and realized that the intense glare from the silver engine cowling coupled with the strain of staring at the blind-flying instruments was tiring my eyes.

Twelve and a half hours out from Africa.... Surely I would see land soon. Vainly I searched the horizon for some sign of the coast. Bending down I switched on to the last petrol-tank. Petrol for only one hour more, and still no sign of land.... Even though I was flying so low, surely I should be within sight of land now.

What was that faint yellow line? Surely my eyes were deceiving me again. No, this time it was real. Land ... land ... I shouted aloud for sheer joy. Nearer and nearer the land drew, until it was possible to distinguish the sand-dunes on the lonely coast of Brazil. Very soon I was within gliding distance of the undulating sandy coast, and at last flew over the long line of foamy white Atlantic rollers sweeping up on to the beach. About half a mile to the north I saw a slight promontory ... a sandy stretch covered with coconut-palm trees.... 'Cape San Roque!' I cried, hardly believing my eyes. It seemed too good to be true that after steering for thirteen hours over almost two thousand miles of ocean I had made landfall within half a mile of the point I had been aiming for. But was it Cape San Roque? My chart showed a lighthouse; there was none to be seen here. Silhouetted against the sandy background I saw the wire framework of a red-painted structure which evidently held the fixed light — a strange, lonely-looking edifice, but nevertheless a lighthouse, I decided. Yes, it was Cape San Roque — an exact likeness of the little photograph in my pocket that I had taken from a book. During the last few months I had looked many times at the lonely palm-fringed point depicted in the photograph, and at the last minute had thrust it into my pocket for a mascot. Now that my position on the Brazilian coast was quite definitely fixed I turned southward for Port Natal. 'Only a few minutes now,' I thought, skimming low along the line of sand-dunes as the sun sank lower in the western sky.

Crossing a hilly part of the coast I suddenly came upon an inlet and a white lighthouse, then saw the buildings of a town. 'Port Natal!' It was like a dream to see real houses and civilization, and passing over the town I gave another shout for joy.

'Aerodrome, 15 kilometres S.S.W. Natal, near Lake Parnamiram,' read my notes, as I steered the machine past the outskirts of the town and over the jungle, where I quickly picked up the large clearing in the dark green tropical vegetation. Having circled the aerodrome, I shut off the engine and glided down to land. Immediately the wheels touched the ground I checked the stop-watch, which registered 13 hours 15 minutes, my time for the flight from Thies Aerodrome. It was exactly 7.45 P.M., G.M.T., on November 13, so my total elapsed time from England to Brazil had been 2 days 13 hours 15 minutes. A wave of pleasure overwhelmed me as I realized I had lowered the record from England by a margin of almost a day, and had also crossed the Atlantic in the fastest time in history.

As I climbed out of the cockpit all my tiredness left me, and I was immediately surrounded by an enthusiastic crowd which had been awaiting my arrival. There were a number of the Air France pilots and mechanics, who warmly shook my hand, and I realized they were genuinely pleased.

Their enthusiasm surprised me. It was not until later, when I had met more French people and had come to love France almost as if it were my own country, that I fully realized what wonderful sports the French are.

In their earnest desire for the advancement of aviation they realize that speed means progress and competition prevents stagnation.

On hearing of my terrible experience in the doldrums when I thought my compass had failed me one of the pilots assured me that in the electrical storms peculiar to that region he had known of similar experiences.

The group of people assembled to welcome me included an English-man and his wife, who were overjoyed at my arrival and invited me to stay with them at their home. 'We had not been out here at the aero-drome very long when we heard the roar of an engine, then suddenly saw your silver aeroplane fly over Natal,' said the Englishwoman. 'It was a wonderful sight,' she kept saying. 'To think that a little over sixty-one hours ago you were in England!' and her eyes glistened at the thought of her beloved country.

'We are very proud that it's a British machine,' put in the Englishman as we walked across to the hangar.

When the refuelling was completed we left the aerodrome and drove towards Natal. The car was well sprung, and sinking deep into the com-fortable seat I breathed a sigh of relief. Closing my eyes, I could still hear the roar of the 200-h.p. engine, and it was difficult to realize that the flight was over and I was really in South America, and not still over the ocean. The terrible storms seemed a long way off now. I must have slept for a few minutes, for on opening my eyes I saw that we were driving along a track above which the dark green trees of the jungle towered like a great arch. The road was not good; it was fortunate that the car was so well sprung. At one stage to pass another car we had to mount the bank by the roadside and drive along at an alarming angle.

'What a terrible road! Is it the main one, and do they drive the air line passengers along this to the aerodrome?' I inquired.

'There aren't any regular passengers,' said my companions. 'You see, the transatlantic 'planes don't take passengers — only mail — and the Clipper ships of the Pan-American Airways are flying-boats, and they land down at the port.'

As we were about to enter Natal the car was stopped by an armed guard. My friends were closely questioned. I was very glad that I had left my revolver at Thies, for in all probability it would have been confiscated. It appeared that special precautions were being taken because of recent

134 Words on wings

trouble and the imminent possibility of a revolution. On being assured that the car contained no firearms the soldiers allowed us to drive on into the town. We stopped outside a large house, and traced our way through a garden the beauty of which I did not realize until next morning, when daylight revealed it in all its glory. After a refreshing bath I changed my flying-suit for a frock and joined my friends, who were genuinely surprised at the sudden transformation of the tired aviator.

'If you listen to the radio you may ear the announcement of your flight being broadcast from London,' said my host, looking at his watch. 'It's just about time for the news broadcast,' he added.

Drawing a chair close up to the radio-set I sat down and listened intently. I could still hear the roar of my engine, which had made me practically deaf.

'There it is,' said my friend, and through the roaring far, far away I heard a voice speaking. There was a pause, then quite clearly the voice came through again: 'Miss Jean Batten successfully completed her flight from England to South America by landing at the aerodrome at Port Natal, Brazil, this afternoon. Her total time for the flight from England was 61 hours 15 minutes, and this lowers by almost a whole day the record previously held by Mr James Mollison.' The voice paused again, then continued broadcasting the rest of the news.

I turned to my host. 'It is wonderful to think that within a few hours of the landing the news is being sent out from London.' Until I heard the voice broadcasting the news it had all seemed unreal and more like a dream, but now the realization that the flight was accomplished came to me, and I experienced once again the greatest and most lasting of joys: the joy of achievement.

It was after a subsequent flight from England to Australia that Jean Batten came home, where she was greeted by huge crowds at Mangere on 16 October 1936.

There was a considerable amount of opposition to my plan to continue the flight through to New Zealand. This was not to be wondered at, however, for the sudden violent storms of the Tasman were well known by all Australians and New Zealanders. No one, however, realized more deeply than I the hazards of this seldom flown sea, for I had carefully studied hydrographic charts of the South Pacific and learned of the high gale frequency and the abnormal number of cyclonic disturbances throughout each year. Several times I had crossed the Tasman by steamer, and had vivid memories of storms when I had awakened in the night and listened almost fascinated to the pounding thuds as tremendous waves shook the ship from stem to stern. Great foaming sheets of spray hurled themselves across the decks, sometimes greedily taking hatches and twisting derricks in their diabolical frenzy.

My fast flight to Australia had been acclaimed with enthusiasm by the Press, and hundreds of telegrams and cables of congratulations were arriving each hour. 'Why not rest on your laurels and stay in Sydney?' a friend had suggested when I received an offer of several thousands of pounds to tour Australia right away instead of flying on. This represented a vast fortune to me, and I spent a long time trying to decide for the best.

'It's all very well for you to talk of linking England and New Zealand and all that sort of thing, but what's your reward?' asked one of the men at a discussion about the offer. 'You can fly the Tasman and risk your life. Then what,' he added, 'are you going to benefit, and what will you gain?'

'The honour of completing the first solo flight from England to New Zealand and linking those two countries in direct flight for the first time in history,' I replied quietly....

On arrival at Richmond Aerodrome the Gull was refuelled and wheeled into the spacious hangar in readiness for the early morning take-off. For the first time in my life I had practically no sleep that night. This was not because of any fears for what the morrow would hold for me, but because of an over-zealous sentry. Just as I was dozing off I heard a sound like thunder, and sitting up listened intently. There it was again, and I recognized the measured tread of a sentry on duty in the corridor. *Tramp, tramp, tramp,* went the heavy footsteps, and at the end of the wooden passage they would halt, right about turn, and return. I tried to shut out the noise by pulling the blanket over my ears, but the even tramp was as clear as a clarion. 'I might ask him to take his boots off and put slippers on,' I thought, but the idea of the sentry in stocking feet was too funny, and I wondered if he carried a bayonet or a gun. At midnight I resolved to ask him, at the risk of offending his sense of duty, to go and find another corridor to parade in. Tiptoeing across the room I cautiously opened the door and looked out. The sentry had reached one end of the long corridor, and at the other an open door revealed a large crowd of reporters and photographers smoking and talking. I had no wish to be photographed in my night attire, so I withdrew my head just as the sentry thundered past again.

At half-past two I rose, and after a light breakfast of tea and toast walked over to the hangars with Group Captain Cole, who was in charge of the base and had been most hospitable, and arranged for every possible facility to be placed at my disposal during my stay at Richmond. The Gull was wheeled on to the tarmac, and as the engine warmed up I said good-bye to the group of friends,… who had motored from Sydney to see the take-off … My small kit was placed in the locker, and the thermos and sandwiches were put in the cockpit. As I sat in the cockpit running the engine up I could see dozens of photographers and news-reel camera men silhouetted against the flares.

The previous night at dinner I happened to mention to one of the officers that before leaving England I thought of taking a lifebelt on the Tasman flight, laughingly remarking that a lifebelt in mid-Tasman would be about as effective as a black cat painted on the rudder. Great was my surprise next morning when one of the officers arrived with a life-saving jacket when I was just about to take off.

'I could never get out of the cockpit wearing this,' I said, as it was slipped over my shoulders and inflated.

'You could if it were half deflated,' one of the officers put in.

'All right, then, just to please you all, but I wouldn't have a chance if the engine failed,' I replied, tying the jacket on securely.

Leaning forward I adjusted the compass, and placed the chart, maps, and log in the leather pockets within easy reach, and switched on the navigation lights. The long line of flares was burning brightly along the runway, and lighting a path in the darkness as I taxied out to the start.

'I want to speak to the Group Captain,' I shouted above the roar of the engine, and a reporter quickly wrote down my message as he hurried forward: 'If I go down in the sea no one must fly out to look for me. I have chosen to make this flight, and I am confident I can make it, but I have no wish to imperil the lives of others or cause trouble and expense to my country.

'Well, good-bye!' I shouted, smiling reassuringly at the sea of tense white faces. 'I'll come back one day.'

Releasing the brakes, I gave the engine full throttle, and the Gull roared along the flare path. The bright line of flares flashed past the left wing, and nearing the last one I gently eased the aeroplane off the ground. Swift as an arrow the Gull climbed through the darkness, circling the aerodrome to gain height, then setting off for Sydney. I flashed my torch across the instrument panel: revs., 2100; oil-pressure, 42 lbs.; altitude, 1500 feet; air speed, 140 m.p.h., etc. All was well, and I breathed a sigh of relief that at last I was on my way.

… The myriad lights of Sydney were slipping beneath the wings, and I picked out the line of lights marking the harbour bridge. The Gull speeded on, and, steering a course for New Plymouth, 1330 miles distant, I left the Australian coast. I felt tremendously lonely as I looked back at Sydney and the land grew fainter, until at last it faded into the distance, and I was left alone to fly hour after hour with only the sky and the vast expanse of ocean to look at.…

On leaving Australia I had encountered scattered alto-cumulus cloud and climbed to 5000 feet. Three hours out I glided down through a gap to check my drift on the white-capped waves. As I flew low over the sea I saw a silvery gleam as a shoal of flying-fish leaped from the sea to speed along, then suddenly plunge in again. I passed just south of a big storm, but dark clouds loomed ahead as storms gathered on the horizon. At this stage, just six hours out from Sydney, I noticed that the gauge on the port centre-section tank registered only four gallons. 'A petrol leak,' I thought immediately, as I had not used any petrol from that tank and the engine had been running on the large rear tank. The leak was not serious, however, as the tanks did not interconnect and there was a very large petrol margin. Nevertheless I switched over to the leaking tank to use up the remaining petrol, to conserve the rest of the supply just in case I might need it all.

I made an entry in the log, and noted the time, which was now 01.00 hours, G.M.T. — just an hour past midnight in England, I thought — and made a note of the date, which was now October 16. There was an object in the sky just ahead, and the machine flashed past an albatross. I had seen no ship nor any sign of life during the seven hours since leaving land, and I was grateful for the sight of the lovely bird. I felt completely isolated from the rest of the world, and the only ties were the thoughts of my friends, who would now be wondering how I was faring over this lonely ocean.

The machine was drifting considerably, and I flew low to check up the rate of the drift. I had just decided to alter course seven degrees to port to compensate when I saw a jet of water rise from the sea just ahead. There it was again, and as I drew near I saw the gigantic form of a whale as the great creature rose to the surface of the sea to spout. Five minutes later I flew over another whale as it came up to spout, then plunged into the depths as the water rose in a spray. The immense back of the whale looked exactly like a reef with waves breaking over it.

Visibility became steadily worse, and storms were drawing in around the machine on all sides. As I plunged into the storm area heavy rain beat down on the Gull, which was tossed about like a feather. The sea was whipped up into a foaming mass, and I lost sight of it completely as I flew blind through storms of such fierce intensity that I almost despaired of ever reaching land. My arm ached trying to steady the machine and steer an accurate compass course. As soon as I would fly through one storm it was only to find curtains of black nimbus heralding another. At times I would fly very low, trying to keep the sea in view, and when it was blotted out I immediately climbed to a safe height and flew on entirely by instruments.

Tired and disheartened I watched the hours pass on as the weather grew even worse, until it became a supreme effort to keep my eyes on the instruments, while sheets of rain beat against the cabin. I realized that the slightest mistake would tip the scales against me and the Gull would go spinning down into the sea. The strain was terrific, and my spirits sank when nine hours out there was no land to be seen, only an occasional glimpse of the sea beneath, when the dark rain-clouds which pressed around the Gull parted for a few minutes.

'If only I could see ahead,' I thought desperately as I switched over to the starboard wing-tank 9 hours 20 minutes out. 'If only I could see land....' Suddenly a dark blur loomed ahead through the rain, and the Gull flashed past a small rocky island.

'Land!' I shouted with joy, recognizing the island as a rock just off the coast. Within a few seconds the Gull swept over New Plymouth, absolutely on its course, 9 hours 29 minutes after leaving Richmond and ten days 23 hours 45 minutes out from England.

It was still raining heavily, and all but the base of Mount Egmont was shrouded in black cloud. My altitude was less than 500 feet as I flew over the town, and I could see people running into the streets waving a welcome.

The weather looked bad to the north, and rain was still streaming down. 'Should I land at New Plymouth?' I wondered, as I flew over the large modern aerodrome and saw the big crowd waiting to see me pass. The Gull had lowered the Tasman record of 11 hours 58 minutes, previously held by Flight-Lieutenant Charles Ulm and his crew. Why not be content and land? After all, I was terribly tired, and it would require a tremendous effort to continue on to Auckland.

Throttling back I glided low over the aerodrome in salute. It would be so easy to land now and be welcomed home by my countrymen, then sleep. 'No,' I thought quickly, remembering my intention before leaving England — to make the first direct flight right through to Auckland, my home town. Opening up the throttle I flew on towards Auckland, 154 weary miles farther north. The weather had been too bad over the Tasman for me to even think of lunch, so I now had time to relax and eat a few sandwiches. Conditions improved, and I feasted my eyes on the mountainous coastline.

An hour later I sighted Auckland, and, escorted by a number of machines from the Aero Club, flew over the aerodrome. The ground was black with people, and hundreds of cars were parked in long lines along the boundary.

I closed the throttle and glided down to a landing, and as the wheels of the Gull came to rest felt a great glow of pleasure and pride. This was really journey's end, and I had flown 14,000 miles to link England, the heart of the Empire, with the city of Auckland, New Zealand, in 11 days 45 minutes, the fastest time in history. With this flight I had realised the ultimate of my ambition, and I fervently hoped that my flight would prove the forerunner of a speedy air service from England.

As I taxied the Gull up to the large reception dais where civic authorities and representatives of the Government and the Services waited to welcome me I was delighted to see my father, and recognized many friends among the crowd. The machine came to rest, and I switched off the engine of my faithful Gull for the last time on that flight and entered up the time of my landing, which was 5.05 A.M., G.M.T.

One last Tasman epic before we close; what was meant to be the first official airmail flight. Again, the pilot was Charles Kingsford Smith but, this time, his biographer and long-time secretary, John Stannage, was also along for the ride.

High Adventure
John Stannage

At 12.25 a.m., all was ready. We had said our farewells and Smithy pushed the three throttles forward to start the heavily-laden old bus with an even bellowing roar down the long flare path.

As soon as we were air-borne, we all set about our accustomed jobs, hardly noticing Sydney's carpet of lights slip beneath our wings. The night was clear and bright as we slipped over the coastline and out to sea, just three men in an infinity of black space controlling a trustworthy machine roaring onwards with a precious cargo of 34,000 letters and some bundles of freight.

The balance of the night was a strange mixture of pitch blackness and moonlight, intense cold and comparative warmth, smooth sailing and atmospheric bumps. The machine ran smoothly, and the radio equipment worked better than any we had ever carried. Even our speed was better than usual, perhaps the newly built streamlined exhausts had something to do with that — why, we were showing just on 105 on the air speed! The signs were good and, believe me, I needed some encouragement.

Amazing how hard it was to resist the call of sleep. The drone of the engines always affected me that way for the first few hours. Smithy appeared to be concerned about the weather ahead. He came back from the cockpit to do a spot of pumping and yelled at me, 'Weather looks dirty ahead, Johnnie!'

Then Bill came back to drop a flare to check drift. Far back under the tail, it leapt into being on the sea, flared, and then died to a pin-point of light.

Dawn came, with a faint suspicion of light far ahead. I left the radio key locked down to go to the cockpit for information on our progress. We were making good time and everything was fine. I was asked to pump some more gas from the main to the gravity tanks. I wiggled the pump until my arms were tired. Three hundred and twenty gallons had to be handed up to the wing tanks; but that would be enough for the moment.

By now it was daylight, and breakfast of some sort became a pressing need. I was just about to open a thermos flask when a terrific vibration shook the whole machine. I went chilly with horror. It seemed as if a giant hand was trying to shake the life out of us. Thoughts flashed through my mind — something dreadful had happened to the starboard motor! What was it? Before I could even guess, Smithy had whipped the machine into a steep stall, cut the starboard engine and swung into a steady turn. I darted up to the cockpit door, and Bill moved aside to show me the shattered end of the starboard propeller pointing raggedly towards me. He explained that part of the new exhaust pipe had broken adrift and been thrown into the whirling wooden blade. So it wasn't a part of the old bus that had failed, but something new, something foreign.

I didn't realise then; but I know now that only the superb airmanship of Smithy saved us from immediate disaster. The starboard motor would have thrashed itself from its moorings, and started the total disintegration of the machine there and then, if it had not been for his split-second realisation of the trouble and immediate application of the remedy. By stalling the machine and killing the motor, he had stopped the broken propeller from spinning and stopped the shattering vibration. At the same moment, he summed up the facts that the weather ahead was bad, a return flight to Australia would give us a helping wind and the distance was slightly less, 590 miles in all against a distance of over 600 to New Plymouth, hence the immediate turn back on our tracks.

I knew the situation was serious. We couldn't keep height with this overload on two motors at full throttle. We had 3000 feet and were slowly losing height. Something must be dumped; obviously fuel. Just then, Bill came back and opened the dump valve on the main tank while I sent our position to Sydney with an account of our plight.

Bill had a difficult job to figure out, with the mental stress he must have been undergoing at that moment, the exact amount of fuel he would be able to dump to lighten the ship and still have sufficient to take us right back to land. Smithy was fully occupied in clutching at the air to make every effort to retain height, while I got to work to heave all our luggage, our freight and spares through the cabin door.

With all the worry, I felt a pang of regret with each parcel I let fall into the sea. This trip meant so much to us all. We had failed to show that a Tasman air service was possible. But, surely, thinking people would understand the difficulties of working with such obsolete equipment and believe that with new modern machines the trip could be done regularly and safely!

As soon as I could, I got back to the radio gear to give our exact position to Sydney, to ask them to see that all ships stood by on 600 metres because I was sure that a distress signal would be the next message I'd have to send. By now, we were down to 500 feet and Smithy was literally holding the machine in the air with his hands and feet, juggling, fighting, and coaxing her to keep flying, getting that extra little bit of response that only he could get from the old machine.

I wonder if it such a fantasy that man-made mechanism can possess a soul? Many men become so attached to their car that the mechanism becomes almost part of them. They are the brains and the machine the body. The dear old 'Southern Cross' always seemed to have something of the human about her. She had never let her master down. She'd been up to her belly in mud and flood water, but somehow struggled out; she had blistered in the hot sun of central Australia where motors were never meant to run, but had still pulled through. Now came this last valiant effort when a newly-made inanimate piece of iron practically tore the vitals from the stout old bus; yet she still staggered on when her master was forced cruelly to spur her, and literally thrash her two remaining power-plants to death. It would be a heart-breaking tragedy to lose the old machine in the sea, a horrible end to a beloved piece of mechanism that had become part of the man who controlled her, obeyed his every command and struggled on to the last gasp.

To me, a well-made and efficiently-working radio set becomes an animate, intimate thing. If one is lonely, and very much afraid the sound of a human voice helps a lot, and the radio equipment provides the equivalent of that helpful human voice.

It seemed pretty certain to me that we would never make land. I could feel those old motors straining and groaning. I was sure they would never stand the racket of 10 hours at full throttle so a landing on the sea seemed inevitable. I called ZLB, at Bluff, N.Z., just for the pleasure of hearing his familiar style of keying, and it cheered me immensely to hear him say, 'Yes, old man, I can hear your every word.'

I was afraid, horribly afraid. Not actually afraid of death; I've been fairly close to it several times in my life; but horribly afraid of intense physical discomfort. I wonder if that is the real foundation of all human fear? No intelligent man is afraid of death; but everyone must be afraid of violent pain, extreme cold or continued physical discomfort of any sort.

I kept the radio equipment going as often as I could. I wanted to be sure that Sydney radio would be listening continuously and that there was no fear of their receivers drifting off our frequency, which might have happened if I had just locked my key and not been sending some interesting material. I felt a little crisp when I read over some of the stuff I sent in newspaper clippings I have beside me now. It's awful drip with a sickly attempt at humour. Of course, I had no idea just then that it would all be published. I figured I was simply talking to a friend at Sydney Radio. Maybe I could be excused by saying my sending those messages was like whistling loudly when walking through a cemetery in the dark. This is the sort of thing I mean …

7.15 … 'Oh why did I ever go to sea? What idiots men are to fly over water, a nice comfy ship will do me if I ever get a chance to use another. The sea's going to be cold when we go in.'

Sickening, isn't it? But, as I say, just whistling in the dark!

Smithy was determined to hold the mail at all costs, and although I didn't like the risk, particularly as I figured that it we were lucky enough to make the land we should probably have to bring the mail over the next day in the 'Faith', I had to agree that we could only dump His Majesty's mail as a very last resort.

I felt a bit cheered next time I went forward to the cockpit when I saw Smithy's cheery old familiar grin and he said, 'Just about holding it now Johnnie. She'll do it if the motors can stand up to full throttle.' I glanced at the throttle levers. They were hard forward which meant we had no margin at all and our altitude was barely 500 feet.

As we made our distance westward, Bill kept a constant check on our track cleverly laying off a course that would keep the southerly breeze on our beam and not allow it to hinder our slow progress. Smithy was in constant trouble. He fought for height every foot of the way. The strain must have been frightful. He had another constant source of worry, too: the shattered propeller kept trying to turn in the slip stream and he had to prevent that at all costs, because even low revolutions would have started that nerve wracking vibration again — so he had to keep a fine balance of air speed without losing that one precious foot of height. Only Smithy could have done it. Bill tried once or twice to cut through the broken blade with a hacksaw by leaning far out of the cockpit into the slip stream; but it was a hopeless job, he couldn't make a mark on the hardened metal leading edge. He had an idea that if he could even up the blades we might get the port motor turning again. That plan he had quickly to abandon.

By about 11 o'clock we had only 360 miles to go, and now the throttles were able to be eased a little as the fuel load lightened and we began to think in terms of reaching land. There was less of that feeling of standing on the edge of a precipice. It didn't last long though. The menace of the sea came closer when I heard that the port-motor oil-pressure was dropping. There was nothing we could do about that…. We'd always had a vague thought that in the event of a forced-landing in the sea, the machine would float on its tank buoyancy for a sufficient time to allow us to cut off one of the wooden wings outside one of the outboard motors; we even carried a wood saw in our demolition tool kit, but what a fragile hope that was! I should not like to risk my life on a frail wooden wing awash in a heaving sea. Now we began to take this forced-landing seriously again. Land seemed far away.

Smithy came back into the cabin to talk to me, whilst Bill struggled to keep the ship in the air. It was good to be near Smithy: his very presence inspired confidence, at least up to the moment he began to divest himself of all his heavy flying gear. He put his thumbs down and said, 'Looks like we've collected it this time Johnnie; port motor won't last another hour. Let's go back into the lavatory and have a smoke.' We did this. A thing that was strictly forbidden ordinarily because of the ever-present petrol fumes. I think that (what we thought) last smoke with Smithy was the most intensely dramatic moment of my life. He pointed to the escape hatch above our heads and said, 'When we go in, you'll have to get out of there quite smartly. Collect all the demolition tools, tie 'em up together and make them fast to a longeron. The tail will be in the air and you'll have to come down and try and dig Bill and me out of the cockpit if we get stuck. We might be a good way under, too. The weight of the engines will drag the nose well down. Take your boots off John. Bad luck isn't it; but we've had fun, haven't we?'

When Smithy went back to the flying, I sat for a while and thought. My thoughts were not pleasant. I wasn't in perfect touch with everything that happened up there in the cockpit. I simply had to take it as final that all hope was lost and that we were due for a forced landing. Well, lots of people would think I was lucky, I suppose, being able to die in such exalted company. Oh rats! Surely there was some way out. I went forward again, and Bill smiled when he said, 'I'm going to have a crack at getting oil out of the tank of the starboard motor. Got anything to put it in?'

Something to get my teeth into at last! I tossed my radio spares and tools from a little leather suitcase, bashed the glass out of a small thermos, and there I had two containers for Bill. When I went back with them, I found Bill had actually made the perilous crossing along an oily aluminium strut from the cockpit to the nacelle of the starboard engine and he was down on his haunches plucking the cowling pins from the cowl panel which would let him get at the oil drain cock. I could see he would want a shifting spanner, so I got mine — and by dint of leaning out into the slipstream managed to pass it to him. Then, when he had unscrewed the plug I handed him the thermos shell. He was able to fill this, and by a little wangling pass it back to me. I then poured it into the suitcase resting on the co-pilot's seat. This passing backwards and forwards went on until I had about a gallon of oil in the case and about two gallons on Bill, on me, and on the side of the machine. The slipstream sucked the oil from the thermos-shell as we passed it across.

Bill then began his passage back to the cockpit. He stood up beside the engine nacelle on the narrow streamlined strut, gripped that with his stockinged feet, shuffled along to the full extent of his arm, then braced the back of his head against the rounded leading edge of the wing for the distance between handgrip on the V engine struts and the side of the cockpit. I watched aghast. I wouldn't have taken the risk for much fame and money. It was the bravest thing I had ever seen done. A mistake meant instant death and all the time the slipstream tugged at his body with a pressure almost great enough to stop him breathing. Underneath was nothing but the mean looking ocean.

With Bill safely back in the cockpit, Smithy changed seats and motioned me to have a shot at putting the oil into the starving motor. I did manage to get out on to that strut on the port side, but with the slipstream of the port and centre motor blasting at me, I had no chance whatever of worming my way along. Anyway my head would not reach the leading edge of the wing. I had nothing to support me during that two foot shuffle between hand grips and the plain truth of it was, I hadn't the courage to try it even if I'd been tall enough, so I crawled back into the cockpit.

With oil pressure now down to about 15 pounds instead of the usual 63, something had to be done about getting oil to that port motor and done quickly! Bill got out on to the spar; but came back again. The slipstream was too strong. He tried again; but had to give it up. Smithy and Bill yelled at each other, and the machine was given full throttle for some extra height; then Smithy eased the port motor to very low revs, and Bill tried again this time with success. He soon had the cowling off to get at the oil filler cap; but had to drape himself along the motor-mount while Smithy opened the throttle wide for height. We had come dangerously low while the motor was throttled back. Soon, we had a sufficient height to allow the easing of the motor again, and I passed Bill a thermos shell full of oil. We lost more of the precious fluid in the slipstream, but as soon as the first pint fed into the tank the pressure gauge crept back to normal. I held thumbs up to Bill and he smiled. We quickly transferred the contents of the suitcase and again with juggling height Bill got back.

This was fine. For the moment we could feel safe. Saved by the sheer bravery and tenacity of Bill Taylor! We all relaxed, and I went back to the radio gear oozing oil all over everything. Again our joy was not long-lived. I heard an uneven throbbing in the port motor. Hell! After all Bill's trouble, was it going to quit? The motor spluttered and slowed. Smithy juggled with the throttle and she roared evenly again; but by now, it was time for another oil transfer and Bill went through all the ghastly business again. This time, when he was out on the port motor and I was handing him the oil, the confounded thing began to play up again. Smithy became very concerned as we dropped nearer and nearer the sea with poor old Bill clinging painfully to the V struts in the dragging rush of the slipstream. Smithy prodded me and said, 'You'll have to dump the mail Johnnie! We can't keep height.' I rushed back into the cabin, cut the ropes holding the mail, and threw 40 or 50 blue air mail bags through the windows in less time than five people could have done the job.

Previously, Smithy had officially radioed the Postmaster-General in Melbourne requesting permission to dump the mail in an extremity. This permission has quickly been forthcoming. Although it was the most shocking thing to have to do, there simply was no alternative. Meantime Bill was still out on the motor struts! I got back as quickly as I could, and we then finished the transfer of the oil. The port motor now ran in spasms; but never completely well, but, with the lightened load of fuel and mail, we were able to keep height with the centre motor full throttle and the port motor helping all it could.

Six times in all, Bill Taylor made that fearful trip to transfer oil. He must have been frozen stiff out there in the slipstream. I was, even in the cockpit, without my flying clothes. Perhaps he didn't notice it. He had too many other physical discomforts to put up with.

Just before 3 o'clock, we sighted a small steamer to port. That sight put new heart into us and I was all for making a forced landing alongside her. Particularly as the centre motor had coughed once or twice. Strangely enough, the port motor was now behaving quite normally. It was the steamer 'Port Waikato.' She had no radio and probably wondered what on earth an aircraft was doing so far out from the Australian coast!

The first faint glimpse of the coast should have brought us utter joy; but we were all too drained of emotion by now to feel even relief. The coast still seemed miles away and came towards us very slowly.

About 30 miles off the coast, the oil pressure began to drop again and Bill had to make another oil transfer. Smithy didn't want him to risk it; but Bill was determined to save the old machine.

When land was actually under our wheels I should have felt happy; but I was too numb to feel any more. When Smithy cut the motors for a landing at Mascot, I slumped in my seat with utter mental and physical exhaustion.

Sleep … the most delicious thing in the world that night.

Next day, Smithy offered to take the few bags of mail that hadn't been dumped, because I had overlooked them in the stern of the machine, across in the 'Faith'; but, thank heavens, that offer was refused by the Commonwealth Authorities. If we had gone in the 'Faith', we should definitely have been lost because during her next five flying hours the

'Faith' blew off a cylinder-head which completely wrecked one motor and she just would not fly on two.

So that was the last flight of the old 'Southern Cross,' except that shortly afterwards Smithy flew her up to Richmond to hand her over to the Government to be preserved as a museum piece as an inspiration to all young Australians.

Smithy made numerous trips to New Zealand, travelling all over the country with hundreds paying for and enjoying a quick ride in the *Southern Cross*. Here's E.F. Harvie's account of one such flying visit, complete with a landing and take-off from Ninety Mile Beach.

Venture the Far Horizon
E. F. Harvie

At that time an Australian racing driver, one Norman 'Wizard' Smith, was up at Ninety Mile Beach intent on bettering the world landspeed record for various distances. Keen to see him in action, we set up camp near the garage in which his car was housed at Hukatere, twenty-six miles north of Ahipara. Whenever the tides were right and Smith was expected to run, visitors arrived from the south and the place became comparatively crowded — a couple of hundred people, perhaps. Even two aeroplanes arrived, one flown by Les Brake with whom I went for a ride up the beach, only to return smothered in castor oil thrown out by its rotary engine. It was among the last of the Avro 504Ks seen in New Zealand.

Some days later G. D. M. Goodwin, Chichester's business partner in Wellington, flew in with Lee Hill, a movie photographer who was anxious to get some shots of Wizard Smith's high-speed runs. They'd come up in the Sports Avian ZK-ACM — a lovely blue and white aircraft that I later owned myself — and were glad of my offer to keep an eye on it and to conceal Hill's gear under a pile of bedding in our tent: there was some doubt whether the movie-making rights were not held by a rival organisation. Nothing was discovered, but one day, just before Smith came hurtling down the beach, the opposition, somehow alerted, rushed from a hideout in the sandhills, sent camera and tripod flying and sailed into the pair with their fists. It was a magnificent display in the course of which Geoffrey Goodwin, unable to get his coat off because his arms had stuck in the sleeves, took a colossal crack in the midriff and went to ground with a grunt. In the Magistrate's Court in Auckland, the episode had its sequel.

In the spring of 1932 the newspapers reported that Sir Charles Kingsford Smith, whose first aerial crossings of the Pacific and later of the Tasman had stirred the imagination of people all over the world, was thinking of bringing the *Southern Cross* to New Zealand again, this time to tour the country and to offer passenger flights to the public. Since this was the kind of lead I had long been waiting for, I acted immediately and a few weeks later received a cheerful letter from Smithy offering me a place in his crew during his coming tour of New Zealand.

Early on the morning of January 11, 1933, the *Southern Cross* left Gerringong Beach, New South Wales, crossed the Tasman in fourteen hours and five minutes and landed at New Plymouth where she was welcomed by an enormous crowd. In the two and a half months that followed, we visited twenty-three places, from Whangarei to Invercargill, carrying thousands of passengers and making a surprising amount of money considering that the Depression was not then over.

Business matters were handled by Smithy's brother, Wilfrid, while the rest of the team included J. T. Pethybridge, co-pilot and engineer; H. C. Affleck, a rigger and fitter; John Stannage, Smithy's radio operator who later became well known in broadcasting circles in New Zealand and Fiji; and Harry Purvis, later co-pilot to Smithy's navigator, Captain P. G. Taylor, on a brilliantly executed survey flight between Australia and South America in the PBY amphibian *Frigate Bird*.

Because the demand for flights in the *Southern Cross* could never be fully met, the overflow of passengers in the North Island was carried by George Bolt in the American Waco biplane ZK-ACV, then without question the finest civil aircraft in the country. Almost every day during the next few weeks George found time to give me a few minutes' dual instruction, best start to a career as a pilot that any man could then possibly have. If he made things seem so easy, he yet impressed on me repeatedly that 'nothing in flying is simple' and insisted that I not merely know *how* to do a particular thing, but far more importantly *why*. And by doing everything the *right* way, one automatically did them the *safe* way. His attitude was scarcely surprising for he was a perfectionist.

In the South Island, George's role was taken over by 'Tiny' White with whom I would occasionally fly in a rather clapped-out Spartan, while Harry Purvis and I, as members of the *Cross*'s crew, were often able to borrow a Moth for nothing and get in some flying practice.

Those Aussies were pretty tough and I found out straight away that no man was accepted as a member of the Boss's crew until he had shown that, entirely unaided, he could start all three engines of the *Southern Cross* in one unbroken session. From the cockpit, Smithy would prime one motor, ready to catch it with the throttle once you had done your job and the thing had begun to fire. Standing beneath the centre Whirlwind first, you pushed a long crank handle up into a starting dog and then began to wind up the flywheel of an inertia starter. It wasn't easy, for a lot of upward pressure was needed and the handle was pretty heavy. At 12,000 r.p.m., reached after more than a minute's winding, you flung the handle away, reached for a toggle and pulled, thereby engaging a clutch which turned the propeller over. If you were lucky, the motor would fire before the flywheel had lost its momentum; otherwise, you had to begin all over again. And it was no help to have the rest of the gang — the examining panel — standing round and offering derisive encouragement.

'Come on, come on!' they would roar, as you wound yourself up in knots. 'That's not fast enough! Speed her up! Give the bastard the works!'

But Smithy always played fair with the throttles. Having got the centre motor cracking, you then had to deal with the port and starboard in turn and by the time all three were booming away, their propellers batting the air, you had just about bust your boiler. Still, you had qualified and from now on it would always be a two-man start. Better than that, though, you could sit alongside Smithy when opportunity offered and have a shot at flying the *Southern Cross* by yourself. And he was pretty generous about that.

I got my first chance one day when we were snoring along somewhere off the Kaikoura coast. The Boss had gone aft for a while, leaving Purvis to see that I didn't get into too much trouble, when Stannage, always the practical joker, came up to the cockpit door and started to do some barracking. Unobtrusively tightening his seat belt, Purvis scribbled a note and passed it across to me: 'Pull up into a climb.' Just as the nose came up he reached suddenly for the stick, gave it the most almighty heave forward. The consequences were fantastic. Stannage sailed straight off the floor, bashed his head on the roof and fell over backwards, while the passengers, instantly thrown off their seats, now grovelled on the floor and fought one another in a swirl of dust and rubbish, as mad as a bunch of bulls. I couldn't see Smithy at first, but he was quick in coming back.

'What the devil's going on?' he demanded, purple in the face. 'Did you have to take a rise out of me?'

Well, we'd done just that — literally. We'd caught him down in the loo.

Little about that tour of the *Southern Cross* will, I suppose, ever be recorded now and even this is hardly the place. But memories remain vivid and some of the things that happened will never happen again. Simple little incidents, I mean, like those great herds of wild horses, each led by a noble white stallion, galloping across the Kaingaroa Plains in full stampede, fearful of the clatter of the J-5s and the shadow of the giant bird that chased them ... The blinding clouds of pumice dust at Bridge Pa, Hastings, every time the *Cross* took off which blew the freshly sown surface to smithereens ... Landing on makeshift paddocks, beaches and estuaries, with tens of thousands of people always in the way and missing the flailing propellers by inches ... Fan mail for the Boss with illuminated addresses, ecstatic poems of welcome, begging letters, offers of inventions ('You could go a 1,000 miles or so,' wrote one chap, 'without such huge amounts of Petrol ... You to arrange finance for Patents, fit up an aeroplane and prove it to the world. If the above appeals to you, I would be pleased to enter an agreement in writing and then divulge the whole idea. As I am a dairy farmer an appointment would have to avoid milking times') ...

Cracking along in a gale thirty feet above the water with the cliffs just visible a hundred feet from the wing tip ... Trapped in a cul-de-sac somewhere in the Kaikouras, unable to climb out ahead and having to do the most fantastic stall turn to achieve a reciprocal course and escape the way we'd come ... A great pile of sandwiches on the compass, Smithy chewing away contentedly, deep in thought ... Forcing the cabin door open in flight and kicking out boxes filled with lunch scraps and an accumulation of other rubbish ... Entertainment everywhere, from parties in gracious homes to clandestine attendances at the kinds of speakeasies you'd inevitably find in places supposed in prohibition days to be 'dry' … The mishap at Palmerston North, with the *Southern Cross* lying crippled in a muddy ditch, starboard wing pointing high into the dreary sky while the sodden clouds wept great torrents all over her … Engine overhauls — grinding in fifty-four valves, slowly, one by one … Rough trips, people sick; one man's head all the way out of a window, another's encased in a huge paper bag pulled right down to his heaving shoulders, hardly a second too soon …

In a technological age when flights across the oceans of the world take place scores of times a day, it is easy enough to forget the pioneer crossings of the past. Not that many remember them, of course; the witnesses were few. Just for a moment, then, let's go back to the 25th of April, in the year 1933, for the *Southern Cross* is now ready to fly to Ninety Mile Beach and return home to Australia.

As soon as the big Fokker had completed her refit at New Plymouth, 'Tiny' White and I flew on ahead and early in the afternoon reached Hukatere, the old Spartan coming to a halt outside the garage that Wizard Smith had used.

'Gee, you fellers come long way,' grinned an old Maori. 'Where t' *Southern Cross*? He come soon, eh?'

A couple of hundred people had already arrived — numbers of white settlers and their families, parties of Maoris who had chugged up the beach in the most ancient of rusty cars or who had ridden tandem on the backs of horses of odd pedigree. Great people, though, overwhelmingly friendly and hospitable, with food and cooking materials, firewood and bedding, all intent on taking good care of Smithy and his crew on their last night in New Zealand.

We marked out the beach with half a mile of white flags and built up a stack of driftwood and brush soaked in engine oil so that the smoke from a fire would give Smithy an idea of the surface wind when he arrived. That done, we could merely wait. Endless lines of surf boomed on to the huge beach from the fullness of the Tasman, the sky was overcast and a light mist beginning to form. Even the baleful eyed gulls seemed excited, nervously expectant. At last we heard her coming; it was 4.12 p.m.

The *Southern Cross* comes in high above the beach. Smithy takes a quick look at the flags and, apparently satisfied, turns and starts to bring her down. In from the south she sweeps, touches the firm sand as lightly as a feather and streams up the beach, the three Whirlwinds thundering defiance at the booming of the sea. We guide her up the beach, beyond high water mark, and she pulls up close to the garage. The crew jump down on to the sand — Smithy, Pethybridge, Taylor, Stannage, Affleck — and with their appearance the excited group of people clustered round the *Cross* puts up a brave little cheer.

'Thanks, people,' Charles says, quietly. 'Nice to be here. Have a good look at her while you can, because we're going to be busy shortly.'

Snatches of quiet conversation among the crew.

'God, what a wonderful beach, Charles. She'd lift ten tons off this place, don't you think?'

'Say, that's a great smell, John. What the devil is it — chicken, or something?'

He'd be surprised if he really knew. It's roast godwit: dozens of them. Everyone eats them around here and they're not going to let Smithy starve.

'Did all that gas arrive, by the way? We'd better start filling up soon. It'll take quite a while, you know. And no smoking. Better tell the crowd now and put up some notices. Keep everyone well back.'

The tide is coming in quite fast now and it won't be long before we are completely cut off from the nearest settlement, twenty-five miles away — unless anyone cares to walk. Down near Kaitaia and Awanui many dairy farms are doubtless deserted and their herds will go unmilked tonight because everyone has decided to come up here and see the *Southern Cross* go. The mist over the beach is slowly thickening and the sun won't be seen again for there's heavy cloud over the Tasman. But it's time to get on with the fuelling — hours of work for us yet.

Eight men gather round the tail of the *Southern Cross*, lift up the back end and slowly manhandle her round till she faces the sea, all set to roll away off the heavy planks under her main wheels in the darkness of early morning. The Boss hopes to get off at 4 a.m., round about low water.

Squadron Leader Isitt, O.C. Hobsonville, and Tom Wilkes, Director of Air Services, come in in an Air Force Moth; there's just enough firm sand to land on. With Smithy they go off in a car and take a look at a mile or two of the beach. The Boss is greatly impressed.

While Pethybridge begins a long final inspection of the motors, the rest of us start to get the fuel aboard. None of your fancy hydrants here, pumping in hundreds of gallons a minute. Instead, every drop must be pumped aboard by hand and the going is mighty slow. Affleck stands on top of the three-foot-deep wing with a nozzle, the others taking turns to wobble the pump handle back and forth ten thousand times while the gas creeps slowly up the hose and into the gulping tanks. There is a whole battery of forty-five-gallon drums to empty at this rate, and even if the load is a good deal less than that carried by the *Cross* on, say, her transatlantic flight, it's formidable enough — 757 gallons in all and thirty-three gallons of oil, sufficient for at least twenty-four hours' flying. And if that seems a long time, just remember that on her last westward crossing of the Tasman she took twenty-three and a half.

After dark work goes on by torchlight, helped by the headlamps of several cars. The *Southern Cross* looks pretty impressive under their bright glare, her name sparkling silver against the dark blue background of her fuselage. Pethybridge checks the fuel gauges, Affleck crawling over, under, round, inside the big ship, having a really close look at every tiny thing. The chink of spanners on lock nuts, the snip of wire-cutters, the deep thrum of a bracing wire somewhere deep in the tail — familiar sounds, perhaps, but strange in a place like this.

Bill Taylor, the navigator, has just anchored his drift sight to its bracket outside one of the cabin windows and his charts are spread out on a small table in the back. Under the feeble glow of the little lights in the cabin roof he gets to work with a ruler and pair of dividers, while his other instruments hang neatly on the wall. There's a powerful smell of petrol, but what else can you expect with that great extra tank in the cabin?

'Heard anything more about the weather?' he asks.

'They said something about a bit of a low halfway across. If there's any strong northerly here at midnight, I don't think Charles will have it on.'

Bill Taylor puts his dividers down slowly and looks up with a bit of a grin. 'Doesn't seem quite real, does it?' he says. 'Away the hell up here. And all this so that we can get home in a hurry. It's sort of man-against-the-sea stuff, in a way. Or having a go at the Almighty. Good fun, though. What about a cup of coffee?'

People are milling around in the garage where these settlers are doing a marvellous job. Enough food and coffee to keep a whole army going for weeks. In one corner Stannage and a couple of men from the Radio Corps have been in continuous touch with Dr Kidson at the Weather Office in Wellington and Stannage has just been talking to the *Monowai*, somewhere out in mid-Tasman.

'Say, this is all right, isn't it?' he says, pulling off his headset. 'Listen to this, Charles.'

The report says nothing about strong westerlies. Only something about local mist and showers, with a slight sea.

'Couple of other ships out there, too,' says Stannage, 'so we're not going to be too lonely.'

Most of those who have come up to see the *Southern Cross* go have decided to get a few hours' rest while they can, and in little groups among the sand dunes have gone to ground, children wrapped up in rugs and tucked away in scooped-out hollows, grown-ups covered with blankets, coats, anything handy. Others are sprawled on the seats of cars. Occasional chatter, the simple laughter and singing of a party of Maoris gathered round a fire some distance down the beach, wisps of wood smoke curling up into the mist. Work on the *Cross* finishes about eleven o'clock and Affleck signs her out. We slip on the engine covers and wrap her up for the brief night ahead. Everything's quiet at last.

11.30 p.m.: Thick fog. Affleck comes up. 'We'll start the motors at four if the forecast's okay. Give you a call about the flares later.'

12.45 a.m.: One of the girls who's been cooking godwits comes up. Doesn't want to go to sleep, she says, in case she misses something. We talk for a while. There's a light easterly now and the mist doesn't seem so thick. Not much of a sea running, either. Soon be time to get the flares ready — sacks soaked in oil and gas.

1.30: Weather synop has just come in: slight seas along the New South Wales coast. *Monowai* has cloud 1,000 to 2,000 feet, some mist and occasional showers. Moderate winds for a start, then 300 miles of northerlies followed by a shift to southerlies. A bit of low cloud and rain over the central Tasman, but nothing to give Smithy any bother.

Can probably get in an hour or so's sleep in the sand grass. What about katipos, though? They live in this sort of stuff.

3.50: A thump on the shoulder from Affleck. 'Come on, give us a hand to get started. You get the covers off.'

For a while we have bother with number one, the centre job, which refuses to fire and takes a lot of repeated cranking. Good Lord, how many more times have we got to bust our guts winding this bloody handle? We change a couple of plugs and she then comes right. All three turning over now, no more trouble.

It's 4.20 and Pethybridge has just seen the Boss.

'Going to wait till it's light,' he says. 'No sense trying to get off in this damned stuff.'

Groups of people have now gathered round the dark shape of the *Southern Cross*. Something's happening soon. The three Whirlwinds keep on churning over at fast r.p.m., gradually building up their oil temperatures, exhausts crackling and spitting out long flashes of blue fire, propellers smacking the air sharply. The old girl is anxious to go.

Just after five a little party assembles in the garage and there's a buzz of talk. It's lighter now and the mist has lifted a bit. A sudden stir. Smithy says he'll give it a go.

'Cheerio, chaps,' he says quietly, 'time to get away.'

Firm handshakes all round, then the crew goes aboard. The crowd claps, Stannage waves and slams the door.

Smithy can be seen settling himself down in the cockpit, checking everything carefully and concentrating on every move he makes. Then a quick glance outside and a brief wave of the hand, the old disarming smile.

The three Whirlwinds let go with a shattering bellow as the Boss opens the throttles and the *Southern Cross* rolls heavily off the planks and on to the firm sand. She's carrying a big load now and is rock steady as she moves down towards the water. Gently, Smithy brings her nose round till it points northward up the beach and she gradually fades from sight into the grey pall of the mist.

There is an interminably long wait and as minute after minute rolls by everyone anxiously watches the beach. Then, unmistakably, and at exactly 5.36 a.m., a faraway roll of thunder comes reverberating out of the fog. The *Cross* has begun her take-off, has started to eat up the beach.

Almost unexpectedly, for it now appears so close, a great grey shape comes racing up out of the mist, forges straight down the beach like an urgently highballing freight. People fling themselves flat on the ground, try to keep her within sight beneath the level of the fog.

She's away none too soon, though, for the tide is well on the way in and an unusually big wave sends a deep carpet of water high up the sand

and right across her path. But the *Southern Cross* can't possibly stop. Straight into it she ploughs, pitches wildly under the braking action of the water against her wheels and disappears in a colossal eruption of spray. God, what a fearful sight! Yet, as people hold their breath, she miraculously lurches out of it and goes charging on down the beach like a mustang that's escaped a corral. Quickly she fades from sight and even the sound of her engines has died.

But she hasn't left the beach and there's another shock in store. For a party of people is at this very moment driving northward in a car, unaware of what's going on. And as the starboard wing of the *Southern Cross* flashes across its roof it is only by the grace of God that those people are still alive.

Not far short of a mile from the point where she began her take-off, she at last comes unstuck and, inch by inch, climbs skyward and leaves the sand streaking away below. Smithy now turns gently out over the water and within seconds the *Southern Cross* has lost all touch with the land.

Back at Hukatere, people turn quietly away, every man lost in his own little emotions. Nothing more to see here, now. Better start heading for home.

Flying Officer A.E. (Alan) Clouston was another 1930s record breaker. In March 1938 he flew a twin-engined De Havilland Comet from England to New Zealand in four days, eight hours and seven minutes, also establishing a new record for the Tasman crossing — seven hours ten minutes.

When not breaking records, Clouston was a test pilot working for the Royal Aircraft Establishment at Farnborough, where he was asked by the boffins to assess the effects on an aircraft of barrage balloon wire.

The Dangerous Skies
Air Commodore A. E. Clouston D.S.O., D.F.C., A.F.C. and Bar

Roxbecox's instructions were simple and to the point:

'All we want you to do,' he said, 'is to fly an aeroplane into the wire and see what happens.'

We decided to begin the tests by flying into strings dangling from a parachute. Initial tests in this manner had already been begun by another test pilot, Pat Fraser (now Air Vice-Marshal Fraser). He had appreciated the danger of the string on the parachute hitting his head and had had a heavy felt crash helmet made. Fortunately it fitted me.

The aeroplane I was given for the tests was a Miles Hawk two-seater trainer, a low-wing, open-cockpit, monoplane of wood and fabric construction. The string was ordinary fishing-line, rolled into a ball and attached to a parachute. I would take the parachute to a height of about five thousand feet and throw it over the side. Whilst the parachute was opening and the string unfurling, I would circle quickly and try to line myself up for a straight run into the dangling cord so that it struck about half-way along the leading edge of the wing.

My first test was made over open country near Farnborough. I tossed the string and parachute over the side, and flew far enough way to circle and line up for my run. The string caught my wing at the correct point. For a few seconds there was no change in the aircraft's behaviour, and then it skidded mildly to starboard away from the direction of the string. The string raced madly over the leading edge of the wing a few feet away from the cockpit, and the friction set up a small trail of smoke. As the last few yards dragged over, the string snaked and lashed back like a whip wrapping itself round the wing. This was due to the ball of fabric, soaked in red paint, which the scientists had attached to the end of the string so that they could determine the behaviour of the string from the position of the paint smears on the wing.

A sharp jerk from the parachute, a slight tug on the aircraft, and before I could apply rudder to correct the swing, the string was free, leaving behind a jagged cut several inches deep in the leading edge of the wing. Continuing test flights proved that the longer the string, the deeper the cut into the wing; and that the damage also varied according to the size of the string, and the speed at which I flew into it. The danger from these tests, I soon discovered, was loss of control not from contact with the string, but from the whipping and curling of the end around the wing.

On one of the early tests I was flying with a new junior scientist, a young fellow making his first flight. We went up to five thousand feet, tossed the string and parachute over the side and made a quick circle. I caught the string on the wing at the correct point, but as the end of the string whipped over the wing it became entangled in the propeller.

The parachute billowed out at the end of several hundred feet of string with the effect of a sea anchor, except that as well as acting as a brake, it was gyrating and kiting, and causing the string to foul the rudder and elevators. The aeroplane pitched and skidded violently. At full throttle I seemed to be holding my height, but I had little control left and there was not a hope of landing the aircraft in such a condition. We would have to jump.

I reached forward and tapped the young scientist on the shoulder. He turned his head.

'Jump,' I yelled, pointing over the side.

He promptly turned his face back to the front and ignored me.

I gave him another punch on the back and he threw me a startled look.

'Jump! Jump!' I shouted at him, again pointing over the side.

The scientist shook his head. No, he wasn't going to jump. Quite unconscious of the fact that there was anything seriously wrong, he certainly could not understand why I should be trying to make him jump out of the aircraft at five thousand feet over Farnborough.

I was growing anxious. Control of the aircraft was getting worse. I knew we would not maintain our altitude indefinitely, and I certainly could not jump myself until my inexperienced passenger was safely clear of the aircraft.

As the Hawk continued to pitch and toss, I scribbled shakily: 'Machine out of control. Unable to attempt a landing. *You must jump.*'

I hit the young fellow on the back once more, and passed him the note. He read it and gave me a very worried look. Then, very slowly, he undid his safety harness, and began climbing out of the cockpit. I managed to turn the aeroplane in a gentle bank to the upwind side of the Farnborough airfield so that he would land on the aerodrome and not in the town.

The scientist put one leg over the side on to the wing. He was moving painfully slowly, as though he had all day to get out of the aircraft. The time he was taking infuriated me. At any moment the aircraft might plunge to earth for all the power I had to control her. I caught sight of his face. It was as white as a sheet and I felt sorry for him. He was hanging on like grim death, otherwise the hundred mile an hour slipstream would have torn him away.

He had his other leg half out of the cockpit when the string holding the kiting parachute broke. The chute flew clear and I had full control of the aircraft again.

I reached forward and grabbed at the scientist. 'Sit down! Sit down!' I yelled, pointing to his cockpit seat.

He stopped, transfixed. He was more out of the cockpit than in it, and he stared at me with an expression of agony. It was easy to read his thoughts. I had told him the aeroplane was out of control and that he must jump. Now that he was almost over the side, I was telling him to return to his seat. Was I raving mad or just pulling his leg?

He made no move, but stayed as he was, trying to make up his mind whether it might not be better under the circumstances to parachute to the ground anyway. Then to my amazement I saw him continue to climb out on to the wing. Fortunately, he was not used to the cumbersome parachute and it caught on the side of the fuselage, delaying him a moment. I released my own safety belt, leant forward, and grabbed him by the back of his flying-suit.

With the slipstream tearing at us, I held on to him, flying the aircraft with the other hand, and nodding towards his cockpit. 'Get back! Get back!' I shouted.

Shaking his head in complete bewilderment, the scientist finally did clamber back into the cockpit.

I landed.

'What the hell goes on?' the scientist demanded furiously as soon as I had switched off the engine.

I explained what had happened.

'Good God!' he exclaimed. 'I knew the aeroplane was a bit frisky, but I thought that was your flying.' ...

I began using thin wire in the same way that we had done with the first string tests in the Hawk. At ten thousand feet I threw out a ball of light wire, several hundred feet long with a parachute attached, and let it unwind. As it descended swinging and swaying from the parachute, I would fly into it. We did these experiments over Salisbury Plain where there was less likelihood of the wire fouling traffic on the ground, or catching in high-tension wires.

The wire was so thin that it was difficult to see. The way it swayed as it descended sometimes made it impossible to catch at the right point on the wing. Often I returned to Farnborough with hundreds of feet of wire hanging from the aircraft. There was nothing I could do about it, but trail the wire across the town as I made my approach to land. Sometimes it fused high-tension lines and caught up telephone wires.

Once I caught some painters on a work-stand. They were painting the senior scientists' mess when the wire whipped round the painters' two-storey-ladder stage, and dragged the entire structure, with paint and painters, to the ground. Fortunately nobody suffered anything worse than bruises and the discomfort of a shower-bath of paint, but thereafter the painters refused to work whenever they knew I was in the air. They would sit around, drinking tea and smoking, and shake their fists at me as I flew in to land. Not until I was back on the ground did they resume their painting.

On another occasion I netted a bicycle rack holding eight or ten bicycles. The wire caught it at the main gate of the R.A.E., and the rack trailed behind me in mid-air, with the bicycles dropping off in a trail across the airfield, as though they were pieces of paper in a paper chase.

The swaying and whipping of the wire made it very much more dangerous than the string. One afternoon I flew at a hundred and seventy miles per hour into the wire, but it caught the tip of the metal propeller and was whirled like a flail round the cockpit, fuselage, and wings. By the time I had throttled back the engine, there were deep gashes several feet long, wherever the wire had struck. One ran down the side of the cockpit and fuselage from a point two inches away from my ear. A fire started behind the propeller, but fortunately went out by itself after a few seconds.

I dragged nearly a thousand feet of wire back to the R.A.E. and landed. The aeroplane looked as if someone had been chopping at it with a giant axe. I noticed another four-foot gash in the cockpit just behind my head; and there was no doubt at all that there was sufficient power in the whip of that wire to cut right through the human body.

It was decided, therefore, that a thick steel-mesh cockpit canopy should be made to protect me from decapitation. Before the R.A.E.

engineers had finished designing this, however, I had another bad experience. This time, as the wire whipped over the wing it tangled in the aileron. The trim tab on the port wing was knocked up to its maximum and the aircraft began to spin horizontally.

I pushed against it with the stick, but the pressure was tremendous. With both hands on the stick I had to exert all my strength to keep the control stick on the left side of the cockpit. I undid my harness, and worked my body round so that I could twist my leg against the stick. With the strength of my thigh added to the force of my arm, I at last managed to keep the aircraft upright. This left me one foot free with which to control the rudder.

By the time I had flown the sixty miles back to Farnborough from Salisbury Plain, my arms and legs were numb and aching. I could not have hung on much longer. Still in an unpleasantly banked position, I landed safely at high speed.

After this, the scientists decided that I should postpone further experiments with the wire until the steel mesh had been fitted to the P.40. They decided also that it would be foolish to risk the lives of two people on such work, and that a scientist would not therefore be carried on future trials! …

For many weeks I tested the P.40 against wire hanging from the balloons. Each time the length of the wire was increased until it started cutting so deeply into the main spar that the tests were really dangerous. We had learned that the last few yards of wire had a habit of whipping in a complete circle round the wing before finally dragging free. If an explosive charge were attached to the end of the wire, it would therefore make contact with the upper surface of the wing and blow the aircraft to pieces. This, however, the scientists were prepared to take for granted without practical demonstration on my part.

Instead, tests now turned to protection against the wire. By putting a sheathing of steel over the leading edge we found that the wire, unless it was very long, did not cut into the wing but made heavy scratches and slid off at the tip. The next step was not just to evade, but to try to eliminate, the hazard of the wire. Special cutters were fitted into the wing tip. As the wire ran along the leading edge of the wing so it slipped into a slot where an automatic knife, fired by a cartridge, severed it.

When war did come, it gave me considerable satisfaction to know that those months of flying into the wire had contributed to the plans for the balloon barrages around the major towns and cities of the United Kingdom. The experiments also led to the fitting of barrage-balloon wire-cutters to the wings of our Lancasters and other bombers for low-level raids on Germany.

In the uneasy calm before Hitler's storm, Alan Clouston was offered an even stranger assignment ...

I was back at Farnborough on test work after the record flight to New Zealand in the Comet when I was called to the telephone. I recognised the caller's name immediately, although I had never had any personal contact with the man.

'I have a proposition I should like to discuss with you privately,' the voice said. 'It concerns the Comet.'

'Can you come down and see me here?' I suggested.

Farnborough in those days was open to the road, not wired off and patrolled by security police as it is today, and the caller was able the drive straight in without anyone challenging him. He was waiting for me outside the flight offices when I landed from an air test. I recognised him at once from pictures I had seen in the newspapers.

'Come in to my office,' I said.

He shook his head. 'It would be better for us to talk about this out in the open, where there is not a chance of anybody overhearing.'

It was not until we were isolated on the Farnborough airfield that he broached his proposition.

'I suppose you have heard of what is happening to the Jews in Germany?' he said.

'I have read reports,' I admitted.

'I am a Jew, as you know,' he said, 'and I assure you that nothing you have read is in the least exaggerated.'

'But what has this to do with me?' I asked.

'We want you to kill Hitler,' he said quietly.

The sun was shining, the surrounding countryside was green, tranquil, at peace. In the distance an engine in one of the test sheds droned like a swarm of bees.

This was fantastic. It was not real. At any moment I would wake up.

I looked back at my visitor. If only he had been some crank! But there was no question of his sanity or seriousness. A man I knew by repute to be a brilliant figure both in industry and in national affairs, he made the statement as if he were at a board meeting discussing the need for new plant. There was nothing dramatic in his voice.

'Don't dismiss the idea as fantastic,' he continued when I said nothing. 'Our plans have been very carefully made. First, though, the question of your fee. You name your own figure.'

'That is ridiculous,' I said. 'I might ask a million pounds. What then?'

'We are prepared to pay a million,' he replied. 'I am not alone in this. I represent a group of wealthy industrialists, and we will put up the money between us. You need have no fear about that. The money will be guaranteed, and can be paid in bearer bonds and notes. Accounts can be opened for you abroad. It will be made available as you wish, and I can assure you that, in our own interests, we shall take every precaution to see that the fee is paid in such a way as to avoid attracting attention or comment.'

Obviously this was no spur-of-the-moment scheme. Details had been worked out with forethought and thoroughness, and I was becoming more and more intrigued. Without committing myself or giving a thought to whether or not I would consider such an assignment, I asked exactly how I was expected to kill Hitler.

My visitor looked at me closely. 'You bomb him from the Comet when he is attending one of his ceremonial parades in Berlin.'

'They would recognise the Comet straight away, and they won't need to put two and two together to guess it is probably Clouston flying it,' I retorted.

'You underestimate us,' he said. 'We have a secret grass landing-strip near the north-east coast. There, we shall disguise the Comet out of all recognition.

'You fly from that aerodrome straight out to sea, keeping well north of the Continent and turn in for Berlin from the Baltic. You come straight down over the Unter den Linden. In these parades Hitler always travels in one of the first two cars. If necessary, you will therefore have to be prepared to bomb both of them, and you will be equipped with two specially manufactured bombs of concentrated high explosive.

'When you have dropped the bombs, you turn immediately north-east for the Baltic again and continue until you are well clear of Germany before setting course for home. There will thus be no indication that you have come from the west. The probability is that the Germans will come to the conclusion that the plane came from the east.'

My visitor paused, but I made no comment.

'By taking this roundabout route,' he continued, 'we have worked out that the whole trip will be something like sixteen hundred miles, which will give you a reserve of two hours' flying. A secret landing-ground is available in Sweden, and in an emergency you can land there as a last resort. But you should have no difficulty in returning to our landing-ground in this country. As soon as you land, the aeroplane will at once be stripped down and overhauled, and altered back to its normal appearance.'

We walked in silence. I turned over the plan in my mind, examining its practicability. I put one or two technical points to him, and he satisfied them without hesitation.

There was no doubt that the whole plan had been thoroughly well prepared, and the more I thought about it the more I came to the conclusion that it was eminently practicable. It should be quite simple to kill Hitler in this way and to get back to the country. But afterwards?

Despite my visitor's assurances of the precautions and protective measures, his schemes for getting me out of the country to South America or anywhere I wished, I could not believe I would not be found out. Our own secret service would, I was convinced, find a lead to me sooner or later. Perhaps they would do nothing about it. But the German Intelligence with their network of agents round the world? I had no illusions about them. They would never close the case on the man who murdered Hitler. If it took years, they would carry on the hunt with ruthless tenacity. No matter where I settled, there would always be the fear of an unexpected bullet or a knife in the back.

More dissuasive than the bullet, however, was the question of my conscience, the fact that I would have to live for the rest of my life with the knowledge that I had murdered a man. For no excuses could disguise the fact that so far as I was concerned, the mission would be no more than one of plain murder for money. I had no personal hatred for Hitler then. I had attended several flying meetings — their 'Rundflugs' — in Germany and, if anything, had a qualified admiration for the way he had built up his nation. Naturally I did not approve of his racial persecutions, but then, with no true knowledge, I was inclined to discount most of the reports as political propaganda.

As if he could read my thoughts, my visitor said: 'This is not a case of murder, you know. It is an execution as just and warranted as any in history. Hitler already has the blood of many innocent men, women, and children on his hands. The man is evil. He is embarked on a fiendish policy of systematic extermination. We are asking you to help us to stop him in the interests of humanity before it is too late. That is not murder.'

I drew him to the side of the Farnborough airfield as a Gladiator came in and touch down. I watched it run over the grass, slow down, and taxi away to the hangar. The noise of the engine faded away, and there was quiet again.

'Suppose,' said my visitor at last, 'they were not Jews but New Zealanders whom Hitler had in his power. Suppose it was your own people, Mr Clouston, upon whom he was committing these atrocities. Would you kill him then if you had the chance? Would you regard that as murder?'

I shook my head. 'That would be different. I should be fighting for my own people like any soldier. Naturally I should have a try and, of course, money would not enter into the matter. The same thing applies here. The one man who can honourably carry out your plan is a Jewish pilot.

'Were I to undertake this mission of yours, I should be a mercenary, not a patriot. I should be killing solely for money. I am being honest with you. And you know what I say is true. That is why you offer a fantastic sum like a million pounds. But it is not a question of bargaining over the money. Half of it would be enough, a quarter of it would tempt me. You can give me more than enough money all right, but the trouble is you would not be able to give me peace of mind.'

'Don't decide now,' he said quickly. 'Why not take a week to think it over?'

I shook my head. 'It won't make any difference.'

'Well, just in case you do change your mind, you can reach me at this number within the next week.'

I walked with him to the car, and watched it purr away down the lane.

Crusader he might be, but there was no evading the ugly fact that I should be no more than a hired assassin. Every man has his price, it is said, and for a week I was acutely conscious of the telephone. If only the offer had been twenty thousand or fifty thousand I could have pondered, perhaps, on what I would have done with the money, and laughed the thing off. But a million!

I have thought about it a lot since. Few men get such an opportunity of learning the truth about themselves. Even so, I have never been really sure how much I was held back by conscience, and how much by fear of being caught.

PART FOUR

WORLD WAR TWO

We can only guess what might have happened if A.E. Clouston had done as he was asked. What we know is that in the years between 1939 and 1945 there was a war that involved nearly every part of the world.

On 3 September 1939 New Zealand declared war on Germany – 12 hours before Britain, in fact, because somebody forgot about GMT. Just over two years later, we were also at war with Japan.

In 1939 the RNZAF had just 97 (mainly) obsolete aircraft; by 1945, its force had grown to 1200.

More planes meant more people, here and overseas. All told, 55,000 New Zealanders served with the RNZAF, the RAF and the Royal Navy's Fleet Air Arm during World War Two. While most aircrew were men, there were a handful of women pilots, including four who served in England as Air Transport Auxiliary ferry pilots. One of these, Jane Winstone, was killed when her Spitfire crashed in 1944.

Casualties were high. In the six years of conflict, 4149 aircrew were killed in action or on active service. All told 3190 New Zealanders lost their lives in Europe – 1680 of those with Bomber Command, and a further 345 airmen died in the Pacific.

By definition, any anthology can only be an imperfect history. There are so many stories; too many to tell here but perhaps these pieces will give a sense of a different time – and different attitudes.

For me, one anecdote precisely measures the distance between then and now. It concerns a group of secondary school students visiting a Battle of Britain museum in the UK. The guide patiently explained when the battle had happened and who'd been involved – for these fortunate recipients of the finest modern education knew neither. A discussion followed about the significance of the combat and the relatively small number of pilots involved. The guide told the pupils how pilots flew several times each day, sometimes being shot down in the morning and flying again after lunch. One student was shocked to hear this. 'You mean,' he asked, incredulous, 'you mean they had to fly again before they'd even had counselling?'

Yes, lad, they did. And some of them did so for five years. Not always in combat, of course. For new recruits, life in the air force began with training. Which had perils of its own although, initially, it seemed like a doddle.

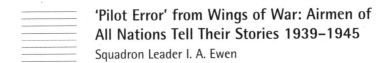

'Pilot Error' from Wings of War: Airmen of All Nations Tell Their Stories 1939–1945
Squadron Leader I. A. Ewen

After finishing my flying training, I was posted as a staff pilot, in 1941, to the Royal New Zealand Air Force Station at Woodbourne, Blenheim. Being a newly commissioned pilot officer, I was detailed, soon after arriving, to ferry a party of ground staff to a drogue-towing airstrip some distance from our base. I had only flown a Vickers Vildebeeste once or twice before.

After taxiing out from the hangar, I was given the green light for take off by the Duty Pilot. Steadily moving the throttle forward, we were soon bowling down the runway and gaining flying speed. Just as I was about to ease the aircraft off the ground, there was a resounding bang; the airscrew quickly stopped turning and we gradually came to a halt, fortunately well short of the end of the runway. Whatever, I wondered could the trouble be?

A ground staff sergeant behind me leant forward. His voice was tactfully quiet. 'Have you switched the petrol on, sir?'

But if training seemed easy at the start, it soon became clear that it wasn't.

The Pitcher and the Well
Edited by J. D. McDonald

In our course, easily the best pupil was Jim. Older than the rest of us, but we all liked him because he was such a good guy and so damned well balanced. Of course I was regarded by the younger pups as coeval with Methuselah and Jim was looked upon as Father Time himself. Perhaps that's why we became, if not friends, at least something more than acquaintances. Apart from our extreme longevity we had nothing in common except our length. Oh yes, Jim was long and thin too. He was as steady as I was mercurial; as unimaginative and pain-staking as I was the reverse. Just plain, solid virtue.

At the Wing's Dance he met a girl. At the beginning of the last dance they announced their engagement. My heart warmed to Jim. How unlike him! And yet, when I think back on it, how like him!

Jim-like, he weighed things up, considered all the factors and came to a firm decision. Jim's decisions were always firm ones. Then he carried it through. Perhaps he'd hurried a little this time but then, life was hurrying us all at the time.

It was the real thing, too. I watched them. She came nearly to his shoulder and, when she got a crick in her neck from looking up at him, she put her palm behind her neck for a little support. The rest of us had somehow merged into the decorations. Who'd have thought that Jim's aged hormones — he was all of thirty — would have acted in such a skittish manner.

The way she looked at him comes back to me with singular vividness as I lie here. I can't see anything to either side except when they move me to work on me but I insist that my position will enable me to see anyone approaching my bed. Nobody does. Nobody except Don.

Naturally, next day Jim went along to beat up her home a little. Now, an aged Vincent, such as he was flying, is the most reliable aircraft in the sky. The Peggy engine which powers it is the most reliable piece of ironmongery in the Air Force, and Jim was reliability personified. Nevertheless he was number three.

We saw the smoke almost as soon as we heard the siren. I heard that Jim had been hanging the old Vincent by her fan in the sky and she'd just slid in. Quite gently. But she burned like all hell nevertheless.

The crash tender was there when we arrived. The N.C.O. in charge in his asbestos suit just walks through the flames and brings the pilot out. Which accounts for the care with which that N.C.O. inspects his suit for defects.

This time he came through empty handed and when they took his helmet off he just looked around him. You see, there was a girl running in aimless circles round the fire. Did I tell you he finished in her father's hay-shed? She was making incoherent noises and that, coupled with the horrible smell of burning meat, made me want to retch.

Her home was about twenty yards away. We tried to take her there but she just stared at us and said very distinctly, 'He called out to me. Honest he did. Do you think it lasted long?' Her rather child-like voice had the horribly serious note of one who is genuinely seeking information.

Then she remembered and began that damned mouthing again. Nothing human, just an un-understanding animal noise like a bereaved beast. She'd stop ... then repeat herself; and so the ghastly cycle went on. Is memory a cyclic or rhythmical thing and not at all continuous as we think? And can shock be discontinuous too?

We were all very cut up about Jim, yet we got over him more easily than we did Ginger. You see, he was number three. That made all the difference.

Later, in England, when accidents were everyday things, we came to regard them as just that, but my earliest memories of them still stick. We grew up tremendously in a very short time. I knew nineteen-year-olds who were almost senile.

As trainee aircrew progressed, their new skills were challenged by the quirks and vices of operational aircraft rushed into production before the bugs had been ironed out. All too often, the results were fatal.

A Noble Chance: One Pilot's Life
Maurice McGreal

It was late in the afternoon and the wintry sun was a red blob in the murk low in the west and ahead of us in the circuit another Wimpy was turning onto final. We watched as he dropped his flaps and then with horror we saw the nose of their aircraft rise rapidly and a wing drop and then the aircraft started to fall like a stone as it stalled. A great burst of black smoke bloomed across the fields near the Cambridge Road and we knew that there would be none that would live from that one.

Bob looked across to me and said, 'The Wimpy flaps have struck again.' He was referring to the problem associated with the way the Wellington changed its longitudinal trim when the flaps were operated. This poor chap had obviously not got his trim wound on quickly enough to maintain control.

A New Zealander and a couple of Canadians were in the crew that died and the Adjutant put up a list setting up a burial party. Grant Jillett,

Bob Fotheringham, Mickey Walker, Ken Manson and I, plus some Canadians and RAF fellows; twelve in all.

In our overcoats with gas masks slung and buttons highly polished we lined the graveside of the four who were to be buried in the Steeple Morden village church yard. The firing squad fired and the bugle blew appropriately. The coffins were lowered, but someone had forgotten to hold the lanyard of the flag on our one and down it went too.

A moment's indecision and then Goldie Webb, a Canadian navigator, jumped down onto the coffin and grabbed the flag in a tidy handful. Bob Fotheringham and Ken Manson gave him a hand and with a spring he was back into line at attention as if nothing untoward had happened.

On the way back we all stopped at the Red Lion in Melbourn Village and had a few pints. Goldie Webb said that he was the only one who could say he had been into the grave and out again. But alas, for most of that burial party few would be alive after three more months had passed.

I never again served on a burial parade; and perhaps the RAF felt that it did little for the morale of the others who were left. 'To go for a *Burton*' was the term the airmen invented to cover death. (*Burtons* was a well known ale) ... a few pints downed by the lads who were left and a jolly ribald song that told of the joy of living and of the pleasures of pretty girls, was often the only obituary most airmen had.

In the early months of the war, British troops and aircrew were once again in France, awaiting the inevitable German assault. Twenty-one years after 'Grid' Caldwell, A.R. Kingsford and 'Mac' McGregor had celebrated Armistice Day, another generation of New Zealand airmen was preparing to fight where they had fought. And when the blitzkrieg struck and the Luftwaffe came, one of those airmen quickly became a celebrity, feted in the press. Officially, this Hawker Hurricane pilot was Edgar, but everybody called him 'Cobber' ... Cobber Kain. He was the first RAF ace of World War Two.

Fighter Squadrons: The Epic Story of Two Hurricane Squadrons in France
Noel Monks

It is Friday, June 7, 1940. On a dusty emergency aerodrome near Blois in France a two-seater Magister communications plane is being loaded up with kit by an orderly. A tin helmet and a gas-mask complete the loading, and the orderly reports to a group of young pilots wearing the uniform of the Royal Air Force chatting gaily a few yards away:

'Gear aboard. Good luck, sir.'

A tall, broad-shouldered, black-haired flying-officer, with the ribbon of the Distinguished Flying Cross newly sewn below his wings, leaves the group and walks to the waiting plane. He calls, 'Cheerio, chaps! Be good,' to his colleagues and climbs into the cockpit, settles himself, gives a mechanic the thumbs-up. The engine roars.

Suddenly a mischievous grin spreads over the face of the young giant in the Magister. On the port side of him he has caught sight of a Hurricane fighter. It is his old ship. Only yesterday at 20,000 feet over Rheims he had 'squeezed the teat' that controlled its eight machine-guns, and down went his twenty-fifth 'Nazi.'

So he uncurls his long legs from the cockpit of the Magister and, going across to the Hurricane, wedges himself into the cockpit. 'One more beat-up, me lads,' he calls, and he is off across the aerodrome in a cloud of hot dust.

The Magister's engine is left ticking over nicely.

With a roar like a thunderclap the Hurricane comes back over the 'drome, above the heads of the little group of officers — only just above their heads, because it is barely twenty feet off the ground, is upside down, and travelling at 350 miles an hour. The boys call this a 'beat-up.'

Still upside down, the Hurricane shoots up to 1500 feet in less than a minute, turns right side up, then starts a series of rolls earthward.

That is just how the young man in the Hurricane is feeling, rolling about in the thin air. The little Magister below, with its engine ticking over nicely, is going to take him home to England for special duties.

Two rolls are completed. The group of R.A.F. officers suddenly stop laughing and chattering. One says anxiously, 'What the hell?' as the

Hurricane goes into a third roll. His experienced eye can see this will bring it mighty close to mother earth.

Then three or four of them yell 'Cobber, Cobber.' They start running. There is a crash. The Hurricane does not quite complete the third roll. Its port wing touches the ground....

The young officers lift their dead comrade from the wreckage. A mechanic climbs into the cockpit of the Magister. He switches off the engine.

That is how Flying-Officer E. J. ('Cobber') Kain, D.F.C., 73 (Fighter) Squadron, Royal Air Force, died. He was the first Ace in the war against Nazidom, and he was the last pilot but one of that squadron's personnel that flew off so gaily to France that autumn to be still on his feet in France on that June day.

Some were in hospital, others were back in England, one, accidentally killed, lay buried at Metz. But not one of that heroic band had been killed in action against the enemy.

Two of the squadron's N.C.O. pilots had been killed in action, but they were avenged many times over — nearly two hundred times over.

I start this story of 'The Two Fighter Squadrons' with Cobber Kain's death because I am sure that is how every officer and man of the two squadrons would want it started....

When young Martin took me out to the airfield on my first visit to 73 Squadron, we stopped before a small bell tent.

'That's the duty tent,' he said; 'we snore off there when we're waiting for an alarm.'

I could hear snores coming from inside, and, pulling back the flap, I peeped in. The floor was ankle-deep in mud, but down the centre some straw had been sprinkled, and sprawled on it, their heads pillowed in their arms, their flying-boots protruding beyond the straw into the mud, were two hulking figures. One was asleep and snoring, and the other was just lying there, shivering as the chill wind rushed in underneath the tent.

The wide-awake one jumped to his feet when he saw me, and kicked his colleague, who groaned, turned over, and sat up.

Straw was sticking from his shock of black hair and his teeth were chattering when he stopped yawning. His feet squelched in the mud as he started to rise, but the tent was too low for him to stand upright, so he bowed his way outside and uncoiled himself.

The young giant was grinning broadly when I shook hands with him. He said, 'Sorry I was asleep, but that's the only way a fella can get warm out here — by going to sleep.'

Boyishly he introduced himself: 'I'm Cobber Kain,' he said.

Some of my English colleagues standing by thought that a strange front name, but, being an Australian, I recognized it as slang out home for chum (England) or buddy (America).

Otherwise it conveyed nothing to me or to my colleagues. Here was just another of the boys. But looking back on that initial meeting with the boy who was to become the first air Ace of World War No. 2, I realize that there was something outstanding about him. Even before Doc Outfin on that first day told me to 'keep my eyes' on Cobber Kain, one sensed that in this towering figure of a lad, with the Rugby shoulders, impish grin, and soft, drawling voice, was the very spirit of the Junior Service, out in France again after twenty-one years to knock the Boche out of the sky, as it did before.

There were many splendid airmen fighters besides Cobber Kain in 1 and 73 Squadrons who were to cover themselves in glory in France, but there was some quality in the whimsical boy from far-off New Zealand that endeared him to every one who came in contact with him, not only as a great destroyer of Huns, but as a natural, friendly, unspoilt young man.

No one ever heard Cobber Kain brag. No one ever knew him to go about with his chin stuck out, looking for trouble. Out of the cockpit of his fighting machine he was mild-mannered and peace-loving. In the cockpit he was a killer.

What more could a nation at war ask of a son?

I was standing beside Cobber that November 8 after he had brought down his first enemy plane. It had been a short, sharp battle.

One minute Cobber was lying on his back on the airfield, scanning the sky through his glasses, as he used to do for hours on end before the show started in a big way, the next he was dashing towards his Hurricane, shouting, 'I've spotted a Jerry.'

He was soon out of sight. It was said of him that whether in the air or on the ground he could spot an enemy plane five minutes ahead of any one else. In a shade under nine minutes Cobber was 20,000 feet above the 'drome, in the wake of the enemy reconnaissance aircraft — a Dornier, that had gamely stuck to its job of getting photographs, though its pilot must have known he was right over a hornet's nest. No one in Germany was ever going to see the pictures taken by that Dornier.

The New Zealander closed in on his prey, gave a slight 'squeeze of the teat,' and the Dornier headed for earth. It crashed at a crossroads just outside a little village near Rouvres, and the plane and the crew were scattered about in pieces all over the place.

The man who had brought it all about reached the spot just as the mess was being cleared up. He stood beside me, still a little out of breath with the excitement of having killed three fellow-humans.

Almost at our feet lay the head of the Nazi pilot, still in its helmet, with wide-open eyes. Cobber did not have any tears to shed. He just said very quietly, 'Well, it was either them or me.' Then he went along to the little home-made bar and served himself a glass of champagne.

When I got to know him better, I asked Cobber about that first victory of his: how he felt when he had the Dornier lined up in his sights. He told me his heart had come right up into his mouth and he was sweating under his helmet. He was scared stiff, he said, at the thought that his mechanic might have forgotten to put ammunition-belts in the eight machine-guns that he was about to let fly at the enemy ahead. Then, while he was in the middle of worrying about it, the Dornier came up in his sights. The Hun's rear gunner must have sighted Cobber at the same moment, for, just as the New Zealander pressed the trigger, a stream of bullets flew past him. The German bullets, far less in number, sped harmlessly by, while the British bullets tore into the unarmoured hull of the Nazi aircraft, and the battle was over.

I asked Cobber later, too, if he had slept that night. He said he went off as soon as he hit the pillow.

Out home in Auckland, New Zealand, Cobber Kain had been flying mad since he was sixteen. His parents put up with him for three years, then gave him his fare and let him go off to England to realize his life's ambition by joining the Royal Air Force.

The day after his ship berthed in London he was up at the Air Ministry to sign on. But bad news was in store for him. He failed in his medical. He was too big for his age, had overgrown his strength. 'Come back in a couple of months,' he was told.

Back he went, and the second time he was accepted.

No more willing pupil ever presented himself for training as an R.A.F. pilot than Cadet E. J. Kain, and for the next two years he went through the mill that turns out the most thoroughly trained airmen in the world. He studied English, world history, aerodynamics, mechanics, maths, mechanical drawing, the construction of engines, airmanship, air navigation, wireless telegraphy.

Then, in his second year, he crammed advanced navigation, armament, flight routine, signals, law, meteorology, the workings of the Army, Navy, and Air Force.

In between all these studies he flew training aircraft. He could already fly when he came to England, but he had to go through it all over again.

Towards the end of his second year Cobber Kain found himself at the Advanced Training Squadron, where he was taught all about Service flying — bombing, fighting, etc., by day and night.

Finally he got his wings. Then one day he turned up at 73 Squadron.

By some, Cobber Kain was thought to be too untidy in appearance to make a good R.A.F. pilot. His hair always wanted cutting, his trousers could always do with a press, and his flying-boots cracked twice as fast as those of his colleagues. In addition, his squadron was at the bottom of the class. But that was in peace-time.

That day out in France, when we stood together looking at his first victim, his trousers not only needed pressing, but they could also have done with several patches. As if that mattered with a £15,000 enemy plane and three enemy airmen in pieces at his feet!

Back at the Lion d'Or in Rheims that night, I jotted down in my notebook: 'Looks like Doc Outfin knew something when he told me to watch Cobber Kain. Got his first Jerry to-day.'

Cobber Kain was the subject of a lot of talk in the War Correspondents' mess at dinner, and next day up at the Press Office. This office, staffed by experienced newspapermen under Wing-Commander Stanley Bishop, was our H.Q.

Every man in the P.O. knew a story when he saw one. All had given up good jobs in Fleet Street at the outbreak of war to 'put over' the R.A.F. to the world. In Cobber Kain they, and we, recognized a potential Ace, and Aces in the last war made world news.

But hardly had we made Cobber into the first Ace of the war — he was the first to shoot down five enemy aircraft — than the Powers That Be closed down on us. We were not to mention his name. Not even his nickname.

The most elaborate Press Section of any of the Forces had been set up by the Air Ministry, and here we were, with a real head-liner in our midst, but on no account must he be mentioned. It all seemed so strange to us.

But we could not help writing about him, though what we wrote was very heavily censored. Every time we visited 73 Squadron after that November 8 there was a story, and plumb in the middle of the story was Cobber Kain. He could not help being there. He was just doing his job, risking his neck for his country. We were just doing ours, telling the world about him.

It seemed that every second story we sent out of France started off with 'A twenty-one-year-old New Zealander,' whom our officers, of course, recognized as Cobber Kain, and they promptly wrote in his name.

The French Press had taken up Cobber in a much bigger way than we were allowed to, and he became France's first hero of the war, too. Several times when I was on leave in Paris with Cobber he was mobbed by French people, male and female, old and young, in the Champs Élysées, or in the foyer of a theatre. 'Monsieur Cob-baire,' they called him. And his exploits were enough to fire any nation on whose side he happened to be fighting.

There was the time when, leading two colleagues on patrol, he ran into twelve Messerschmitts; he shouted into his radio, 'Get going, chaps. Make for the middle of 'em.' He got one enemy plane in the first fifty seconds, then, when one of his colleagues got put out of action and had to make a forced landing, he saw the other member of his flight hard-pressed by five Messerschmitts directly above him. He did a rocket loop and came up underneath the M.E.'s, his eight guns blazing away in a mad fury. One fell in flames, narrowly missing Cobber as he shot upwards. Then he found himself sandwiched between two M.E.'s, one in front, one on his tail. Shells and bullets whistled all round him. His own ammunition had run out, but did he put his wheels down in token of surrender? Not Cobber Kain. The colleague whom he had rescued knocked off the M.E. from Cobber's tail just as a shell landed in his engine.

The New Zealander's Hurricane burst into flames, as, shouting profanities into his radio, Cobber made a wild effort to ram the Messerschmitt ahead of him; but his engine was dying on him and he lost speed.

He was at 20,000 feet, and, choked by smoke, Cobber reluctantly put his nose down. The flames got worse, and he pushed back the hood to bale out, but remembered, just as he was about to jump, that he had not fixed his parachute properly when he had taken off.

So he returned to the cockpit, into which oil and smoke were pouring, although the downrush of air had put out the flames. The engine had begun to tick over again, and the lion-hearted Cobber, instead of looking for a place to crash, looked for a spot to land.

He just made the edge of the Metz aerodrome. He climbed out of the badly damaged Hurricane, then, in his own words, 'I fell flat on my bloody face: passed right out, like a cissy-boy.'

Back at 73 every one was sitting round the mess, waiting to hear the worst. Cobber had been seen shooting down in flames, well over the line. He had probably wiped himself off.

After about two hours' suspense the phone rang. It was Cobber, from a French hospital. He wanted to speak to Doc Outfin. He said, 'These big-hearted "froggies" think I'm dying or something. They've taken my clothes away and bloody well won't let me get up. For Pete's sake talk to the surgeon and have him send me back to Rouvres.'

Doc Outfin helped out, and Cobber arrived back at Rouvres, to the cheers of the villagers and his colleagues. This time his trousers were far beyond the patching stage.

We did not want to get Cobber court-martialled, so instead of saying he had not adjusted his parachute, we said it had come unhooked.

Cobber had another glass of champagne, but only one. He was on patrol in the morning.

Not long after that, Cobber was 'grounded' by Red Knox for some minor breach of discipline and was sent to take charge of the operations tent. To a man like Cobber Kain that tent was a torture-chamber. He had only to switch on the radio, and he could hear the boys on patrol chatting to each other as they spotted some Huns, or went into action, or just coasted along.

By the second day he was like a caged lion. Then, on March 26, after the early patrols had got into several scraps, word came through that there was much enemy activity over the line. That was more than flesh and blood could stand. Cobber bolted from the operations tent and was 30 miles away in five minutes.

Here is his official combat report of what followed:

'In the Luxemburg corner I saw a number of enemy aircraft and proceeded to investigate at 2.30 p.m. I gave a message on R/T (radio telephone) to Flying-Officer Perry and Sgt.-Pilot Pyne, who were with me, "Enemy aircraft ahead," and proceeded to attack. I turned into the enemy, which had started to climb and gave a burst at the leader, who pulled up, turned on his back, and spun away in flames. I then noticed 5 more M.E. 109's working round behind me, so I turned hard right and took a sight on the near machine. I fired a burst at him; he dived away and I took three deflection shots at another M.E. 109 which was slowly turning ahead of me. I got behind this aircraft and gave it a burst. He turned on his starboard side and dived right down towards earth.

'I then took observations about me. The sky seemed to clear, so I looked for my other two machines. I was just turning south when my cockpit was hit by a cannon shell, while another hit my gravity tank.

'The explosion on the hood of my cockpit rendered me unconscious, but I came to diving steeply. After a while I managed to pull out of the dive and tried to bend down and turn off the petrol, but the flames burnt my face. I headed towards France to gain as much ground as possible, and when the flames got too intense I decided to abandon my aircraft.

'I got out from the port side and pulled my ripcord at 12,000 feet. I came out of the cloud at 10,000 feet. It was all very still, and I thought I was in heaven. Believing I was now near the frontier, I started slipping the air to get down quickly. I landed at Ritzing, near a wood. I gathered in my parachute and scrambled towards the wood, where I hid it. I then headed in a southerly direction, as I didn't know whether I was in France or not. A French Captain challenged me at pistol point and asked my nationality, and we set off for Evendorff. The Captain told me I had landed in no-man's-land. I was burnt and my leg was hurting, so I received medical attention, and was sent back to Rouvres in a French staff car.'

Cobber arrived back at the mess in a terrible state. Not only his trousers, but his tunic also was in tatters now. It took Doc Outfin two hours to get the New Zealander washed. Then, in answer to the Doc's query as to what was wrong with his leg, Cobber said casually, 'I dunno, Doc. Some things went in and I don't think they came out.' He fainted then, and Doc Outfin extracted twenty-one pieces of shrapnel from his leg.

Two hours later, when the officers were leaving to attend the sergeants' mess as guests for dinner, Cobber started yelling for someone to come and get him. His shouts went echoing round the village.

Doc Outfin and one of the medical orderlies carried Cobber up the village road to the sergeants' mess on a stretcher, and after he had drunk their health and they had drunk his, he was carried back to this billet, satisfied. He was asleep as the faithful Medical Officer walked at the head of the stretcher.

Quite a day for a boy of twenty-one.

When he had recovered sufficiently, Cobber Kain was granted ten days' leave in London, during which he got engaged to a charming girl, and collected his Distinguished Flying Cross, the first to be awarded to any British airman in France.

Some of the recklessness had gone from him when he returned to the squadron on April 11. He had a keener desire to live than before, but he had only about seven weeks to go.

From then until he was killed Cobber Kain was a hard-working pilot, along with the other British fighter pilots in France. Hardly a day passed that he was not in action, and his bag of enemy aircraft had jumped up to a round dozen.

Soon after he resumed duty from his last leave, Cobber Kain lost his lucky charm that he always wore round his neck. It was a green jade Maori *tiki*, or god, given him by his sister. He told me later that he was sure that every patrol he did without it was going to be his last. The first time he noticed his *tiki* was missing was when he was about to engage a Messerschmitt 110 at 20,000 feet. He had his prey all lined up in his sights and reached inside his jacket to pat his *tiki* before 'squeezing the teat,' when he found it was missing. In the split second that his mind drifted off the work in hand and got on to worrying about where he had left his *tiki*, the enemy had dived, and the next thing Cobber knew was that bullets were thudding into his plane from the rear. He had a narrow escape. He found his charm two days later. A mechanic had picked it up on the 'drome. The day Cobber got it back he shot down a Dornier and a Heinkel.

Before he returned from his last leave in London, Cobber Kain went into the Air Ministry and said he was worried about all the publicity he had been receiving. He said that in fairness to his colleagues, he should be moved to another squadron.

That was a typical gesture of that generous-hearted flier.

From May 10 until he died, Cobber did not get much time to think about anything but shooting Huns out of the sky. And this he did practically every time he went up in those crazy days when the whole world, or at least the air world, went mad.

On the night of May 15 Cobber Kain walked slowly into my room at the Lion d'Or looking as though he had just come up out of a coal-pit. I turned the bath on while he peeled off his clothes — for the first time in six days. He was dazed for want of sleep, and said hardly a word. He got into the bath, stretched out, and sighed loudly with the luxury of it. I started brushing the dirt from his tattered uniform, and rang for some hot coffee. After about ten minutes I looked into the bathroom which went with Room 124, and Cobber Kain's chin was resting on his massive chest, his eyes were closed, and the water was making little bubbles as he snored a fraction of an inch from the surface.

When the coffee arrived I woke him up, and he cried, 'Blow me down! I dreamt I was back home in dear old N.Z.'

Cobber told me before he left to go back to Rheims-Champagne aerodrome that the Nazis were expected to drop parachute troops all round us that night. Officers and N.C.O.'s of 1 and 73 Squadrons, he told me, were standing by with rifles all night, ready to do battle with the parachutists.

As if chasing about in the skies all day was not enough for a pilot, without having to sit up all night with a rifle in his hand!

At 4 a.m. I was hauled out of my bed and told to get going in the clothes I stood up in. I must not take so much as a typewriter. We were to make for Château Thierry and wait there for orders.

The room that had been my home for nearly eight months was left with all my war-kit and bits and pieces in it.

A miniature of my wife and an autographed photo of Cobber Kain were all that I brought with me.

Bombs were dropping all over the place as we left Rheims. I wondered how the Hurricane boys and the bomber boys were doing — if they were still alive.

The Hurricane boys were, but not many of the bomber boys who had been based in Rheims were left.

My next meeting with Cobber Kain was in Paris on May 31. We had a slap-up dinner-party at Maxim's, now patronized daily by Göring. Doc Outfin was on his way home, and Cobber and Willie Williams, equipment officer, were up to see him off. There were only three of the original squadron left now.

It was a grand party. Word got round that 'Monsieur Cob-baire' was dining in the famous restaurant, and people crowded in off the streets to cheer the young Ace.

It was to be the last time I was ever to see Cobber, and he had never looked fitter or neater. He had on a brand-new uniform; his hair was cut short; his eyes were clear. He looked just what he was: an officer and a gentleman. The blue-and-white ribbon on his breast told, also, that he was a hero.

My wife, an American, had heard much of Cobber Kain, but had never met him until that night. Because of all she had heard, she was afraid she would be disappointed in this great hero. But she was not. She, not long from the land of football and baseball heroes, had never met such a 'guy' in her whole life as this six-foot, genial sky-boy.

When we parted that night for the last time on this earth, Cobber Kain said to me, 'Don't put it in the paper — but I'm getting married next week. That is, if I'm alive.'

A week to the day, and he was dead.

Fifteen days later, the French surrendered. In less than two months, German forces had done what they could not do in four years of fighting during World War One. During the Battle of France, 915 RAF aircrew, including 534 pilots, were killed, wounded or missing and 386 Hurricanes and 67 Spitfires were shot down or destroyed on the ground. It had been a costly battle.

But some 340,000 troops had been evacuated at Dunkirk. And when the anticipated invasion of England came, they would be able to fight. So would the pilots who had provided air cover over the beaches. One of those was

Alan Deere. He was shot down near Dunkirk, hitched rides to the beach and returned to Dover in a Royal Navy destroyer carrying evacuated troops.

At Dover, he caught a train to London (at one point he was asked to get off because he didn't have a ticket), then took the tube back to his squadron at Hornchurch airfield. Mind you, that was just the beginning. In the next few months, during the Battle of Britain, Al Deere was involved in two mid-air collisions, one with an enemy fighter, and one with a 'particularly promising' trainee pilot on his final test. He was also shot down three times (once by another Spitfire), either crash landing or bailing out; and, on another occasion, didn't even get airborne before a bomb destroyed his aircraft during take-off. No wonder he called his book *Nine Lives*.

Nine Lives
Group Captain Alan C. Deere D.S.O., O.B.E., D.F.C.

About 3,000 yards directly ahead of me, and at the same level, a Hun was just completing a turn preparatory to re-entering the fray. He saw me almost immediately and rolled out of his turn towards me so that a head-on attack became inevitable. Using both hands on the control column to steady the aircraft and thus keep my aim steady, I peered through the reflector sight at the rapidly closing enemy aircraft. We opened fire together, and immediately a hail of lead thudded into my Spitfire. One moment the Messerschmitt was a clearly defined shape, its wingspan nicely enclosed within the circle of my reflector sight, and the next it was on top of me, a terrifying blur which blotted out the sky ahead. Then we hit.

The force of the impact pitched me violently forward on to my cockpit harness, the straps of which bit viciously into my shoulders. At the same moment, the control column was snatched abruptly from my gripping fingers by a momentary, but powerful, reversal of elevator load. In a flash it was over; there was clear sky ahead of me, and I was still alive. But smoke and flame were pouring from the engine which began to vibrate, slowly at first but with increasing momentum causing the now regained control column to jump back and forwards in my hand. Hastily I closed the throttle and reached forward to flick off the ignition switches, but before I could do so the engine seized and the airscrew stopped abruptly. I saw with amazement that the blades had been bent almost double with the impact of the collision; the Messerschmitt must have been just that fraction above me as we hit.

With smoke now pouring into the cockpit I reached blindly forward for the hood release toggle and tugged at it violently. There was no welcoming and expected rush of air to denote that the hood had been jettisoned. Again and again I pulled at the toggle but there was no response. In desperation I turned to the normal release catch and exerting my full strength endeavoured to slide back the hood. It refused to budge; I was trapped. There was only one thing to do; try to keep the aircraft under control and head for the nearby coast. The speed had by now dropped off considerably, and with full backward pressure on the stick I was just able to keep a reasonable gliding altitude. If only I could be lucky enough to hit in open country where there was a small chance that I might get away with it.

Frantically I peered through the smoke and flame enveloping the engine, seeking with streaming eyes for what lay ahead. There could be no question of turning; I had no idea what damage had been done to the fuselage and tail of my aircraft, although the mainplanes appeared to be undamaged, and I daren't risk even a small turn at low level, even if I could have seen to turn.

Through a miasmatic cloud of flame and smoke the ground suddenly appeared ahead of me. The next moment a post flashed by my wingtip and then the aircraft struck the ground and ricocheted into the air again finally returning to earth with a jarring impact, and once again I was jerked forward on to my harness. Fortunately the straps held fast and continued to do so as the aircraft ploughed its way through a succession of splintering posts before finally coming to a halt on the edge of a cornfield. Half blinded by smoke and frantic with fear I tore at my harness release pin. And then with my bare hands wielding the strength of desperation, I battered at the perspex hood which entombed me. With a splintering crash it finally cracked open, thus enabling me to scramble from the cockpit to the safety of the surrounding field.

At a safe distance from the aircraft I sat down to observe the damage to person and property. My hands were cut and bleeding; my eyebrows were singed; both knees were badly bruised; and blood trickled into my mouth from a slightly cut lip. But I was alive! I learned later from the technical officer who examined the wreckage after the fire had been put out, that the seat had broken free from the lower retaining bar thus pivoting upwards, and so throwing my knees against the lower part of the dashboard.

The aircraft had ploughed a passage through three fields, studded with anti-invasion posts erected to prevent enemy gliders from landing, and bits of aircraft and posts were strewn along the three hundred yards of its path. My Spitfire was now a blazing mass of metal from which a series of explosions denoted that the heat was igniting the unused ammunition, to the consternation of a knot of onlookers who had by now collected at the scene of the crash.

A woman, whom I had observed coming from a nearby farm-house, approached me and said:

'I have telephoned Manston airfield and they say that an ambulance and fire engine are already on the way. Won't you come in and have a cup of tea?' ...

The morning of August 31st was strangely and ominously quiet in the Hornchurch Sector, particularly in view of the good weather, and it was not until about midday that the squadron received the order to scramble. We had just taxied into position for take-off, and were all lined up ready to go, when a counter order was passed over the R/T. No sooner were we again parked in dispersal with engines stopped than a wildly gesticulating telephone orderly indicated that we were to start up again. In a matter of seconds all twelve aircraft were again taxi-ing to the take-off end urged on by the Controller's now near-hysterical voice shouting over the R/T 'Hornet aircraft get airborne as quickly as you can, enemy in the immediate vicinity.'

Hurriedly, desperately, for I had no wish to be caught taking off, I swung my aircraft into wind only to find my take-off run blocked by a Spitfire, the pilot of which was looking vaguely around for his position in the formation. 'Get to hell out of the way, Red Two,' I bellowed, recognising my number two from the letters on his aircraft. It was a second or two before he made up his mind to move; immediately he did so I opened the throttle and careered across the airfield in pursuit of the squadron which had by now cleared the far hedge, and with wheels retracting was turning and climbing away from the airfield.

I was not quite airborne when a bomb burst on the airfield, ahead of me and to my left. 'Good, I've made it,' I thought. To this day I am not clear exactly what happened next; all I can remember is that a tremendous blast of air, carrying showers of earth, struck me in the face and the next moment thinking vaguely that I was upside down. What I do remember is the impact with the ground and a terrifying period of ploughing along the airfield upside down, still firmly strapped in the cockpit. Stones and dirt were thrown into my face and my helmet was torn by the stony ground against which my head was firmly pressed.

Finally the aircraft stopped its mad upside-down dash leaving me trapped in the cockpit, in almost total darkness, and breathing petrol fumes, the smell of which was overpowering. Bombs were still exploding outside, but this was not as frightening as the thought of fire from the petrol now seeping into the ground around my head. One spark and I would be engulfed in flames.

'Al, Al, are you alive?' urgently, and to me miraculously, the voice of Pilot Officer Eric Edsall, who had been my number three in the section, penetrated to my dazed senses.

'Yes, but barely. For God's sake get me out of here quickly,' I answered breathlessly, desperately afraid of fire.

'Can you reach the release wire on the door, Al? If you can free the catches I might be able to lever it open; I can just get my hand underneath it.'

It was no mean feat to move my right hand across, locate the spring-loaded release wire, and exert sufficient pressure to free the locks. But somehow I managed it, and after a tremendous struggle Eric managed to force the door outwards. The next problem was to free myself from the parachute harness, as the small aperture created by the now opened cockpit door was barely large enough to wriggle through, without the extra impediment of a parachute pack. This too I eventually managed, and after a frantic struggle squeezed my way out into the blessed fresh air.

In his book, *A Clasp For The Few*, Kenneth Wynn provides brief biographies of the 129 New Zealand pilots and aircrew who flew with the RAF during the Battle of Britain. One of them, Donald Cobden, had been selected as an

All Black in 1937 and chosen to play as wing three-quarter in the first test against the Springboks in August of that year, but was injured during the first half and had to leave the field. (For the record, the All Blacks won.)

Later that year, Donald Cobden worked his passage to the UK, where he subsequently joined the RAF. By 1940, he was a member of 74 Squadron, flying Spitfires over France and then in the defence of England. On 11 August, 'the squadron was patrolling a convoy in the English Channel when 40 Me 110s were sighted and attacked.' According to Kenneth Wynn, 'a fierce combat ensued, during which Donald Cobden was shot down and killed. He crashed into the sea a few minutes after mid-day on his twenty-sixth birthday.'

Throughout the months of fighting, another New Zealander played a crucial role — not in the air, but on the ground. Keith Park was the commander of 11 Group, Fighter Command, which bore the brunt of German attacks. In 1947 Lord Tedder, marshal of the RAF, said: 'If ever one man won the Battle of Britain, he did.' But it was a close thing. On 15 September, 1940, with massive Luftwaffe raids in progress, Winston Churchill visited Keith Park's headquarters. Noting that more German aircraft were coming in, Churchill asked Park: 'What other reserves have we?'

'There are none,' said Park.

Mercilfully, none were needed. By the end of that month, it was clear the Luftwaffe could not sustain its daylight assaults — there would be no invasion.

Along with Al Deere (and Donald Cobden), Bob Spurdle was another New Zealand pilot who fought in the Battle of Britain. Later in the war, he also served in the Pacific, before returning to Europe to fly Typhoons. His war ended on the ground with the army, but we'll come to that later. For now, it's 1940 and the last stages of the crucial struggle to control England's skies.

The Blue Arena
Squadron Leader Bob Spurdle DFC and Bar

October 22nd, Sortie 37. 1.00 hour duration: Sure enough, the warning wailed as we tore back to dispersal after a hasty lunch. George put his foot down and the brake (station-wagon) howled round the perimeter track. We could see the white fog-trailers weaving fantastic clouds far above us. I felt dead inside; it was our third scramble of the day. How long would this keep up? How long could it? ...

Up, up, oxygen on, air-scuttle open — slowly the blue sky turned a deeper shade and at 23,000 feet faint trails of mist formed, thickened, and streamed behind like comet's tails. The radio nattered away to our supporting squadron.

Far off we could see the trails of our adversaries — a 50-plus, according to Operations. They were higher than we, but ten to fifteen miles away. We climbed on, up and up in great circles. Now we were above them. We spread out into sections of two in a rough line abreast, and we could see their yellow spinners, Me 109's. One after another the yellow-nosed devils were peeling off for the deck — the usual trap for the unwary; follow one down and five more are on your tail.

A strange thing happened. We were some 700 yards away and closing head-on at astounding speed, when a cloud of them rolled for the deck. Then it turned to a rout without, as yet, a shot being fired. 109's are breaking in all directions. I see Johnny* and his section zoom up after seven or eight which climb towards the sun. Steve is whipping round after a dozen or so that overshot to starboard. It is going to be a real show.

I hared after Malan. A tinkling sound of empty cartridge cases bounced off my bullet-proof screen and scratched 'R's' wings. Malan had fired at a 109 crossing ahead of us. It rolled lazily on its back, hung there dead and ugly. A puff of black smoke and it disappeared in a dull, red flash. Something black and ragged fell away. Someone calls 'Look out!' Malan's 'Shut up, you fool!' brings a warped grin. I am sweating hot, choking with a dry mouth; I don't seem to be getting enough air to breathe. I feel reckless; impatient to squirt at one of the devils. Suddenly there doesn't seem to be a machine in the air; just a lot of crazy white cords lacing the blue above. 'R's' belly seems horribly bare and vulnerable and I roll her on to her back and look down. Sure enough, there's a 109 just a couple of hundred feet below. It sees me and rolls on its back and dives.

I heave the stick back, black out, recover and glance at the altimeter

---

* Flight Lieutenant J. C. Freeborn, DFC and Bar.

— 27,000 feet. The 109 is diving vertically. I grin and open 'R' full-bore, fine pitch. The airspeed winds up and up; it passes 450 miles an hour and the red patch denoting top permissible speed. I can't depress the nose sufficiently to get the gunsight on the Hun. I push at the stick and wind on full tail-trim. I go into coarse pitch and 'R' drops like a rock.

Now the slip stream is screaming and wailing past the hood and the whole machine is taut and quivering like a violin. Three hundred yards in front, the black 109 seems to be slowly drifting up as the Spitfire draws down on it. Below, the chequered fields are lazily expanding. I swallow and my ears clear with a popping sensation. The 109 is beneath me somewhere still dropping vertically. I twist the stock over and roll 'R' round her axis and the 109 reappears and the joy-stick starts to judder in my grasp.

Wham! A blast of solid air hits me. 'R' flicks into a spin like a mad thing and the stick threshes round the cockpit. Everything goes brown but I can see a horrible space where the starboard wing should be. I heave and struggle to get out. Something is holding me down. My hands are cut and bleeding where I grip the shattered perspex above my head. I pull and pray. Then, with fatalistic calm, I remember the Sutton-harness. I pull out the pin. Now for my helmet. I fumble with my neck-strap. Get out! Get out!

How much height is there left; but I can't read the blur which is the 'unwinding' altimeter. My hands are sticky with a clear slime which evaporates as I gape at them. A whining scream. Something black whips past and I struggle up and, like a cork, popped out of the spinning wreck.

I'm out! It isn't like falling at all. I shut my eyes and there is no rush of air, no sound. It is quite peaceful — a marvellous sensation. I take several deep breaths. 'Count ten', someone had said. Count ten! My fingers scrabbled at the rip-cord's metal ring and I heaved it out a foot. Nothing happened! Mad prayers and thoughts. I opened my eyes. Then my head nearly snapped off and pain cut into me; I felt as if I'd been knackered.

There is a white saucer hanging over me and everything is quiet except for little rustling sounds as the silk flexes. The chute seems small, the silken cords rigid, straining. I look at the ground. How funny not to have wings and a motor and yet fly! I look for the other boys but can't turn my head as the harness, loose in rest, has ridden up and squeezed my

Mae West about my ears. I 'brown out' through lack of oxygen. It is dreadfully cold; my boots have been torn off. The ill-fitting harness is cutting into my crotch. One of my testicles is being crushed. I wound the steel rip-cord round the harness release-plate and put my foot in the ring, standing up. If I can't release my weight and free the leg-straps, I'll go crazy; the pain is so awful. The cord, slipping, unwinds the quick-release plate. I sag back into the harness. The freed cord falls, turns slowly, grows smaller and smaller and vanishes thousands of feet below. I hang there in terrible fear; the slightest jar on the release-plate and the chute will be jettisoned and I'd fall like a stone.

Something whining shrilly streamed past and I saw strange twisted white lines drawn as into infinity. More of them and weird rushing sounds. I appeared to be the centre of a mad, wind-blown spider's web. Amazed, I heard the crackling, tearing sound of cannon-fire like a giant ripping canvas, and then a high whistling shriek. Something big and black tore past me — a 109E.

It climbed right in front of me, turning for another go. I cursed and wriggled frantically in the harness trying to draw my revolver.

There was a deep purring roar and 'P' flashed by, followed by 'S' — Steve and Wally! I laughed with relief. Boy, what a grandstand view! The Hun half-rolled for the deck and I watched him twist and turn beneath me. It seems that I can almost stand on him and my chute above quivered and rippled at the machine-gun fire as Steve got on the Hun's tail. Now he's had it! Ha! Ha! A good show, Steve! The Jerry staggered, slipped and fell, crippled and smoking into a wood. I saw rows of hop poles below me and the ground seemed to have come much closer all of a sudden. It is rushing to meet me. I got ready for the landing and fiddled with the harness but I can't make the thing manoeuvre as I want to. Loud and clear the rending crash of the Hun came up. Serve the bastard right!

The landing was just as they said — like a jump off a 12-foot wall. I lay still on the warm earth and panted. Hell, but my hands and feet were cold! I shut my eyes and relaxed on the soft ploughed field. I heard someone running over the uneven ground and next minute a huge farm-labourer threw himself on me.

'Get off me, you stupid oaf!' I screamed. 'Don't you know your own bloody side?'

He reared up, fist ready to smash me down. Slowly comprehension dawned.

'Be you RAF, sor?' he asked with a beaming smile.

'No, I'm a fucking angel! For Christ's sake, let me up!'

Then he heaved me to my feet and put his arms around me. I suppose I look scared, shivering with fear, and he started to pat me and tried to stroke my head.

'I'm not scared! I'm bloody cold. Leave me alone!'

By this time there must have been a hundred hop-pickers surrounding us, all smiling and shouting, 'Good for you.' 'Up the RAF' and other nice things; all, under the circumstances, quite inappropriate. There were some real dolls among the crowd and, looking into their eyes, I could see I had it made. But what can you do in socks, ankle deep in ploughed soil, bloody hands and a whole village watching?

A bobby arrived and took me in charge. Soon I was having tea and cake in his kitchen while he phoned Biggin for a squadron car to pick me up.

Three scraps in one day, a parachute descent and Peter St John dead. It had been quite a day....

October 30th, Sortie 38. 1.30 hrs duration
It was my first flight since baling out eight days before. We were at 33,000 feet and the cold made my left leg ache. The bruising was one long hurt from knee to ankle. The Jerry formation of some fifty fighters was about a mile ahead and we had the jump on them by several hundred feet. About a dozen peeled off and dived away towards the Channel. Then twenty or more! It's a rout!

Malan called, 'Get them boys! Go!' and 74 split into pairs, spread out and were after them in whistling curves.

I couldn't follow. Suddenly I was a mass of jangling nerves hanging in space. A terrible fear gripped me. Oh God! I'm stuck up here! I can't go down! And now about five of the yellow-nosed 109s were heading for me, guns flashing and sparkling. I was on my own.

'What happened, Spurdle?' Malan had me in his office. 'I couldn't dive, sir. I found I couldn't follow the others down! Then the Jerry top cover got stick into me and by the time I'd shaken them off, I was down to about 16,000 feet. I was all right then.'

'Did you get any of them?'

'No, sir. I know I damaged one, but every time I got into a firing position, two or more would latch on to me.'

'Well, you'll just have to get over it. Keep your eyes fixed on your No 1 and stick with him.'

The same day, Sortie No 39. 1.25 hours duration
Again, the bell rang. We'd only been down an hour! The boys started and looked towards the phone. The telephone orderly grabbed the receiver. Breaths were held — a faint pop-pop-popping of a battery charger in the distance — the orderly stiffened and looked towards us, repeating, '30 plus Maidstone — 20,000 feet'; the klaxon blared out. There was a great rush for the door, the boys grabbing their helmets and gloves as they dashed out. The inevitable stooge pilots started their loud barracking and bemoaning at being left out of the flap. Pure bluff!

Outside mechanics and riggers, warned by the scramble buzzer, race madly towards their machines. Soon the roar of the first Spitfire shatters the warm afternoon sky and rooks rise raucous from the field. Harnesses are being buckled and chocks being pulled away; over at the far edge of the dispersal some unfortunate pilot is still running to his kite — his is the furthest from our hut; he must have missed the brake. Christ! He'll get a trimming from the CO.

Now they are taxiing away from the field towards the runway, the ground crews standing by their fallen chocks, their arms slack and a strange longing in their eyes. Poor devils! They don't know the thrill of smooth wings, the rushing eager power surging underhand. No Merlin's magic for them — theirs is the dirt and the mud and the oil; they can only trudge away, forgotten until the machines return.

Thirty-plus Maidstone — 20,000 feet: the familiar old call! My breath came hard as I reached my kite. Christ, but I'm out of training! My knees were weak and trembling — silly jerky thoughts flashed through my mind — Florance — Mum — my tiki. I patted it. Practised hands and automatic movements got me into the cockpit. 'Good luck,' the rigger shouted — I grinned in answer, but the face mask gave him no sign. I waved and winked; Bill saluted — a good guy — the sucker wants to be a pilot.

Air pressure low — never mind, it'll build up, radiator temp — oil pressure, gun sight, tail and rudder trim — OK. Oxygen? Christ, it's a bad bottle — I'll kill the bastard. The radio warming — crackled into life — yes, old slow coach is copping it — Jeez, but the CO is tough. Must remember that phrase — nice for a party. I'm scared I'll funk diving again when the time comes. God! It's an awful feeling.

The ground bumped by under my wheels, and some oaf in 'T' nearly ran into me. That's Nelson — or is it? I can't remember but cursed him mentally. I saw the first three zoom off down the runway — that's the CO with his Nos 2 and 3. Now I see Mungo taxiing into position, his 2 and 3 rolling either side of him. With a bellow and a cloud of dust they roared down the path and slid into the blue. I rapidly rechecked my taps — everything is OK and I saw Steve with his head in the office. The two No 4s take off — I hear Malan cursing his for being slow. There is no reply — they are bucking in the second section's slip stream — guess he can't take his hand off the stick to answer.

Where the hell's our No 2? I slewed my kite around a bit and saw him. He'd pranged it — he'd gone over on his nose; must have struck a soft patch in an old bomb hole. He is green — I laughed into my mask — he's a cocky oaf. Steve has seen it and raised his hand — thumbs up! I raised mine and opened the throttle as Steve's kite rolled forward. We tore down the runway and staggered into the air.

Up wheels, alter pitch, I felt the Rotol bite and throttled back a bit, the rev counter was fluctuating — I'll kill that mechanic! He hasn't checked that cable yet! The CO was calling up Ops and getting more dope. Curse 2 — now I'll have to take zero* and miss some of the gen. Christ! 30-plus at 25,000 feet and another bunch of 20-plus crossing the coast. I glanced around the squadron — 11-strong. I wriggled in my harness and could just see the tips of 4's wings. With Steve's 2 missing, that left only three in our section. I decided to stick to Steve; become his No 2 and 4 can tag along.

Higher and higher we climbed. The glycol temperature was nearing the danger point, the oil temperature was off the clock. Air pressure built up OK. Somebody's wheels weren't fully retracted and I called him up. I

---

* The Identification 'Friend or Foe' radio device was switched on by one of the No 2's detailed for this chore. Its intermittent transmission interfered with RT use in that particular machine.

raised my voice till it sounded like the fancy boys at the Running Horse — (it transmits better) — 'Your wheels are down, Yellow leader' — and I heard a laconic 'OK'. The wheels jerked into place.

Altimeter reads 13,000 feet — time to turn on the oxygen — Jeez, but it's foul — must be an old bottle — Hallo, Ops say, the Jerries have joined together and are heading north. 50-plus to 11! Bloody hell, just about the decent proportion for a good scrap. I switched on my gun sight and firing button. I closed the radiator flap a little. Oil from Steve's kite was spotting my windshield and I slid out to one side a bit. Hell's Bells, but it's getting cold — the fresh air scoop was jammed open and I cursed myself for not reporting it before. It's always the same and each time I come down I forget it. My mirror was streaky too — that's a kick for someone!

Things were now getting pretty active; above us, and towards the afternoon sun, there were several vapour tadpoles forming, streaking along — they were turning towards us too. All of a sudden the few turned into a cloud. Tallyho! someone shouted and Malan turned the squadron towards them. As usual my mouth dried, the palms of my hands got moist, the engine seemed to lose its power and the CO seemed to have gone off his head. He's turning towards the Jerries' port. They were about ten miles away and in two minutes we will be into them. I prayed and cursed silently. Christ! We haven't got a show. There seemed to be hundreds above us too — the CO is mad — why doesn't he climb? Now we were going at full bore just to their port. I saw long trails of white vapour streaming back from the black devils. We were smoking too and I slid under Steve's tail and got between his smoke trail and the Jerries — I can watch the bastards better. Hell! they're streaking along — I can see their yellow noses now. Oh! cunning move, cunning move! The CO turned left sharply, says, 'Oh Kayee!' We split up into pairs. I stuck to Steve's tail like a leech — God! If I could only have a drink!

Something's died in my mouth. Fierce thoughts surged through my mind and black thoughts too: Ricalton and Hastings, my friends — dead. Kirkie too — all right you bastards! Come and get it! Steve headed straight towards a bunch, a dozen or so, and I slid further out to one side and drew a bead on the starboard 109. I saw great streams of tracer twisting past my port wing and could see the wicked flashes from the

Jerry's guns. My eight Brownings spluttered and the acrid smell of burnt charges filtered into my mask.

I held my fire on the Hun as we screamed together and, as I saw the smaller details of his kite, I heaved back the stick. My face dragged down and strained on my skull, my mask nearly broke my nose and everything went brown, then black. I held the stick hard back and kicked on left rudder — that'll fox the bastard. The black-out faded and for an ageless instant I couldn't see a single plane. Then I spotted a lone Me 109 on my starboard and, turning towards him, I crouched forward, my head nearly touching the gunsight. Damn my harness — now, now — allow plenty of deflection, quick now, give him a squirt. Hell, the bugger is made of iron! No! I've hit him — yes, yes — Christ! What's that? A great stream of solid fire screamed past my hood, I got a fleeting glimpse of four 109s diving on me, a wicked flickering at their ugly noses. Right rudder — stick hard over, and full back — I half rolled and cut in underneath them. Something flashed past my nose — I flew through the black cloud that streamed behind. Hell, that was close, my heart was banging and I felt sick. Nearly collided. I gave myself more oxygen. Thwack! Something hit my kite. Sounded like a pickaxe on an iron roof. Hell, I've been hit — my hand shot up in reflex to the hood release and with a curse I realised that I'm a fool — bail out indeed! Yes, there's a hole down by the starboard flap panel. Where are those bastards? I turned and twisted — couldn't see a thing. I felt as big as a glasshouse. There's a hell of a lot of nattering on the RT — sounds like a party — Christ! If only I had a drink.

I saw Jerries below and again it happened! Again I couldn't dive. I was trapped in a ghastly nightmare, a horrible mental barrier which held me in the icy blue while far below the others caught up to and mixed with the enemy. Again there were a few Huns who stayed high and now turned their attention onto me. This time it's almost a relief not to have to face descending. In a few moments we met head-on, to zoom past and climb again. Aircraft are very sloppy in the rarefied air and tight turns are impossible. The 109s with their higher wingloading couldn't out-turn me. This time they stuck together more or less and soon I managed to get on one's tail.

The Teutonic twit! The fool's changed his mind! He's turned the other way and now I've got him in my sights. A quick glance at the others now away to my port. They're diving for the ground.

The Brownings spluttered and white tracer bullets screamed away in thread-like spirals to envelop the enemy. De Wildes sparkled in tiny red flashes on fuselage and wings. He's lifting, lifting; white contrails streamed back from both wing tips as his G's built up into a high speed stall and he fell onto his back. More strikes! A puff of black smoke from his motor. A shower of sparkling perspex and he flashed past and behind, spinning down inverted.

I can't descend! In a trembling fury I watched the German plane falling out of sight still inverted, until he disappeared through a thin cloud layer miles below. Now the problem of getting down to the 'drome. Obviously I'll come down when the gas runs out.

Slowly, oh! so slowly, I throttled back, turning the plane to head north away from the battle area. By watching the far horizon and just keeping the plane above stalling speed, we sank down to the warmer air below. By watching the horizon and not my instruments or the land beneath slowly expanding, an illusion of level flying was achieved. Gradually confidence came back and soon I managed to pull myself together and began to fly again. The old Spitfire magic returned and we became as one.

I wondered how the boys were — was that a Spit I saw go screaming past me with the fiery tail? God, I hope not. Ah! here's the 'drome — funny how easily I've found it — usually have to fly north a bit, then west a bit along the railway till I come to the quarry. Silly half-lucid thoughts flocked through my mind. Has the wind changed? Where's that kite of Johnny's? Oh yes; there are the obstruction flags around it.

My knees were trembling with reaction and my mouth was like a kiln. That's funny, I was one of the last back — seemed to be a lot of kites at our dispersal. I shoved the radiator lever open and dropped the undercart. The wheels locked with a solid click and I saw the green lights flash on. Down flaps then alter pitch, then turn in to land. A little motor, steady; now she drops down with a last rush and sigh, and I'm braking her to a roll. Up flaps and I taxied around to my pen.

Bill rushed up as I switched off. The Merlin coughed to an unwilling stop and I heard the faint clicking of the cooling exhaust manifolds. Funny how quickly they lose their heat.

'Any luck, sir?' Bill asked the old question. I could see he was thinking of a Swastika on F's sleek sides. I pointed to the hole in the wing; Bill grinned, the mechanics clambered up and started to re-fuel; armourers already creeping over the wings, unscrewing gun panels with practised speed.

My hands trembled, fumbled and Bill pulled out the harness-locking pin. It dropped away and I lifted myself out. The oxygen hissed loudly in the empty cockpit; I let it run out — Christ it was foul!

'Any luck sir? Any luck?' I shook my head. 'Can't say; I damaged one, probably won't get home. We'll soon find out — maybe damaged a second.'

The armourer corporal called out, 'Your starboard No 2 gun has a stoppage, sir.' 'OK let's know the cause.' No more holes in my kite, thank goodness, so I climbed into the waiting brake. 'Any luck, sir?' 'Hallo George. Yes, I think I got one, damaged a second.' 'What were they?' '109 bastards.' 'Oh.' 'Is everyone back yet, George?' 'No. Mr Nelson and Mr Chesters and Sergeant Soars aren't — you're one of the last, sir.' 'OK.'...

November 1st 1940: Five scrambles today! Five hours thirty-five minutes in the air with three wild mixed-up whirling scraps from thirty-odd thousand feet to the deck. No real results for me. I've decided to discount the odd hit on a E/A. Unless something actually comes off it or there is smoke or fumes, it wasn't damaged as far as I was concerned. However, each flight brought more confidence and a more professional outlook. Flying became absolutely second nature, my machine and I one entity. But each take-off was a conscious effort of the will, the butterflies in my stomach not at all keen to formate and fly.

Sergeant Soars got shot again and, this time, wounded; he should be taken off fighters. Flight Sergeant Burnard, too, but my friend Flying Officer Nelson got himself killed. He was an American who had joined the RCAF for the adventure. Or was it because he believed in freedom and was prepared to die for it?

November 2nd, Sortie 45. 1.20 hours duration: I got him! I got him! The Spy has just told me my Jerry crashed near Ashford and is woodened out. Oh! Bloody marvellous! My first Hun confirmed and my lost Spitfire now squared for. There were eleven of us and over forty of them. My confirmation as Flying Officer came through and to round off a wonderful,

wonderful day, I scooted off to the Mess bar, shouted the crowd, had a batman sew on my new broad rings and drank myself unconscious.

Wizard! The Spy has got me souvenirs from my Hun — he was Oberfeldwebel* Fritz Noller, holder of the 1939 Iron Cross 2nd Class. On him was the photo of a woman, ration coupons, odds and ends.

The photo gave me pause, but Wally got quite excited, waved it about and shouted, 'Don't be a clot! It could have been you! Remember! They started it! Now let's get onto the plonk and it's your round.'

But on other days, after other battles, there were different reactions.

 ## Typhoon Pilot
Desmond Scott

Once we had cut our Typhoon squadron formations down to eight aircraft, although we may not have qualified for a prewar Hendon air display, I don't think any squadron in the RAF could have outshone us in the art of formation flying and all-round manoeuvrability. The New Zealand pilots would cling to me like leeches, whether I was upside down, on my nose, or standing on my tail. We did a number of offensive missions, mostly escorting Bombphoons — a Typhoon loaded with bombs — which was good experience for many of the new pilots.

However, it was not until the morning of 14 April that I had the opportunity of firing my cannons again. Group Captain McGregor, who was about to relinquish command of the station, phoned from the sector operations room to report enemy plots on the board, midway between Cherbourg Peninsula and Le Havre. They were flying low and appeared to be escorting a surface vessel of some sort, since the plots were moving slowly across the board.

I took Pilot Officer R. Fitzgibbon, a Culverden boy, as my number two, and invited Harvey Sweetman to come along with one of his B Flight pilots. We hurried off across a glass-like sea, and as we approached the

---

* Equivalent rank to our warrant officer.

general area of our search we came upon patches of heavy sea mist and began a square search. After about five minutes we flew directly over what appeared to be an E-boat. Pulling up sharply over its deck, and expecting a shower of flak, I caught a quick glimpse of some of its crew, who looked as surprised as I felt. The mist, which was really more of a thick haze, was only about 200 ft deep, and as I broke into the clear sky above it, I came almost directly under and behind a pair of Me 109s in wide search formation.

My zoom-up from below closed me in so rapidly to the 109 nearest me that I had to open fire almost immediately. I got in quite a decent burst and bits flew off him in all directions, including what appeared to be his canopy, which flashed past my own cockpit by inches. I was forced to pull quickly away to starboard, otherwise my propeller would have minced off his tail and we would both have been in a similar predicament.

I could see he was in real trouble. His propeller began to windmill and short sharp bursts of black and white smoke began leaving his exhaust; but I could see no fire. I looked around for Sweetman. He had apparently followed the other 109 down into the haze as it fled quickly for France. Fitz was still with me and had taken a shot at our 109 directly after I had pulled out to the starboard.

As our victim dropped his nose into a slow shallow dive towards the sea, I throttled back in formation with him. He was trying to climb out of his cockpit and I could see quite clearly the terrified expression on his round young face. You do see things when your blood is up and your heart is pounding that you would not do under normal circumstances. I followed him down in the direction of a reasonably clear patch of sea, where I thought he was going to attempt a ditching, but he must have changed his mind, or was perhaps injured. Still clinging to the side of his cockpit, he pulled himself out on to the starboard wing when only 100 ft above the water. For reasons which I have never been able to analyse, I pressed the firing button again, and he and his aircraft hit the sea almost simultaneously in a fountain of spray, framed only by the pattern of my own cannon fire.

As Fitz came alongside me while I was turning for home, he gave me the thumbs up sign. I buried my head in the cockpit and was suddenly overcome with a feeling of deep remorse. When you shoot down an aircraft, you don't normally think of its pilot. But in this case we had come face to face, the victor and the vanquished. Why had I fired that last burst? It had not been necessary. I tried to console myself in the fact that he was the author of his own destruction, and had been far too low to bale out. Yet why could I have not kept my bloody fingers out of his final moment? The passing years have not erased the magnitude of this brief encounter. I often see him looking back at me — and well may he ask 'Who won?'

There were times when each could say they had.

New Zealanders with the Royal Air Force Volume II: European Theatre, January 1943–May 1945
Wing Commander H. L. Thompson

Squadron Leader Checketts had a remarkable experience early in September. He was leading the squadron as high cover to Marauders bombing ammunition supplies in the marshalling yards at Cambrai, and the Spitfires were just swinging away from the target when some twenty Focke-Wulfs dived on them from out of the sun. The squadron broke up and dogfighting began. Checketts records how he sent one of the German machines down in flames and then:

Suddenly flak bursts appeared all around me and I started to weave and twist to avoid them. I then saw 5 F.W.190s at three o'clock above me coming down to attack us and called my No. 2 to break. We fought for altitude and finally got it, when to my surprise saw two more F.W.190s above me. One of them came for me in a port turn, the same as mine, and the other took the other turn and attacked head on. The first enemy aircraft could not get me and I thought the other one could not either. His first attack was miles out and I thought I would get a shot at him next time round, but we both missed. His third attack was terrific and I saw all his cannon firing, also his spinner and engine cowlings. There was a terrific explosion at my feet and my cockpit filled with flames. I frantically clutched my hood release and dragged the hood open. The flames gushed round my face and I released my harness and stood out into the slipstream. The stench of burning flesh was sickening and I seemed to be hours trying to escape this inferno. At last my body was wholly out but the toe of my flying boot caught on my windscreen catch and I was being dragged swiftly down; a terrific kick and I was hurtling head over heels down and down. I clutched my ripcord and pulled and a hard jerk stayed my fall. The F.W.190 flew close to me and I was terribly afraid — would he shoot me? No. I saw my No. 2 fly away home to dinner as I drifted slowly down with the white canopy billowing above me and my friend the enemy watching me.

Pilots knew that friendship was a fleeting thing. Combat meant casualties — some of them close friends. In 1941, a pilot called Dick Bruin joined Des Scott's squadron. Born in Hokitika, Scott had the unique distinction of being decorated by every country in which he served — England, France, Belgium and Holland. At 25, he also became the youngest group captain in the RAF.

One More Hour
Desmond Scott DSO, OBE, DFC and Bar

Six feet tall, large grey eyes and ears that stuck out like a pair of ping-pong bats. He arrived in at A-Flight dispersal after spending the best part of two days on a train. This hadn't improved his looks, nor had it helped to tidy up his uniform. In fact the CO said he looked as rough as a badger's arse and that he would be happy if Dick rejoined the squadron after a hot bath. Dick was to share my billet in a brick bungalow just over the boundary fence from Martlesham Heath. I had shared this room with a number of pilots before Dick arrived. Some lasted a week or two, some a day or two — and there was a boy from Dundee who was killed before he could unpack his bag.

Before Dick arrived I was beginning to feel the room had a jinx on it and was about to shift into a small cubicle below the stairs. But there was something about this new arrival that impressed me. He seemed a bit different from the general run of sergeant pilots. He was one of the silent

types, although I was soon to learn that this was not a natural characteristic. He had lost his upper front teeth in a training accident and the
resultant wide gap made speaking difficult. He would lisp, and when
swearing, had trouble pronouncing his f's.

Dick's lanky frame and casual air added a further dimension to the
ranks of the squadron sergeant pilots, and although we were soon to
share a number of close calls while escorting Blenheims over France, we
managed to stay alive and thus widen our flying experience. Most days
we would team up as a pair and take turns at leading. Our CO, the
bewhiskered Gibb, was a real 'press on' type, and a man of very firm
convictions. 'Always stay with your leader Scott. Stick with him no matter
what happens. If he flies into a mountain — I expect you to follow suit.'
It didn't sound like good arithmetic to me, but with Gibb sergeant pilots
were sergeant pilots who, like children, should be seen and not heard.

Dick was always hungry and like most tall boys, never in a hurry. In
fact the only time I saw him get a move on was during an air raid. My
own bed was set in an alcove or what would be better described as a
three-windowed bay, and Dick's was against the wall on the other side of
the room. These low-set windows opened out onto a small garden and
lawn at the rear of the house. On being requisitioned by the RAF at the
start of the war the house had been stripped of all furnishings including
the floor coverings, and a slit trench dug in the centre of the lawn.

To prepare ourselves for any night emergency we would normally
leave one of the smaller windows open and during the day practise
vaulting off the end of my bed through the opening and onto the grass.
Five or six yards on all fours and we were into the slit trench. We soon
had it down to a fine art. About four seconds flat. One night we had
hardly fallen into our beds after visiting a local pub when the uneven
beat of enemy bombers began reverberating overhead. As our ack-ack
guns thundered into action so too did the enemy. A stick of bombs
exploded some distance away, followed quickly by another that shook
the walls of our billet and rattled the roof tiles. I was just about to suggest
to Dick it was time we got out when he beat me to it. I could hear his
bare feet pounding on the floor boards. There was a bounce on the end
of my bed and a crash of breaking glass. Hurriedly pulling on my flying
boots, I ran down the passage way and left the house by the side door.
Dick had made the trench and seemed oblivious of the fact that he had

just sailed through a plate glass window. All he said was 'Jeeze that was a close one'. Once the 'All clear' had been sounded we picked our way back to our room, cold but sober. I thought Dick might have been cut to pieces but the only damage resulting from our forgetfulness was a few superficial scratches and a broken window.

About a week after this incident Dick and I were detailed for a two plane dawn patrol — 10000 feet between Orfordness and Harwich. I had flown it many times before, but in this instance I had an uneasy feeling that something out of the ordinary was about to happen, and was ready and waiting well before take-off time. As I paced the dispersal floor, a mug of tea in one hand and a cigarette in the other, Dick sat on the floor casually pulling on a second pair of grey socks. I asked him sarcastically if we were going to the Pole. He didn't answer immediately. After slowly pulling on his flying boots he stood up, pointed a finger at the ceiling and said 'Scottie, I told my old Mum who knitted these socks if ever I am shot down I will die with my feet warm.' His grin resembled a hangar with both doors fully open. I followed him outside towards our Hurricanes, barely discernible in the pre-dawn light. We were soon off, tucking in our wheels as we flew over Chesley Petersen's sleeping Eagle Squadron. To the east the sky was beginning to light the way for a new day and I could see to starboard the balloon barrage above Harwich — steel-grey and daunting.

Dick was leading and I positioned myself close in to his starboard wing to be ready for the unbroken cloud that covered the sky above. We entered this at 2000 feet and climbed through into a crisp morning sky at about 6000. The sun was still under the horizon but as we climbed higher the cloud below began taking on the appearance of a white blanket, and the higher we climbed the lighter it became. At 10000 feet the sun burst from below the horizon like a huge ball of fire. I looked across at Dick. His Hurricane was illuminated as if by an orange searchlight. It stood out sharp and clear against the dark blue of the western sky, a picture of vitality and beauty. I could even see Dick's oxygen mask. It was hanging free and waggled from side to side as he looked from left to right. As we dropped our port wings and began wheeling to the south-west, a sharp staccato voice came crackling through my earphones. It was the voice of the Hornchurch controller. 'Hello Spartan leader. Eighty plus bandits approaching you from zero-nine-zero Angels 25. Do you read me. Over.' Dick's Midland accent quickly replied 'Roger Clapshaw. Message received. Proceeding Angels 25 on zero-nine-zero.' We opened our throttles to full boost and began climbing hard, the hair on the back of my neck rising to the occasion.

Other friendly aircraft were quickly airborne for I could hear the ground controller vectoring them towards the approaching raiders. We were soon above 20000 feet and heading for Angels 25. While nearing our target height a squadron of Spitfires appeared. Twelve of them. It was a beautiful sight and I was more than thankful for their climbing superiority as they cruised up ahead of us. Dick began a turn to port and as I followed — my eyes on swivels — I caught sight of a swarm of aircraft directly above us but much higher. The Spitfires must have seen them at the same time for my radio became alive with many voices, some high pitched and anxious, others low and authoritative. I tried unsuccessfully to chip in and warn Dick but he slowly turned on his back and went rocketing downwards. I thought he must have seen some enemy aircraft below us and was racing in to attack. I made sure my gun button was in the 'fire' position and did my best to stay with him. His throttle must have been through the 'gate' for he was going straight down like a bullet. As the cloud raced up towards us my Hurricane began to shudder and I realised we had reached a critical speed and that Dick too must be in trouble. By this time he was well ahead of me and Gibbs' instructions began ringing in my ears. They came through loud and clear like a hammer on an anvil. 'Stick with him.'

My desire to live suddenly took hold of me and in sheer panic I closed the throttle and wrenched back on the control column with both arms, my feet pressed against the rudder pedals. A great weight pressed down on me, forcing my face into my chest and exploding my eyes into what seemed like a million stars. I regained consciousness with my aircraft rolling towards the heavens, the altimeter reading 12000 feet and climbing rapidly. Although tail-heavy there was no difficulty in bringing her onto an even keel and as soon as I was able to re-set the artificial horizon I asked Debden control for a vector to base. On receiving a course to steer I let down through cloud, breaking out a little to the east of Bawdsey. Martlesham Heath soon came into view and I wasted no time in lowering the undercarriage and dropping onto the grass. While taxying towards our dispersal hut I noticed my aircraft's wings had taken a severe beating. Many rivets had popped and there was a pronounced gap between the wings and the fuselage stub fairings. They had been stretched to their limits. There was no sign of Dick's aircraft and my anxiety was soon confirmed. The observer corps had already reported by phone that an aircraft had crashed at very high speed into the sea off Harwich.

I went back to the bungalow and faced again the agony of losing another room-mate. The broken window had been replaced some days earlier but there were signs of Dick everywhere. His best boots were lying where he had left them in the middle of the floor. There was a half eaten NAAFI bun on a bedside locker and his shaving brush — still white from the morning shave — stood upright in the basin. Apart from a leaking cistern in the bathroom the house was as quiet as a tomb. I sat on my bed as thoughts of the past hour began crowding in on me. The hard climb through the dark cloud into the blazing sunrise. The heavenly arena and its ghostly voices. The screaming power-dive and the rapid return to the silence of a dull grey room. I began wondering what I was going to say to his mother. She had written to him every day.

At that moment a Rolls Royce Merlin cruised overhead. As it whispered its way into the distance a shaft of sunlight crept through the window. Warm, comforting, like a friendly arm upon my shoulders. It bowed my head and turned my eyes into pools of rain. I had some difficulty finding my forage cap, made my way out the side entrance and quietly closed the door....

By 1941, with the Luftwaffe forced to abandon its assault on England, a stalemate of sorts set in, at least in southern Europe. The major battles were fought elsewhere; in the Mediterranean, in Russia and on the Eastern Front and also in the Pacific. On 7 December 1941, Japanese aircraft attacked Pearl Harbour and invaded northern Malaya.

Suddenly, the great British bastion of Singapore was under threat. Since the end of World War One, Singapore (and the reassuring presence of a major British force there) had been a key part of New Zealand's defence strategy. It was seen as a major obstacle to any aggressor coming south.

Not any more. The island was ill equipped to repel an attack from the north. Its fortifications and defences had been organised on the assumption that any attack would come from the sea. Moreover, the aircraft supposed to defend it were obsolete — including the American Brewster Buffalo fighters of No. 488 Squadron, RNZAF.

Through December and January 1942, RNZAF and RAF pilots did their best to repel Japanese air attacks. But it was an uneven contest. There was a desperate need for better aircraft and more pilots.

Ian Newlands was one of them. After training in New Zealand and Canada he'd been stationed in England until the last weeks of 1941. His squadron then left for North Africa but, whilst at sea, there was an urgent change of plan.

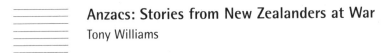

Anzacs: Stories from New Zealanders at War
Tony Williams

On the way to Singapore, they told us on the carrier that the Japanese only had obsolete biplanes. Later on we found out that the obsolete biplanes belonged to New Zealand. We were hopelessly outnumbered all the way through.

On 5 February, we were flown up to Singapore and they wanted half a dozen pilots to go up to bring back some planes that had been repaired in the jungle. We were told to leave all our kit bags behind, except for our toothbrush, razor and wallet. I never saw my kit again and I lost my log book, which had all my details in it.

We climbed into two Lockheed Hudsons belonging to the Australian Air Force and had several Hurricanes as escorts and flew up to Singapore. But we didn't know that the northern dromes were under shell fire.

The first one we landed on was Sembawang. We were getting hungry. It was lunch time so we went into the cookhouse and there wasn't a soul anywhere. The place was deserted, but we could smell the dinner cooking in the oven. We opened the door and there was roast beef and roast potato … so we grabbed a plate each and piled it up. We were half way through this meal, when … *boom, boom, boom* … we were under shell fire. We all rushed out and got in the plane again and took off to Tengah. The wheels had only just left the ground and shells were exploding everywhere.

We had to operate from Kallang, which is near Singapore City, on the south side of the island. All we could see when we first landed there were great big oil tanks on islands outside of town. All of them were on fire. The smoke was going straight up in spirals and spreading out in a black cloud about 2000 feet up. And the *Empress of Asia* was lying on the bottom and smoking away but still upright. It hadn't sunk because it wasn't sitting in enough water to sink.

Kallang was just a grass aerodrome, no concrete runways. It had already been badly bombed. The whole aerodrome was just a mass of craters about 20 feet in diameter and about eight feet deep. You couldn't land in a straight line anywhere, you had to zig-zag. It was more or less a banana-shaped runway.

The control tower and administration building had been flattened by bombs and we operated from a little old bungalow on the side of the drome. It had a telephone and the barest necessities. We thought this was a bit of fun.

At the most, we had about 10 or 12 operational Hurricanes and there were a couple of little Buffalo fighters there. There were plenty of pilots because a lot of them had gone direct on a different ship, getting there before us. There were a few New Zealand pilots and a few odds-and-sods Australian ones.

The Japanese bombers would come over about every hour, starting in the morning from about 8am until about lunch time. They seemed to pack up in the afternoon. By the time we got warning, we couldn't catch them up. They would overfly the drome and come back and drop their bombs as they were heading back to base up north in Malaysia.

We had trouble with their Zero fighters, which were one of the best fighters produced in the war. Our Hurricanes had inline motors and had glycol for the cooling system whereas the Japanese had radial motors and relied on air cooling. They didn't have the same problems as we did with overheating on the ground. The Zeros were very light. They had only about two cannons and two machine-guns as armament. The Japanese were very accurate and very capable pilots — again, the opposite to what we'd been told when we first came out to Singapore.

It was very hot on the ground. We just had shorts and short-sleeved shirts, and miniature pith helmets, narrow-brimmed to keep the sun off our heads, and goggles and New Zealand flying boots. Once you got up to 25,000 feet, it would be freezing cold. Then you'd come down. You'd spend the day being hot, cold, hot, cold, which was very exhausting. You'd do two shifts a day — one in the morning, the other in the afternoon.

We only had a few planes and after a few days there weren't that many left. Our Commanding Officer crashed on take-off one day, as he flew off from the drome. He had to fly over a little basin in Kallang harbour and he hit the mast of a barge. That was the end of him and the plane.

A lot of the pilots were shot down but managed to get away with it. The planes had masses of cannon holes in them. Then we started running out of different stuff. Oxygen (for the oxygen masks that pilots wore) was pretty scarce and we were short of hydraulic fluid. When we ran out of glycol some of them filled the radiators with water in the finish. Anything would make do.

As a sergeant pilot, I usually got called up as a number two and would guard the tail for the number one, the officer. On 6 February 1942, I was up at 25,000 feet with a New Zealand pilot. We spotted this lone dive bomber flying along and we opened up our special throttle (called 'pulling the tit') that for about 10 minutes allowed the engine to go over the designated spot before it would overheat.

We went hell for leather for this Japanese. We got up pretty close to him. My engine was overheating a bit, but I had to have a crack at him. I pushed the machine-gun button and at the same time I could hear, *Rat-a-tat-tat*, and I thought, 'Is that me?'

I was getting hit myself from somewhere. He got me through the motor and the engine seized up. I couldn't see who hit me, and I had no motor, so I couldn't get away. I had to make out I was dead. I rolled the plane over and put it into a spin and dived for the deck from about 25,000 feet.

The cockpit got full of fumes and I was going that fast that the speed dial indicator went hard over right and jammed. I think I was almost approaching the speed of sound. I started to pull out of the spin at about 6000 feet, because I thought, 'He's not behind me now.'

But the plane wouldn't pull out of it. I was going straight down.

The day before, my friend W.A. Moodie had gone straight in, crashed and died.

All I could do was to use the control trim. You should never do that because it puts your tail into a high-speed spin. But I had no other option. About 500 feet off the ground, I managed to pull it out of the spin.

But I still had to land and I had no engine, no air speed and no hydraulics — nothing was working — you have to have your motor working to give your hydraulics pressure. I don't know whether I had brakes or not. Luckily I was above the aerodrome. I circled the aerodrome until my speed got down and I managed to get down on the ground and headed for the boundary fence.

I can still see the fence today. It had six wires and upright posts like a normal New Zealand fence.

The wheels weren't down, which they call a belly landing. I knew I wouldn't be able to pull up because it wasn't a very big aerodrome. I selected the wheels down and kept pumping the hydraulics as I was coming over the fence. I just cleared the fence by inches and waited for the crunch.

Nothing happened. I just kept going along and I thought, 'Oh the wheels are down.'

Now I had to keep steering in between all these bomb craters and the next boundary fence was coming up and I had hardly any brake pressure again. I pulled up about 20 yards from the far fence, in one piece and not on fire. The fire engine came out and all they could find was two bullet holes in the side of my motor on the right-hand side.

They send you straight up again to get your nerves right. I went up by myself and I was really mad that I had been shot down. I was looking around for anything to fire at. If I couldn't find an aeroplane, I'd always fire guns off into the jungle somewhere on the far side of the strait where it was lousy with Japanese troops. You couldn't see them for the trees.

[A book, *The Bloody Shambles*, was later written about the Japanese invasion of the Dutch East Indies, Malaysia and the Philippines and the pilot is mentioned in it who shot Ian down. He claimed a kill for the incident.]

A few days later, on the 11th, I went up with a New Zealand officer to bring back two planes from a northern aerodrome that was under shell fire. In order to get the planes off the ground to avoid shell fire, we had to get them off first thing.

We had an early cup of tea and were told we would be back in time for breakfast. We went up by taxi. It was pitch black and we were driving this taxi down a jungle road and every now and again there were Gurkhas on guard. They'd stop us and we asked if it was still okay to go on and they'd say, 'Yep, away you go.'

Finally we got to the drome and they showed us the two planes to take.

Sometimes they are buggers to start, but mine started quickly and I couldn't be bothered waiting for the officer, so I took off.

The Japs had an observation balloon on a wire and they could see all over the island. I thought I'd see if I could find that bloody thing and shoot it down for a start. My guns were fully loaded, but I couldn't find the balloon. I strafed all around the perimeter until the guns were empty. I landed back at the drome again. My mate got back eventually and I didn't tell him I'd been strafing. I used to keep it all to myself.

I went back to the hotel for breakfast and to spruce up. Then back down to the drome again ... well, there wasn't a bloody plane left ... they'd all gone off. All the officers had gone and left us sergeants there.

We knew we had to get out of Singapore. One of my mates had been shot down and bailed out in his parachute and landed in the Johore Strait, between Singapore and Malaysia, and the bloody Japs had machine-gunned him when he was in his rubber dinghy. He was fairly badly knocked about. We knew he was up in the local hospital, so we thought we'd see how he was getting on and if he could come with us.

So we went up there and, 'Oh what a mess.' The whole hospital had people lying in bandages, all down the corridors, doctors stepping over them. There were not enough beds for everybody.

Our friend didn't look too good. We thought we should leave him there. He was saying, 'Take me, take me.' But we thought he'd be well enough and left him there. It turned out later that the Japs visited that hospital and bayoneted half the people there, nurses and all. But he got out. We saw him later on.

As we had no planes to get out with, we had to get out by ship. We looked for a ship or a barge to pinch — anything that would float to get out of the country. We got a hacksaw and a hammer and other tools in case we needed to cut an anchor chain or something.

We were having trouble finding a boat, when someone said, 'There's an RAF boat going out tonight, the *Empire Star*.' It was a big cargo boat. We got on that, it was full of nurses, RAF, and a few Aussies.... I think unofficially they were deserting — they'd had a gutsful. The conditions were awful.

We didn't sail until almost daybreak. Ridiculous! We were waiting for the last people I suppose. We got out and had a bit of a convoy with a lot of smaller ships.

For four long hours we were hit continuously by 54 Jap bombers and fighter escorts. The Japanese bombers always came over in formations of 27 in a V formation and they let all their bombs go at once.

They scored three direct hits on our boat and the cruiser got three hits. There were about 18 men killed on the *Empire Star*. There would have been a few more because some of them would have got blown straight over the edge.

I only had my shirt and shorts on in Singapore and one of the officers left his jacket behind with a couple of stripes on it and I put it on, not thinking. When I was on the ship everybody thought I was an officer. I was screaming around the deck, saying, 'Clear the decks, man the guns, and fire!' and all this. That's the way I used to carry on anyhow.

I was behind a wire which held the mast up and had a revolver firing at this dive bomber coming down. All the deck was like a hail storm, bullets hitting the steel and bouncing up. One bomb had gone through the deck into a cabin and hit eight Air Force officers playing cards. Another bomb, just a little bit away, landed right in the middle of the spare propeller and didn't do any damage.

Everything was covered in this yellow dye. Another bomb hit on the side of the boat on the front of it and knocked out a big hole and I was shouting, 'Man the fire hoses.'

I looked down and all the fire hoses had been knocked out and all the pipes were hanging over the side. Luckily, the fire was put out.

There was an officer next to me and a blast knocked his vision out. I don't know if he recovered. I took him down below to the cookhouse for someone else to look after him and went back up the top for more action.

There was a naval officer marching along in his white uniform, looking immaculate and swinging his one arm like he was on the parade ground. He wasn't swinging the other arm because he'd been shot through the elbow. There was only a bit of skin attaching it. He was going down to the sick bay. I thought it was pretty cool the way he was doing it.

The deck I was standing on had bodies on it everywhere. I was dragging them over to one side, clearing the decks. The cabins were behind the bridge and I didn't want the women and children to see all that.

First of all, the three dive bombers had come over and made three direct hits, then they came over in formations of 27. They were dropping their bombs from about 12,000 feet. Twenty-seven bombers would come from one direction and the other 27 would go from the other direction at a higher altitude. One lot would let their bombs go then the other lot would let their bombs go. That was 54 lots of bombs coming down, four or five bombs from each plane.

It would take about a minute for the bombs to reach us after being dropped. The captain of the ship would do a hard left to avoid them, so there were quite a few that missed. Then he just cut the motors and stopped the boat. He just sat there and of course they were expecting us to zig-zag. Not one bomb hit and the whole ship lifted up in the water and down again from all the explosions as the bombs hit the water.

The next day was Friday the 13th. We thought it was going to be a black day, but nothing happened on that day. By then, about 90 per cent of that convoy had been sunk. Afterwards, we heard all sorts of stories about that convoy, like some of them got on an island and the Japs found them and shot them. Some of them got over to Sumatra. There were very few survivors.

We got to Batavia and funnily enough the bloke who owned the blooming jacket was there waiting and he got his jacket back. I ended up with just a shirt and a pair of shorts again. On the 20th, I was flying again.

There were a few dog fights, then we heard that they had captured an aerodrome about a hundred miles east of us so we went over and strafed that. They were strafing us and we were strafing them back.

We moved to Bandoeng, which is halfway across the mountains. We'd no sooner landed there than they were bombing us again. We scrambled and I had only just got airborne and the whole cockpit was full of white fumes. I couldn't see a thing.

I almost did a circuit around the aerodrome boundary fence to land again. I swung around and landed in the bomb bays, which were big piles of dirt to park up to four planes in.

I slammed the brakes on, jumped out and dived into the nearest slit trench, which was about six foot deep. I could see the four bombers coming straight at me. They'd seen me park the plane. And on the other side of the bomb bay was a Dutch bomber.

The Japs started dropping their bombs … one two, three … I could hear them coming and the next one went over my head and landed on the bomb bay a few feet from me. Then the Dutch bomber caught fire and activated the machine-guns. Every time I poked my head out of the trench, the machine-guns were firing right over my head. I was trapped there for about half an hour.

When I finally got out, I was able to find out what had happened. The fitter had left the cap off my glycol tank. If I'd caught him I would have shot him on the spot with the revolver that I always carried around with me. I was that mad. Mad as maggots.

We evacuated from there the next day but the Japs were right down the island by then. There was only one plane left to fly and they asked everybody who wanted to fly it because there was a Dutch pilot in a Buffalo who wanted to go and strafe some of these Japs in a gorge where they were fighting the Dutch troops. They wanted someone to go and cover for him. I thought, 'Oh, the bloody plane would hardly fly....'

A few minutes later, they capitulated and waved the white flag. They told us to hand over our guns, no escaping and all that jazz which was confirmed by the CO. If anybody escaped, 10 or 20 other prisoners would be shot and also tortured.

I was now a POW.

Today, one of the most controversial aspects of World War Two is the saturation bombing of cities and civilians. Such things do not sit well with us. The modern mind does not accept the view expressed by Lord Kitchener during World War One: 'We cannot make war as we ought; we can only make it as we can.'

There's an additional argument, best summed up in this quote from a 1961 speech given by a British military historian: 'The great immorality open to us in 1940 and 1941 was to lose the war against Hitler's Germany. To have abandoned the only means of direct attack which we had at our disposal would have been a long step in that direction.' Whatever the final verdict, the courage of those who flew on bombing raids cannot be denied.

A Noble Chance: One Pilot's Life
Maurice McGreal

With a call of 'Here we go' Frank pushed open the throttles and we raced across the darkened aerodrome along the line of lights and climbed into the air. It was a heavy aircraft tonight, with fuel for six hours or more and the bomb bay filled with 500 pound bombs and canisters of strange, hexagonal black sticks about the size of a relay runner's baton that were deadly phosphorus incendiaries.

I took over the climb towards 12,000 feet, or higher if we could make it, and Frank stood watching all about, from the vantage point in the astro dome. The crew were all silent and there was no useless chattering over the intercom. They all seemed to know that this trip was the first real one.

The coastline of Holland showed up ahead and Frank climbed back into the seat while I took up station in the astro dome. We were going to go in a bit north of the target and then turn due south for the bombing run; at least that was the plan. The stars were bright and the pale light of the half moon showed scattered cloud far below.

Davie, as the bomb aimer, was now prone by the bombsight and with his carefully shaded torch was trying to match faint ground features with the map of Rotterdam he held.

A searchlight came on away to the right and after waving about for a few seconds went out. Suddenly in the same place six beams sprang to life and then swung swiftly into a cone. Flak bursts filled the apex but we could see no sign of an aircraft in the trap — there were eighteen others with us here tonight and perhaps it was one of them.

We droned on towards the target now on the southerly leg. Suddenly close beside us two strong beams sprang to life and wavered about for a few moments and then went out. Then like six great angry fingers, a cluster came swiftly to life in the same place and in one sweep they all moved across to us and locked on.

The cockpit was filled with harsh bluish light: the shadows were all hard edged and sharp. In a flash it was the morning NFT all over again and Frank's force on the controls was brutal. He dived and rolled away from the cone; then with a savage heave, dragged the aircraft over and away from the lights as they strove to follow. Then a fierce pull up and a corkscrew down again. Anything to try and baffle the radar and the trackers down below.

I worked my way forward to be close to Frank in case he should be hit and as I reached him he threw the aircraft into another steep dive and I floated weightless; but we had achieved the blessed darkness of the nameless world outside the cone of light. Frank climbed again to regain some of our lost height and turned in towards the target in reply to Davie's calls.

There were no tired eyes now. No one said they were sick; no funny remarks came across the intercom.

Davie asked if anyone could pick up the bend in the river we had been briefed to look out for as a guide to the aiming point. From his voice you'd have thought that nothing had happened over the past few minutes. The routine drawl of his Toronto accent came clear through to our ears as we all stared down into the darkness to try and identify something that would be of help. There were glimpses of the river water in the reflected light of the moon.

We were certainly over Rotterdam, but just where?

Frank pulled his mask-mike aside from his mouth and shouted, 'Go back and stick a flare out. It should show us something.'

I gave him a thumb and turned to go aft and at once my head jerked sideways as the oxygen tube and the intercom connections, which I had forgotten to unplug, nearly dragged my head from my shoulders. Syd looked up from his dimly lighted desk at the radio as, clumsy in my multiple layers of clothing, I struggled across the main wing spar and aft along the catwalk to where the flares were stacked in racks on the starboard side.

Each was about four foot long; black and finned at the end. The trigger end had a safety fork and a lanyard that tugged the safety fork clear and armed the thing as it slid out of the chute. I unclipped one from the rack and in the dim light slid it into the chute and engaged the trigger. It was minus 15 °C outside and my thick gloved fingers were clumsy, and by the time I had got all this done I was gasping for lack of oxygen. The outlet point nearby was just too far away and I had unplugged while I was working on the flare.

I flicked my torch to check that all was right and from the front I could see that Frank was getting more and more shirty. I fumbled with the trigger but the beast would not slide free. I rechecked the small lanyard and wriggled the trigger.

Syd Parrott, the wireless operator, was now making all the signs to show just how irritable Frank was becoming. Their fingers and the signs were saying, 'Get the bloody thing away.'

I pulled again and again. Suddenly in the darkness I saw, and my subconscious knew, that I had triggered the thing. It was doing a sort of fizzing and that meant that we had something like a minute to go before a million candlepower bomb burst in the aircraft. It would be worse than being hit by flak. It would be the end.

But none of these thoughts really went through my mind and with the numbness of shock I kept on blindly fiddling with the mechanism that would free it. Then suddenly it was no longer there and only the bit of cord that was the lanyard was dangling by the tube. Almost at once there was a great flash close below the aircraft and MacMillan the tail gunner called that he thought we'd been hit.

I flopped down beside the intercom and oxygen point and plugged in. Les Gore in the front turret said that it was pretty close, whatever it was, and Frank reassured them when he said that as far as he could feel from the controls all was well with the aircraft.

I added nothing to assist their guessing.

Then again came the calm voice of Davie Florence saying, 'I think I have picked up the pinpoint we were looking for.' He told Frank to turn left about thirty degrees for a run into the target. Then, 'No; that was the wrong river bend.' Once more we began the search for something that would give us a lead.

I had recovered now and rejoined the front end team. Les Gore said he had a view from the turret of something and Davie couldn't pick up what part of the terrain he was talking about.

Frank called up on the intercom, 'For God's sake, Davie, get a move on, I've had a gutsful of this bit of sky for tonight.' Then Les Gore again had a helpful sighting from the front turret.

This time Davie recognised it and started his monotonous litany of the bombaimer as he guided Frank on the final leg of the bombing run. 'Left. Left. R I G H T, … STEADY; Left. Left; STEADY.'

Then the words that all bomber crews greet with joy … 'Bombs away.'

Almost before the words had left his lips Frank was diving and turning away to the west and heading for home.

Mac from the tail turret said he believed he had seen them hit the target area and that some fires had been started. Davie was very unhappy because he had intended the bombs to go in two separate sticks with a canister of incendiaries after each, but Frank had reselected his 'mickey mouse' (bomb selector) and dropped the whole load in one go.

They exchanged views on this for a few moments and then Frank told everyone to shut up because this was just the time that the night fighters attacked — in the post bombing euphoria when crews were heading home. I thought it was not the time to tell them about the chaos associated with the flare drop earlier on. It wouldn't have helped.

The Pitcher and the Well is a fascinating and poignant book. Published in 1961, it is a collection of letters written to — and edited by — J.D. McDonald. The letters came from an unnamed New Zealand navigator and were written in a German prison hospital where, finally, the writer died of burns received when his aircraft was shot down. This extract describes two raids (one in daylight) and a first-hand account of the effects of the Blitz on London.

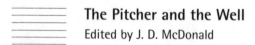

The Pitcher and the Well
Edited by J. D. McDonald

We went to Hamburg one night. The briefing was more detailed than usual and the out and alternative home routes were to be decided just before take-off, as the false prophets were a little discouraged by the outcome of their previous night's forecasts. This means more work for the wretched navigators. Two sets of data to be prepared in time hardly adequate for one.

The gen about Hamburg is always a little intimidating. And this is early days, remember. The searchlight coverage is very good, so is the gun co-ordination. Besides which, there are God knows how many fighter stations ringed around it and, just to pile on the agony the radio location is invariably very good too. Thorough! Fritz's way. Rumour spoke of 9.2s being used and of cones of hundreds of searchlights. Coming along?

There are two major routes of approach, one over land, the other a sea route. Naturally we prefer the sea route. Wouldn't you?

The final briefing ordered the land route and the target was indicated with extreme precision. What do they think we are? A curse on all this pinpoint stuff. Pip's view was (a) it was a lot of bloody lunacy, (b) it wasn't possible and (c) too many aircraft and their excellent crews are lost on propaganda stunts featuring precision bombing.

Still, a job is a job. I go along for a last minute nav. check. Pip does the same for the old girl. It's a chastening thought that for half the eight hundred miles we're going to be over enemy territory and he's bound to resent it all.

And so for the usual, after dusk final touches. Look around for extra food (you never know your luck), leave behind all papers and documents other than those essential for the business of the evening, check watches, check all necessary equipment; and then there's always that deadly few minutes as one waits. Remember that bit in *Journey's End* where the older man engages the younger in a discussion about tea versus cocoa as beverages while they talk away the last few moments of their lives? We usually wait in silence.

This night we joke a little — you know how it is — every aircrew has its family jokes. Once in the aircraft things are much easier. We all have work to do but yet I'll swear we all have our ears cocked, as we always do, for the first coughing bellow as the cylinders fire. That always means, 'This is it!' The engines stutter a bit as they clear themselves and then settle down. They are warmed beforehand and we sense the familiar change in their notes as Pip gently pushes the throttles forward. Soon the old girl shakes with their thunder as one after another they are raced at full throttle. The test satisfactory, the nose dies to the pleasant sound aircraft engines make as they tick over. Strange how very vividly all this fills my mind.

Anyway, they quicken a little and we lurch forward and waddle towards the runway to await our turn to take off. There's always a moment or two. Then all hell bursts loose as our horses roar their heads off. The bumps and jolts gradually decrease as the tail comes up, the acceleration seems to be forcing one backwards through the back of the seat. Now we are riding with only an occasional solid bump. Soon these too cease and the second dicky is winding up the legs.

Why am I giving you all this gen? To show I know the pilot ju-ju? Do I need to convince *you*? Or just because, when I write to you, by Christ, I'm right there, riding.

I verify time with Pip and get on with my job. You were right. A navigator *is* the loneliest soul under heaven. But who would swop that loneliness, that power?

Soon we cross the coast and there is our sea, steely grey below us in what remains of the light. I love the pattern wind lanes make on it but tonight is no time for admiration even if they were visible. The leader of a formation is as busy as a one-armed paperhanger. This is no time for beauty, unless it be the cold beauty of exact calculation. This is our first big do, the first time we lead.

I swear to myself that Pip will make a name as a leader if navigational skill can do anything for him. All the other required qualities he has in full measure, pressed down and running over. Just quietly, I have all he lacks, however little that may be. Still and all, he does *lack* it, and he does *need* it.

I begin making checks at absurdly short intervals. They were largely unnecessary as the Met gen is good all the way. See how I'm identifying myself with Pip. We're giving nothing away, Pip and I.

Dead on E.T.A.* (Coast) there's the dirty line of the Dutch dunes and we spend a little time later dodging searchlights near Ostend. This valuable time has to be allowed for and made up later. Our escort turned back and we edged through the gaps. By this time all is known and the radio location boys down below are on the job, and there's some evidence of perturbation around the fighter aerodromes down below. It's not possible to operate an aerodrome in complete concealment but what interests me isn't what fun and games goes on below, but whether the bright R.L. boys have stacked up a squadron of fighters above us. Actually we see a couple of Me 109s looking for trouble and one or two of those little Heinkels with the inverted gull wings. They're so small they look farther away than they are — you'd swear they're much farther away — an error of that kind can be fatal, as the little bastards carry cannon.

We slipped the first lot. We're a much better viewing platform than

---

* Estimated time of arrival.

they are, but our luck runs out a little later. The formation tightened up but the last kite didn't weave as much as the tail guard should; result: they beat him up and he begins to fall back. We throttle back for him, to protect him as well as we can until the fighters run out of fuel or ammo or both, thereby imperilling our own E.T.A. already knocked about a bit by the delay at the coast. Even a well closed up formation is no guarantee of safety when the fighters have cannon. Soon it's obvious we have to leave him. The heroic picture in the gutter press of a formation fighting its way through swarms of fighters is just childish. It's not a battle on even terms, or why design fighters? No bomber *wants* to fight. If they have a really worth-while bomb load they can't bristle with defensive armament. That's what's wrong with the so-called 'Flying Fortresses.' They fly high, they fly fast, they fly far, they're very well defended, but they don't carry any bombs. Or, at least, none to speak of. Did you hear the story of the one which made a forced landing on one of our fields, and before it could be checked, an absent-minded bombing crew had included it in the bomb load of one of our Stirlings? A wicked, wicked lie, no doubt.

Our straggler dropped still further back. Pip's decision, now. Not mine, thank God. He jettisoned his cargo, but we know how it is. The fighters all leave us to concentrate on him. We imperil the whole show if we don't get moving. He knows how it is too.

The fighters curve in leisurely fashion, or so it looks to us; actually they're most business-like. They'll concentrate on the tail gunner and do him in, and when he's gone for a Burton, they'll chew the aircraft up with stern attacks.

That's just what they do. As soon as the gap opens they blow his fin off and he goes into the deck on fire. Only Alec, in our tail turret, sees it, and he doesn't see any brollies in the hazy, shifting moonlight. Perhaps we're too far away. And we're in a hell of a hurry. The fighters want to do him in as speedily as possible so as to have fuel and ammo to tackle the exposed tail end. We want to put as much distance between us as possible, now that we are buying time and distance with aircraft and crews.

We by-pass Bremen, almost as tough a nut as Hamburg, but we lose another poor sod to a fighter attack just to the south. He seemed intact too; just slowly banked to starboard and went into a spin. Perhaps they got the pilots.

Hamburg is better camouflaged than anyone can imagine but I think we are on the target. Nothing compares remotely with Hamburg. Brest, Bremen, Berlin, the three B.s are nothing to it.

Yet miracles are with us. First in, and nobody cares. The lights cone behind us, the fighters quarter behind us too. How come we are missed? And for how long?

What's the use of talking. It was bloody hell I tell you. No man should be asked to go to Hamburg....

Do you remember the kid's song 'London Town is Burning Down'? I've seen it happen. And I don't even know how to begin to tell you about it. It's true I didn't see the worst — or, at least, so I am told. What I saw was bad enough.

One feels one ought to do something, that a mere bystander is really as much a nuisance as the fire-fighters say; yet, what to do? Actually there's so little one can do, and the helplessness is the worst feeling.

Little narrow streets aflame from end to end, blocked with rubble, criss-crossed by firemen's hoses; thin shrieks sound so unimportant beside the roar of the fire and the thunder of the water, yet they probably indicate that some poor devil is being burned alive. Hear me! For Christ's sake, hear me! Someone being burned alive! Do something if it's the last thing you ever do.

I hate to think of anyone pinned down in the path of the fire.

Can you hear me? Pinned down in the path of the fire, I say. Don't let's think any further of that! You were in the Napier earthquake. Try to imagine Auckland like that. It won't give you much idea of even a minor blitz, but it'll help.

Bombed people are numbed unless actually injured. They don't cry out or have hysterics but they do revert to the most primitive of all instincts and aimlessly huddle together. Only the animal instinct left.

The discipline of the firemen carries them through. They're a great bunch. Their frantic efficiency contrasts oddly with the stupidity of the victims of the 'incident'. Isn't that a lovely word? The bombed folk know it's no 'incident'. Does the fat boy really think that calling bombing by another name actually alters it? He's such a windy spouter of words that he may think they do alter things.

Yes, bombed folk say, and mean 'bombed'. They know war's war. It's to be expected, now that the civvies are the front line, that they'll catch it — apart from those in bomb-proof underground towns of course. Meaning the politicians.

A man is safer in the Air Force than in London. Statistics prove it.

Pip and I watched. Clumsily we try to help but our lack of training is a hindrance. Little acts of heroism occur all around us, and, of course, the inevitable bathos as well. One old lady looked at a house where the side had been torn out and then said with obvious satisfaction, 'I always knew the Robinsons didn't have a sideboard.'

Actually the material damage is dwarfed by the human wreckage. A fragment of a woman is a horrifying and sickening sight. Intestinal and head wounds are very common but less so than broken limbs. The latter are so normal looking that Pip and I look at them for relief from pity and fear. Was it Aristotle who said that tragedy, by pity and terror, purged the emotions? It does. On recollection later. At the time, one is as numbed as the poor wretched beings dug out of the remains of their homes.

It's always the working class that gets it worst; not that the uninjured are unduly worried. Their lives are so near the subsistence level that a little worsening is scarcely noticed. They double up a little more and share with a cheerful improvidence.

Fire is the real terror! A broken gas main, a shower of incendiaries, an overturned stove, and panic begins: Fire! Fire! Even those dazed by blast show signs of fear at the mention of fire. And Christ alone knows what would happen in New Zealand. We'd all bloody well panic too. Yet one wonders what there is to burn in some of the English homes, they're so poverty-stricken.

As I lie here, in retrospect, I can see clearly those hellish nights, but at the time, I saw things and nothing registered. I was numbed, too. It was like turning over the pages of a book of horrors and having fleeting glimpses of each thing more terrifying than the last until there's only a vast confusion of terror. Later, my memory would toss up vivid little pictures. There was a living child with half a face; a terrified pet dog; fireman cursing as a leaky lead shot water all over him; a warden primly taking notes before the dust of the debris had settled; a whistling bomb with its devil's shriek; the vague shafts of the searchlights. And, over all, the stench and the roar of the fire.

You may ask what the hell I am doing here. I'm with Pip. He wants to see for himself.

I'm afraid. Afraid of being killed. But afraid, most of all, of what I may see.

We saw plenty. What is courage anyway? The chap who doesn't know what fear is isn't brave: merely insensitive.

Next morning the fight is still on, the peril to London is even greater, the pall of smoke denser, but somehow everything is less terrifying. We can see the damage more clearly. It is indescribably bad, but it's not as bad as it had felt at night, when we guessed at unmentionable horrors. Perhaps the real terror is the night. Man has always feared what the night may hold. A vast fire doesn't turn night into day. It merely makes the night more evil.

Yet, when it was scorching our faces Pip turned to me and said very quietly 'I suppose Hamburg is something like this.' Neither of us spoke again for a long time....

'Bring back a photograph of this and this.' You know the idea. Navigation to yards. Altitude to inches. All in lovely crystal daylight. Off with the wheels brushing away the morning dew just as the first light flushes the buildings around the field. Off with the bloody early birds. (In passing, city birds haunt airfields. All the original rural avian inhabitants have beat it to Central Africa or the North Pole. Perhaps the city sparrows like the familiar invigorating smell of petrol.)

A lovely day, my hearties. Specially laid on by the false prophets for your exclusive use. As we climb the sun pops up in a most sprightly manner. He was rising anyway but our rapid ascent brought him up more rapidly too. The aircraft looks somehow less warlike than usual, no one would say beautiful, but perhaps a little less out of keeping with the loveliness of nature.

We climb. And after that we climb. We've been using oxygen since take-off as this is a special job and everyone has to have his wits about him. Because of the oxygen everything looks mighty fine. Except that, as we climb I get an attack of gas in the guts. Excruciating while it lasts. I double up in agony and it's misery to try to do my job. Geoff's sorry for me and mumbles something about diet. Curse his sympathetic soul! 'Try a good fart,' says Alec, seizing essentials as always. 'Nothing like it to clear

the arse.' The others don't phrase it that way but they're, well —
expectant. If I can't do my job we put back. All for want of a fart.

Presently it comes, a real arse-splitter. I'm on top of my job again
within a minute. I tell Geoff so, but he suggests mildly that the information is superfluous.

Still we climb. The sky looks more black than blue now. We're above
even the feathery cirrus clouds and the little spicules of ice in them glitter
in a most attractive way. The world below is spread out like a carpet. I kid
myself I can see the curvature of the earth. The occasional cirrus drifting
below merely serves to veil the broad expanse. Oh, lovely day!

Higher still. The old girl is about at her ceiling. It's true she's new and
specially tuned for this job so we must be nearly as high as my altimeter
incredulously indicates. Which is bloody high, my braves.

Then along comes the first faint breath of trouble. Calmly as ever,
Don mentions difficulty in moving his joints. The old girl wallows with
soggy controls at her ceiling and Don has 'the bends'. How to save him?

Then, God in Heaven! Vapour trails! And above us! Above us, I tell
you. How the hell can fighters get up there without pressure cabins? I
can't see them against the sun but the trails show they're there. 'Run
down the line of the vapour trail,' I yell to Alec; 'you'll see them with your
slit goggles.' Back comes a sort of croak from Alec. Oh God, if I can't hear
him he can't hear me. Is the intercom defective?

But trust Alec. He's spotted the situation but they're out of his range
and of course the little bastards have cannon. Now Geoff, it's over to you.
Your front gunner is crippled and the fighters up there must be the latest
thing or they'd never get so high and there are at least three of them. Why
hasn't the second dicky relieved Don? Pettishly, I clip on my portable
oxygen bottle and scramble along to edge out Don. He looks bad, what
I can see of him. It's the heavy, thick-set guys like Don who get the
'bends'. Not skinny fellows like Geoff and Alec and me. But wait a
minute, Pip was stones heavier than Don, he flew at the ceiling of every
aircraft he was ever in, and with duff oxygen gear too, yet he never got
the 'bends'. Perhaps that's part of being Pip. Eh, Don?

Don's resolution is remarkable. It must be agony for him to move a
joint. I settle in his place while Geoff circles to try to get the fighters out
of the sun. There's an unreal air about the proceedings. It's too high and
too lovely a morning for sudden death.

The first whumps of the cannon shells soon dispel that one. In they come, the little sods, one broadside, one on the tail. And more to follow. Why the bloody circus? Why all the protection? What has Fritz that he doesn't want to be photographed?

The broadside guy slips underneath and turns back for another pass. Geoff deftly sets him up for me but in so doing buggers up my shot at the tail bloke who has just overshot us forward. I loose off a burst but don't need to look at Don to know I've mucked it up. Curse it, deflection shooting is for gunners. Geoff should have let me have the other one.

But here he comes again and this time I think I get one or two aboard. The old girl shudders with the cannon shells but nothing vital is hit. Jesus, how I sweat.

Geoff tries to set him up for Alec but the old girl is mighty soggy on the controls and in any case we're not built to turn with fighters. All the same, Alec does some damage. His bird straightens and flies off on the level but too far away for Alec to get him.

All this time we'd forgotten there were more than two. The other fellow dropped neatly on top, right out of the sun's eye, and this time is it. The old girl begins to spin, quite slowly. There's no sense in husbanding ammo now, so we let them have it at extreme range. That keeps them off momentarily. How long? Another minute. In they come. Alec does some more damage. Not vital, just enough to make him break off. I spray the upper atmosphere, to Don's misery and reproach. Then miraculously, the undamaged guy draws off.

Can't he see we're crippled? Or is he out of fuel or ammo? Was it a trap up there and were we late? Were they up there longer than they bargained?

Slowly we spin. Not a word over the intercom. Just a strange mumble. All the same, I'll bet all hands have grabbed their 'brolly bottles' — the little oxygen bottle which will keep you alive for a quarter of an hour until you reach breathable air. At least, that's the idea. But what about the frost-bite? And how'll we get Don out?

The second dicky shows up. He's only been with us twice. 'Can't we hear Geoff on the intercom, or won't we answer? Are we all clots? Don't we know the doors are down and the hydraulic gear has gone to hell? Come along and wind up with the hand gear.' I leave Don and crawl after him. My portable oxygen bottle bobs along with me. Lord God, how we worked! If we can get the doors up she may come out of the spin. It's a slow spin but the extra G's force me to crawl and fight my way along. If the spin quickens or tightens we'll black out. Geoff, Geoff — don't let her!

The doors take an eternity. The second dicky's on the outer, harder side and he works like a fiend. Presently he passes out. Stupidly, I stared at him. Automatically I notice that his oxygen is O.K. Why did he pass out then? Hit? In a moment or two he recovers but passes out again as soon as he tries to work. I see it now and so, very cautiously, I finish my share and edge by him to tie off the job as well as my clumsiness will permit. He recovers again in a minute or two and crawls shakily forward. The spin slowly stops and we are flying straight and level. It's only then that I notice that one fan is feathered. On the outside of the spin, thank God. That must have helped. A lovely feeling to be flying straight and level.

The sound of Alec's guns soon ends that pipe dream. Like a fool, I leap for my place and the whole aircraft slowly greys and spins around me. When I come to, seconds later, I crawl slowly to Don's place and, in spite of the urgency of the guns, I plug into the main oxygen supply at once. The whole world clears up and I can see the little bastard slipping into position again but out of range. All the same I can see him clearly and without the furry outlines of oxygen lack and there's something odd about the tips of his wings. I've never seen that shape before and so he must be new! New fighter, eh?

Jesus! What's that? A great gaping hole appears near me and a noise like all the thunder that ever was. Through the hole I can see an outer engine go to hell. The little cloud of ice that suddenly appears could be Geoff jettisoning petrol, I suppose. And the cold! The blasting wind and the bitter, bitter cold.

Slowly the nose comes over and down, down, down we go. The speed must be terrific and I doubt whether we're stressed to take it in our damaged condition. And how do you get out of an aircraft at this speed? If we try the escape hatches the blast will pin us there. Stupidly and dully, I find myself cursing, of all people, Don and his bloody bends. I make no move to save myself or Don; the hole gapes in front of me, its jagged edges will cut us to pieces; no escape there. Stupidly, I just won't think of anything else. I put the idea of getting out away from my mind, rather petulantly and feebly return to cursing Don and his bends.

Seconds only, I suppose, but what an eternity of time a second is. The main oxygen supply goes and I recognize, stupidly again, that Don has passed out and I feel, in some infantile way, that it serves him right. All the same, after I'd plugged myself into my portable supply, with fingers like sausages, I do the same for him. How is Alec doing? The intercom is still mumbling.

Still going down, and then, ever so slowly, we come out of it. Ever so slowly so the damaged wing won't leave us. Ever so slowly, Geoff. Oh yes, we come out of it. Not much to spare that time, Geoff. The countryside is distressingly close. However, now that we're almost on top of it I know exactly where we are and, forgetting the intercom can only mumble, I pass the news on to Geoff. His reply comes crisp and clear. Now, how do you like that?

Then it dawned on me, or at least I think it dawned on me. Up where we were there wasn't enough air to carry our voices from the vocal cords. Hence the mumbles. That's my explanation, anyway.

And now for the chastening look around, because it's my job to check on damage. The old girl, without a load, will hold her altitude on what she has according to her manufacturers, but she's been specially tuned for high-altitude stuff and she's not so good down here. Besides two engines are not intended to maintain altitude on a flying colander.

Very slowly we slip downhill. To my surprise we lose height at about the rate the land is falling away from us and so we keep the same relative distance above it. That's fine until we meet the sea which doesn't fall away.

Count your blessings. The fans are turning. The crew is intact, or is it? What a sod you are for flesh wounds, Alec. The doors are up but one wheel is down in the way it always happens, the flaps are jammed but that's not present trouble. That'll worry Geoff at let-down (if there is a let-down). Crowning blessing — the bloody big hole is where the old girl is strong. And how wonderful to breathe without a mask that hurts the face, constricts vision; the way a mask cuts into vision has to be experienced to be believed. The biting wind still pours through that cracking big hole and God knows how many smaller ones, but my thermometer has tentatively moved back on to the scale so the temperature must be rising. My apprehension keeps me warm, anyway.

Rather ashamedly I sneak a look at Don. He must have gone through bloody hell in that dive. A sort of decompression chamber back to front. It's a wonder the gas bubbles didn't find his heart and kill him. The sweat pouring from him is freezing around his mask. I pull the mask off and try to make him comfortable. Slowly he revives.

Still we drift downhill. Good for you, Geoff. Hold her off the carpet.

Why is nobody interested in us? We're a sitting duck. Fighter defence knows all about us. Fritz has some bloody good reason for not wanting the area photographed and he has every reason to do us in. It'll be easy too. Geoff has a little lateral control but hardly any possibility of vertical manoeuvre. What are you waiting for, Fritz?

By way of comfort, I do know exactly where we are. I also know our parody of a ground-speed. And I know we're on the best possible course for the next possible airfield. And it's a lovely, crystal-clear day.

I need hardly mention that the radio has gone to hell. It always does. The multiplying of gadgetry to compensate for the inferior man is plain folly. An aircraft has a surprisingly small vital area, and without half the junk would be mighty hard to clean up.

Bless me, here's the sea, the very abode of Britannia herself. Just a little dirty smear on the horizon but the veritable ocean nevertheless. Geoff, how far above the little wavelets can you hold the old buzzard?

But as we approach the sea the crystal air changes a little and distressing signs of thermals appear. The damned cumulus begin to gather.

The old girl is working a bit already. You can feel her coming apart. She's in no shape to take a beating from the cumulus.

Behind us the land is bright, ahead the sea is as grim as the gates of Hell. What has happened to our lovely day?

Still no fighters? Why not? Why are they letting us off the hook? They can't want us to get away.

The light fades. Somehow, the murky light makes it feel colder. Perhaps it is. The first clumsy lurch tells us the old girl doesn't like the turbulence. I don't blame her. Neither do I.

Very slowly Don flexes his arms, and then with a sort of wonder, his legs too. The nitrogen bubbles are clearing themselves and if he doesn't have a seizure we're all right.

What did I say? At this very moment the old girl drops sickeningly. I know it's only a bump but the falling seems to be going on for ever as though there were no end to it and the sudden reversal seems to be driving one's guts through one's mouth. God, I feel bad. Surely not going to be airsick before we get home. I grab some of the dope we were issued as an experiment, and then force another couple on Don.

Almost, it seems, methodically, the elements proceed to batter the old girl to pieces. Quite slowly. There's plenty of time. But, little by little, she's coming apart beneath us under the savage hammering of the cumulus and it won't be long now, my friends. The second dicky is violently airsick and I get some satisfaction out of that. He's no help at all. He just stays in his place and spews straight ahead while his whole body is convulsed with great shudders. Geoff signalled to me and I fed him some of the dope, hoping frantically that I was in time because if there's one thing will make you airsick quicker than anything else it's the smell of spew and the horrible retching sounds that go with it. Perhaps you're surprised to know it can be heard?

Going back, I took a bad toss. My own fault, of course, but it shook the wits out of me for a bit, partly because it was so near that bloody hole. Then it dawned on me that the course for a clear, cloudless, windless morning at thirty thousand feet mightn't be necessarily right for the bashing we were getting at under two thousand. The Met report was all to hell. But wait a minute, that report was for our operational height and it said nothing about conditions at sea-level. Very well then, what are these conditions at sea-level? Occasionally I can see the waves so I have a line on the wind. It's all over the place. Mostly upward. Lovely!

Can you keep her airborne for another couple of hours, Geoff? Geoff thinks not. But Geoff had better think again. I tell him he has no option unless he wants to ditch the flying junk heap we're in. He gets the point. He always does. So minute by aching minute we sit and ride her with Geoff. I try not to look at the altimeter every ten seconds and I try even harder not to think of the two remaining engines, tuned for high-altitude stuff and now certainly running hot at this level. Only two of them and the old girl as full of holes as a bath sponge.

Minute by minute she settles. Minute by minute we inch closer to England. Strain your eyes through the gaps. Never mind the jolts and the breaking up of the old bus. Of course she's breaking up but it's broad daylight and soon, in the gaps between the cumulus, you'll see England. So strain your eyes until they start out of your heads. Anything that doesn't alter position is England.

And so it is. We slither over the coast-line and I notice for the first time how roughly the engines are running. Geoff does his stuff. He doesn't put the other leg down until the last possible moment at the first possible airfield. All at crash stations. This is it!

We could almost put a hand on the ground at this stage but from my position I couldn't even see the little field we sat down on. All I know is that the other leg did go down and that we hit with a God-awful crack, kangarooed along for a bit, dipped to one side as the leg on that side collapsed, swung in a wide circle, the wing-tip touched and disintegrated, we half rolled over and then, suddenly … unbelievably … stopped!

With me still alive. With us all still alive.

The blood cart and the crash wagon made a dead heat of it.

We stood and looked at the write-off. Geoff turned to me, 'And we didn't get those photographs either.'

That's just what I was thinking.

In 1943, on the island of Guadalcanal. Bob Spurdle was flying in the Pacific — not in bombers, but escorting them.

The Blue Arena
Squadron Leader Bob Spurdle DFC and Bar

Two small islands to our left were curiously uniform in size and I banked the Kittyhawk in a gentle turn towards them. A hundred yards to the right and slightly astern Sergeant Pirie slid across behind me and over to port. At 200 feet I tightened the turn and, banking steeply, we stared down at the green islets.

They were elongated and covered in thick foliage. They had no fringing coral shallows but lay in the dark blue of deep water. It was awkward to fly close to them in the narrow bay because of steep hillsides and I kept an eye out for steel cables. Sometimes the wily Japanese slung them between high points to snare the unwary.

There was something wrong and when up-sun of the two islands, I saw it. Saw the flash of sun from glass hidden beneath the leaves. Then I spotted a third boat nosed in against the shore.

'Drop tanks! Line astern — go!'

The 90 gallon belly tanks fell tumbling end over end to burst open on the sea. Pirie dropped behind me as I turned away and flew up-sun to position for attack. A wide turn and —

'Nine-degrees to port — go!'

We dived almost line abreast towards the mystery boats.

'You take the port one.'

'Roger!'

And in we went in a shallow dive. The .50 calibre machine guns chattered out their deadly hail and tracers spiralled away trailing white smoke to lash the sea to foam or be absorbed in the green foliage. Ricochets could be seen caroming away to skitter across the water in leaping splashes. Leaves and branches were smashed and blown away to reveal the motor torpedo boats beneath.

Three of them, and each about 70 feet long! Gyoraitaeis! The grey devils that plagued the Yank navy in the close in-fighting of night actions among the reefs and islands of New Georgia.

With steep climbing turns to avoid the jungle slopes of Choiseul, we flew back down again and again, diving within feet of the sea's surface to flatten out and spray the boats. The sun shone hot into our cockpits and the saliva dried in our mouths with the excitement of the kill. No need for instructions; no need for words. We flew as one.

The damn things must sink or burn soon, I thought, and then decided to fly away for a spell and try to trap any Japs that might attempt salvage. The boats must be riddled and sinking!

For some twenty minutes we scouted the coastline for more game and then returned to the attack. Fuel covered the sea in iridescent swirls but there was no sign of life. Japanese gunners were probably shooting at us from shelter of trees along the shoreline but we couldn't care less in the thrill of the moment. At the first burst of our fire, the tracers ignited the fuel and the 'islands' burst into flames which ran down and spread across the water. Oily smoke roiled and billowed into the blue sky.

It was time to leave — the black column and raging fires would be visible for miles and we were much too near Jap fighter bases for comfort. We turned towards Kolombangara Island and fled the scene like naughty schoolboys, fearful of punishment.

Thirty miles away, looking back over my shoulder and by slewing the tail of the Kittyhawk to one side, I could see the smoke like a huge exclamation mark. Kolombangara's high peaks were dead ahead and in the shallows of the eastern coast Japanese flak gunners like to bathe. We caught a few in bursts of fire before climbing for the sky and on back towards our base at Guadalcanal. The Kittyhawk might have been a poor fighter but as a ground strafer it was magnificent with its six .5's, lots of ammo and long range.

More and more the RNZAF were relied on to give close support. US bomber squadrons made a point of asking for us to be detailed to this onerous chore. Their own fighters had the much more pleasant (and safe) job swanning around free-ranging the target area. We were used as close front lower cover — the most difficult, exacting and dangerous position.

The Jap's method of attacking the B24's was a quick dive for speed and to climb up, firing continually, in a head-on approach. Then a half-roll diving away out to one side to climb, head off the bomber stream and repeat the performance. Each attack, made usually in threes and fours, meant that we defending escort-fighters had to dip our noses to meet them and so lost height. Then we'd climb as fast as we could back into position.

At over 20,000 feet, a B24, having dropped its bombs, could actually climb away from a Kittyhawk and on this next occasion this is exactly what happened.

August 28th, Sortie No 445: Close escort for 27 B24 bombers attacking Kahili air-field with 40 USAAF supporting fighters. Kiwis — three.

'Ma port outer generator's gone blooey — am returning to Cactus,' and a bomber peeled off.

'Having trouble with waist gunner's oxygen' and another went. Some just turned away without even an excuse. As we neared Bougainville, more and more chickened out until, out of twenty-seven big four-engined bombers, only fifteen remained. In excellent close formation these stalwarts forged on. Of the forty US fighters, only eleven remained.

You could feel the tension mounting as we droned up the Slot climbing to 21,000 feet. To my surprise the bomber leader kept on until about five miles inland when we wheeled around in a great curve and began the bombing run. The big boys were going fast and we could not weave. The bomb door opened. Flight Sergeant Pirie took the starboard side and I the port with Flight Sergeant Laurie as my No 2.

Flak started to burst amongst us and then I saw Jap fighters coming up in a quarter attack from four o'clock at Pirie. I warned him and at the same time saw a mix of eight or nine Zekes and Haps at ten o'clock at our level. These machines bored in, in a semi head-on attack on the bombers behind us. I fired on the leader seeing a few strikes. At 400 yards, he started to fire and, rolling on his back, continued to fire but at nothing — he was stuffed. Black smoke from both wing-roots poured out and, looking back and below, I saw the thing falling in a ball of flames.

Up above us the B24 gunners were hosing away, their white tracer smoke streaming out in great arcs. Away over to starboard I saw Pirie get a Jap which burst into flames and hurtle down. Sergeant Laurie had somehow gone over to him and now he came back to drive a Zero off my tail. The fight got very confused, the only focal point being the great bombers which released their eggs in long streams to plaster the airfield below. With bombs away, the squadron commander gave his turbo-charged Pratt and Whitney's the gas. His whole squadron began to climb! We couldn't stay with them; our Allison motors wheezing away in the thin air. Steadily the gap between us widened and now the three of us were left to the Jap hornet swarm. Soon we three became separated in the mêlée. There was only one thing to do and that was put our noses down in screaming dives for the deck.

Ahead of me, I saw a lone Kitty and drew up to it — Noel Pirie. My No 3 Flight Sergeant Laurie had got lost somewhere (he proceeded on his own and landed OK).

Formatting together, Noel and I flew along at fifteen thousand feet below the B24's. Above them several Japs were milling about and suddenly there was a huge white bomb-burst near the bombers. Great streaming tentacles of white smoke hung down from the central cloud. The buggers were trying to bomb the big planes with some new sort of weapon! And again but not so close! There was nothing we could do about it and heading towards a patch of smoke on the NE tip of Ganongga Island, we found two small ships on fire. We tickled them up a bit with the last of our ammo. These boats were the craft Jack Day and his boys had attacked shortly beforehand on shipping recce to Buri Village.

Back at base more criticism for 'having left the bombers.'

'Left them? Left them? The bastards left us!!'

What a bloody lovely situation.

I was having a miserable time at night with pain from the injuries received on my bale-out from the crippled Spitfire in England. The heat and sweat were making my life hell and the continual frustrations of these pitiful bomber escort jobs became more than I could bear.

One pilot who never left the European theatre was Johnny Checketts. Born in Invercargill, he joined the RNZAF in 1939 and remained with the force after the war, leaving in 1955 to start a topdressing company. Shot down in September 1943 he was able to avoid capture with the help of the French Resistance. Earlier that year (while Bob Spurdle was enduring the heat on Guadalcanal) Johnny Checketts wrote of a typical day at the famous RAF base, Biggin Hill.

New Zealanders with the Royal Air Force Volume II: European Theatre, January 1943–May 1945
Wing Commander H. L. Thompson

We had been wakened at 0500 hours, and though I had stolen an extra forty winks, I felt really contented as I walked from the mess to hear 'Al's' briefing at 0600 hours.

The early morning sun gives promise of a sweltering day, and the ground mists are dispersing leaving the hawthorn and blackberries smelling fresh and clean. Even the sombre buildings look fresh — it is grand to be alive. The 'show' looks interesting and we expect some fights. Al decides not to fly with my squadron and as we go to dispersal he gives me a few final instructions. The Spitfires look sleek and pretty and my pilots were all happy and contented. I had had mail from home yesterday so had every reason to be pleased with life. As the pilots changed into flying kit I watched them and marvelled at their laughter, jokes and perfect fellowship. I only hoped I could always be in such grand company.

As we were strapping in, Al gave me a cheery grin and we waited for him to start up. I get a little tense at this time, because the minutes drag so slowly. At last his Spitfire starts and I start mine as all the others stir to life. I have a lovely aircraft and the huge engine splutters to life as Al taxies out for take off. I wait until his pilots form up on him and my boys taxi out behind me and form up on me in one long line abreast. As Al takes his aircraft off I glance over my pilots and take a last look at dispersal. Doc., Spy and Adj. are there waving, as well as our ground crews. I wave my hand as a signal to start and open my throttle. The long line of Spitfires slowly gather way, faster and faster until they become airborne. Wheels tuck away like birds and the aircraft take formation positions. I can see Al about three miles ahead and take up position behind and down sun as we slowly climb over the beautiful Kentish countryside. I look round at my squadron and then check my gunsight, gun safety catches, oxygen, wireless and petrol. England is away below and looks so fresh and green. I can see the channel from Ostend to Le Havre very blue and calm.

The bombers loom up in the distance like a swarm of bees and as they approach we take position and set course towards France, leaving England's white cliffs behind us. I open the boys to battle formation and search up sun for Huns. Controller reports Huns away to south-east and the formation looks in that direction as we drone on our way. Al's voice calls a turn to starboard and we approach our target unmolested. The bombers drop their eggs slap on the target and the huge bursts throw up immense clouds of dust and smoke — I'm glad I'm not there. Up comes the flak at us now, as well as at the bombers, and I hastily clamber for altitude because some bursts are pretty near. Al calls another turn, and so far no Hun aircraft has come near us though many are reported inland. The flak and target fall behind and we approach the French coast on the way home. Jerry seems to be crazy this morning because the flak is all round the bombers and the ominous black blobs speck the sky behind and to one side of us. As we cross out, Al's radio comes in and his voice asks if we are all alright. I answer that we are O.K. and we carry on.

England is awake now — I can see the smoke before we cross over her white coasts. We leave the bombers over the friendly coast and dive swiftly towards home and another breakfast. I hear Al break his squadron into sections for landing and watch the fours landing with swift efficiency. I break my boys and lead my four down, lower our wheels, open hoods, lower flaps and sweep in on to a lovely smooth aerodrome. I watch my other sections come in like graceful birds, as I taxi in.

As I clamber out my ground crew rush up and help me and say, 'No luck, Sir?'. They always look at our gun patches and can tell whether we fired or not before we finish taxying in. I grin and say, 'Not today boys.' Al looks over the bay from his machine and hurries me up for a late breakfast. I grab my shoes, tie and dog and clamber into his car.

We laze in the mid-morning sun discussing the show and chattering away like children. Some of the pilots sleep, some just sit and generally take life easily. I inspect my tomatoes and chase my dog away from the precious plants, much to the amusement of the boys, because 'Winkle' won't wander far from me at any rate.

We are expecting another 'show' and Al calls us up for briefing at noon. This is to be a good one we hope. Al and Spy are still busy when I look in at 11.30 hours and so I keep out of their way and try to get the 'gen' on where we are going. Sailor comes in and we chat about aircraft, the morning's show and shotguns. He is a very fine fighter pilot and a jolly fine friend. Everyone files in and Al starts his briefing. He is very thorough and explains what he wants done and finally decides he is going to fly with my squadron. The pilots' voices murmur and finally break out into hopes that Jerry will come and fight. I like watching their expressions during briefings; some are sombre, some are keen, some express nothing, but I know that they are all keyed up and really anxious to get to grips with the Hun. They are all good boys and I think an awful lot of them. We have been briefed early so that we can have lunch without having to hurry.

The boys all go to the bar and have a beer before lunch. I am very dry and like to take my drink out on to the lawn and gaze across the valley into the soft green woods on the other side, with the nice white house in its very tidy grounds. The weather is too hot for lunch, but I have a little and go off to dispersal. We still have 15 minutes and the pilots are not all here yet. The radio-gram plays some new records and everything is peaceful. As the pilots come along we start to change into flying gear; Al arrives and everything is now very busy. Ground crews rushing here and there, pilots looking for gear and studying maps, phones buzzing and innumerable last minute questions and hustle. There are some disappointed faces too; pilots cannot go on every show and the boys hate to miss one.

Al and I sit in the sun as the pilots go to their machines and we finally have to stir. I am leading one of the sections and Al is leading the squadron. The old tense minute arrives, how I hate it too. I check and recheck my cockpit instruments and controls, but Al's motor bursts into life and I come back to earth with a start. As we taxi out I see the heatwaves from the ground rising like fire; the old Spits. are very warm and we are really hot until we get airborne. The squadron forms on Al and off we go. Wheels tuck away and we meet the bombers and evade the coastal flak and approach the target. There is not much flak and I'm not certain where Jerry is. He is being reported near us to the south-east and south-west but we have not made contact. I'm disappointed but the bombs make a nice big mess on a Hun target. It's awfully hot and I'm glad when Al calls his turn. I have a look at my No. 2 and he is busy looking above and behind. I nearly look there too but check myself and look elsewhere — it is no good wasting a pair of eyes. In spite of our keenness we see no Huns and dodge the flak on our way over the coast and land at base after an uneventful show. The pilots cluster around Al and discuss the sweep and chatter away as they usually do, discussing tactics and all the things that happen on such a mission.

We don't think there is going to be another sweep today and the boys settle down to station duties and their non-flying tasks, or just read and write in the sun. I have a little office work to do and when I've finished we have a clay bird shoot and lay small wagers on our own skill at this fascinating training. At 1600 hours the pilots start to go up to the mess for tea and we have a nice hot cup and lie on the grass or go swimming. I think I shall change and go out this evening, so after a shower I don uniform but am informed of another sweep — we are to be briefed immediately. The pilots rush off to briefing but we don't expect to see anything on this sweep. Al is not flying with my squadron this time. Take off is 1800 hours and Al and I yarn about tactics on the way to dispersal. The boys are all there changing and the usual pre-sweep bustle is noticeable. I think it is a shame to fight on such a beautiful day and wonder what Jerry thinks about it. Winkle is very hot and just pants in the shade; he is very keen to retrieve a stick though and I think of home and the happy days I had duck shooting in the estuary, and of my friends who are now on all the English battlefronts.

We are assisted into the Spits. and Al starts up. I feel the heat very much and my Spit. is hot to touch; she is a perfect machine though and

I've had some good fights in her. These aircraft are beautiful and sleek; I love flying them and playing in the cloud valleys and tail chasing with my boys. The slipstream rushes the cool air into my cockpit as I taxi out after Al and watch his pilots form up on him and take off. My pilots form on me. I watched them take off one day when Sailor would not let me fly and they looked beautiful tearing down the aerodrome in formation and tucking their wheels away, almost like birds. I open my throttle and the Spits. gather way slowly then terribly fast and at last leave the ground. Al is a bit further away than I expected and we close to position just as we meet the bombers. One big circuit and we cross the coast over the blue water. The sun is strong and at 22,000 feet the bright sky is hard to look into. As we cross the French coast Huns are reported east, south, and west of us but as yet we don't see them. Al has called a turn to port 30 degrees as we approach the target and the Huns are still nowhere near us. The bombers drone on very steadily and look like big moths, only very sinister. We sweep slightly south of them and I watch their bombs burst on the target with a terrific upheaval. I carry on on my present course and call my turn to the pilots. We turn 120 degrees to port. The Huns should be in sight soon. Those specks over to port look suspicious and I finally identify 14 F.W. 190s approaching slightly below and above two miles away. I swing to attack and call Al on my radio that I've made contact and am attacking. My Spitfire is tearing along and I can see Jerry trying to get at the bombers. Suddenly we are among them, black crosses and sinister aircraft dart everywhere. I get on the tail of one and my sight picks him out. A two second squirt, the cannon vibrate my aircraft and cowlings, smoke and flame gush from the unfortunate Hun; anyway the only good Hun is a dead one. His leader rolls over and goes vertically down and I chase after him, closing the range slightly because of my superior speed; we tear down at a terrific speed and every time I fire my Spitfire judders to cannon recoil. At last I hit him on the starboard wing and close the range to 200 yards as he levels out at ground level. This Hun heads for the south taking my No. 2 and myself inland as fast as he can go. On my next squirt a cannon stops and I close the range to 100 yards and let him have machine gun only. I can see my bullets striking but he won't go down. At last a thin white trail of smoke, gradually turning black. I have to leave him because we are too far inland. We break to port and set course for England at ground level. The French peasants wave to us and I find I am wet with perspiration but the fascination of flying over enemy territory at zero feet, seeing people, towns, harvesting and the thrill I got out of one Hun destroyed and another probably destroyed make my wet clothing seem as nothing.

We maintain full speed and I look about for more Spitfires but there is only my No. 2 and myself so we fly back towards the coast and home.

As the highest award given to British (and Commonwealth) forces, the Victoria Cross is a medal rarely bestowed. Three New Zealand airmen were awarded the Victoria Cross during World War Two. One of these three was a Coastal Command pilot from North Auckland who'd been farming before the war.

When Lloyd Trigg said goodbye to his wife and two small sons, then aged about two and one, he gave an assurance that he '... would do the best job he could but that he wouldn't go looking for decorations'. Yet he got one, anyway. As cited by Alan W. Mitchell in his book *New Zealanders in the Air War*: 'There must be very few cases in which the award of the Victoria Cross has been based entirely on evidence supplied by the enemy ... He won the decoration, and lost his life, when he attacked a German U-boat off the West African coast.

Having previously flown twin-engined Hudsons, Lloyd Trigg and his six crew members were on their first operation in a larger, four-engined Liberator. After eight hours on patrol, someone spotted a U-boat on the surface, and as Mitchell describes: ' ... Trigg immediately prepared to attack it. In the face of intense anti-aircraft fire from the U-boat's forward gun, the New Zealander made two runs across the target, straddling it with bombs but receiving direct hits which caused the Liberator to catch fire.'

According to the official citation: 'That moment was critical. Flying Officer Trigg could have broken off the engagement and made a forced landing in the sea.' If he continued to attack, his aircraft would be a no deflection target 'and every second spent in the air would increase the extent and the intensity of the flames and diminish his chance of survival'. Yet, despite 'the already precarious condition of the aircraft' he attacked again. 'Skimming over the U-boat at less than fifty feet with anti-aircraft fire entering the open bomb-doors', he dropped his depth charges 'with devastating effect. A short distance further on the Liberator dived into the sea with her gallant captain and crew.'

All this only became known after seven members of the U-boat's crew (including the captain) were found two days later, drifting in the Liberator's dinghy and rescued by a Bavy corvette. It was the German survivors' generous praise for the captain and crew of the Liberator that earned Lloyd Trigg his VC. He had flown 46 operational missions.

The first New Zealander to receive a Victoria Cross was a co-pilot from Wanganui. This is his story ...

Great New Zealand Adventures
Brian Joyce

Jimmy Ward completed the cockpit check as the Wellington bomber lumbered out of the dispersal point, with the sound of its engines shattering the peace of a lovely English evening. As they rolled towards the main runway, Ward could pick out by their navigation lights the other eleven bombers that were taking part in the night raid. One by one they turned onto the runway, paused to clear their engines, switched propellers to fine pitch and applied full power. Gradually the runway lights blurred together as Ward and his crew roared down the tarmac. There was a bump, then a lurch and they were airborne and the base faded away below the wings. Under full power they climbed steadily, banking to the east — towards the North Sea and Germany beyond. The year was 1941 and Britain and her allies were at war. Jimmy Ward was a New Zealander.

Britain was fighting for its survival in March 1941 when Ward, a twenty-two-year-old school teacher, arrived in England. For a year-and-a-half the war had gone badly for the Allies, with Hitler sweeping all before him — Poland, Holland, Belgium and France. However, the English Channel and RAF fighter squadrons confined him to Western Europe.

It was decided that Bomber Command would strike at Germany. As the Wellington bombers crossed the coast of England on the night of 7 July 1941 they set course for Munster, an important river and rail centre. Damage to Munster would hinder Germany's attack on Russia, which had been launched the month before.

Ward and his captain, Squadron Leader R. P. Widdowson of Winnipeg, had developed a smooth partnership. The Canadian, himself a skilled and experienced aviator, admired Ward's coolness when they were under fire. His quiet, unassuming second pilot had proved himself in his first month during raids on Kiel, Dusseldorf, Cologne, Mannheim, Brest and Hamburg. Then it was Munster's turn. All eyes in the bomber swept the dark skies, alert for signs of night fighters which the Germans had recently pressed into service. Below, the North Sea shimmered in the bright starlight as they climbed to 4,000 metres above sea level. Ward probably had time to reflect back over his short life to date.

Flying bombers for the famous No.75 (New Zealand) Squadron was a far cry from the quiet provincial life of Wanganui, where Ward was born in 1919. Apart from a spell at Wellington Teachers Training College, he had always been a Wanganui man. He was in his first year as a teacher at Castlecliff School in 1939 when Hitler's actions changed the course of his life. Now, he found himself in a freezing-cold plane passing over the coast of Holland. For the next few hours, during the flight to Munster and back, enemy fighters could be a problem. As the plane roared over Holland a full moon began to rise in the east giving the bomber crew a spectacular view of Western Europe. Only a year before, in July 1940, Ward had had his first flying instruction at Taieri Aerodrome near Dunedin.

In September 1940 Ward had gone on to Wigram at Christchurch to complete his service training, receive his flying badge and promotion to Flight Sergeant. Then in January 1941 he had boarded the *Aorangi*, bound for England. Events had moved quickly when he arrived on 6 March at the Bournemouth reception centre — a few days later he was sent to train on Wellington bombers at Lossiemouth in Scotland and then he was given a posting to No.75 Squadron. He had experienced a hectic month over the night skies of Germany since his arrival at Feltwell, and, like all other members of bomber crews, he would never forget his first experience of flak, or anti-aircraft fire.

As they approached Munster visibility was perfect under the bright moonlight — perfect for identifying targets and shooting down bombers. Their luck was in that night as they started the bombing run with only sporadic bursts of flak punctuating the sky. With their bombs falling towards the ground, Widdowson circled the city to gauge the night's work. Heavy damage appeared to have been done to the railhead and the river port. Then, with Munster and the worst danger behind them, they turned west for the long flight home. However, they could not afford to relax so soon. Even over England the enemy could still be a menace. Night intruders had shot down bombers while they made their approach to base, perfectly silhouetted against the runway lights. They knew they must keep their eyes moving, searching, always searching, for the enemy.

Holland crept past below their wings as the plane lumbered along at 275 kilometres/hour. Ahead they could see the coast and the glimmering waters of the Zuider Zee and the North Sea beyond. In the distance the dark mass of England, home, Feltwell, loomed large, almost close enough to touch, it seemed. Ward had enjoyed his month at Feltwell, home of a collection of buildings and tarmac camouflaged to blend into the shape and colour of the East Anglian countryside. Perhaps he liked it because of the many New Zealand aircrew on the squadron or perhaps because the East Anglian farmers were so friendly and hospitable, just like those in Wanganui.

If the good weather held over Europe they would be sent out flying again the next night. The day would follow the old familiar pattern — night-flying 'tests' in the morning and, after lunch, servicing, refuelling and 'bombing up' or loading the planes ready for the night's raid. All aircrew would be in the crew room for final briefing in the late afternoon. There would be an evening meal, a short rest, then take-off for an uncertain evening over Germany once again.

As the Wellington droned out over the Zuider Zee the crew searched the skies ceaselessly. But they didn't see the danger until it was too late. A Messerschmitt 110 night fighter fell on them from above with its machine-guns blazing. They were sitting ducks. Cannon fire and heavy calibre incendiary shells raked the plane from nose to tail, making her stagger in the air. Only the rear gunner, Sergeant R. J. R. Box of Auckland, had a chance to hit back. As the fighter flashed past his turret he raked the enemy with a point-blank burst of machine-gun fire. A moment later one of its engines caught fire, it stalled and dived steeply towards the sea.

Back in the Wellington, Widdowson and Ward regained control of the bomber after the staggering burst of enemy fire. She was flying, but only just. Out on the starboard wing the motor was vibrating, threatening to shake the plane apart. They closed down the motor and the vibration ceased. One shell had severed a hydraulic line and the bomb bay door hung down, slowing their speed. Miraculously, only one member of the crew was hurt — the front gunner, Sergeant D. H. V. Evans of Wales, who had been hit in the leg. Another shell had destroyed the wireless and the intercom. Then what all aviators fear most happened. Fire broke out on the starboard wing, close to the useless motor. Fed by escaping petrol, the flames flared and seemed likely to engulf the entire wing.

Ward and the navigator, Sergeant L. R. Lawton of Auckland, hacked a hole in the side of the fuselage and tried to attack the flames with fire extinguishers. The slipstream whipped the foam away uselessly into the night. They even tried pouring coffee from their vacuum flasks along the leading edge of the wing, hoping it would run along to the seat of the fire, which was a large shell hole over a metre away. It, too, was blown away into the night.

Widdowson passed a message to all members of the crew: 'Prepare to abandon aircraft.' Baling out of the bomber would be a death sentence for some, if not all, of them. They were over the North Sea, too far from England and unable to alert rescue services of their plight because of the damaged wireless. Those that survived the cold water would almost certainly end up in a prisoner-of-war camp.

Ward tried one last, desperate gamble. He suggested that he climb out on the wing and try to put the fire out. Widdowson initially refused, but there was no other hope. Already the bomber was almost uncontrollable. If more wing fabric was destroyed they would soon plunge into the sea in a ball of fire. Reluctantly, Widdowson agreed. Ward wanted to leave his parachute behind because its bulk would hamper him, but the others persuaded him to use it in case he was torn from the plane by the slipstream.

Quickly they tied a rope around him from the aircraft's dinghy and opened the narrow astro-hatch in the top of the fuselage. Ward squirmed his body half through, gasping as the wind whistled past. Lawton shoved from below as Ward eased himself out onto the top of the fuselage, with his body spreadeagled. The wind threatened to tear him from the plane, so he quickly kicked holes in the aircraft's thin skin.

Then came the first tricky part. Lawton passed Ward's parachute up through the hatch. Holding on with one hand, he grabbed it, but when Lawton let go it was almost torn from his grasp. With the slipstream almost sucking him from the plane, Ward somehow managed to wriggle into the harness and fasten it. Next, Lawton passed through a canvas engine cover which one of the crew had been using as a cushion. Ward hoped to smother the fire with it. He slipped it under his body, but the wind kept getting under it and lifting his body off the fuselage. It almost caused his death.

Ward never wrote about his doubts and fears of being outside an aircraft high over the North Sea, his only link with the bomber being a slender dinghy painter. He said afterwards: 'It was just a matter of getting something to hang on to. It was like being in a terrific gale, only worse than any gale I've ever known.'

Widdowson did everything he could to stop his second pilot vanishing down into the sea below. The one remaining engine was throttled back to almost stalling speed and the propellers on the damaged engine were feathered to reduce turbulence.

Using all his strength, Ward inched down the curved side of the fuselage, kicking footholds as he went. It was a drop of little more than a metre down onto the Wellington's broad wing, yet it must have felt like a kilometre. Several times the slipstream lifted the canvas cover and Ward with it, then slammed him back down again onto the hard fuselage. Ward hung on until he felt the wing under a probing foot. A German shell hole gave him a secure hold as he dropped onto the wing. The slipstream was stronger now because of the aerofoil shape of the wing. His eyes were streaming and he shivered in the icy blast. However, the light from the fire helped him as he edged sideways towards the blaze. Finally, he was within reach of the flames streaming from the cannon shell hole. He saw that a fractured fuel line was feeding the fire.

Ward dragged the canvas cover from under his body. The slipstream caught it and almost hurled him into the night. But he clung on and got it under control before stuffing it into the hole and partially smothering the flames. Then, the slipstream whisked the cover away. Ward beat at the fire with his arms and hands until there seemed little danger of it spreading. By then he was exhausted. The return trip up the fuselage threatened to be a nightmare, although he did not have to contend with the canvas cover or kick footholds. Lawton kept the tension on the rope as Ward clawed his way back along the wing, then up the fuselage to the hatch. Off came the parachute and Lawton dragged Ward inside. He slumped against the fuselage and they plied him with warm coffee and chocolate.

A few minutes later the fire on the wing went out and they concentrated on getting home. Feltwell was too far away so Widdowson chose Newmarket aerodrome instead. Tension mounted as the one good engine droned on over the English coast towards London. With no radio direction fixes because of the damaged wireless, Lawton was thankful for a clear night as he conned them in on astro-navigation alone.

A few kilometres short of the runway, flames flared up again in the starboard wing, presumably fed by a residue of petrol. They watched anxiously, but the fire soon died out. Now, all hopes were on Widdowson as he brought the crippled bomber into the unfamiliar aerodrome. Lower and lower they sank, with the bomb doors hanging down, then with a screech of rubber against tar-seal the plane landed safely. This was no mean feat because the Wellington was so badly damaged it never flew again.

Two days later Jimmy Ward was awarded the Victoria Cross — the highest award for valour in the British Empire — for his courage on the wing of the Wellington. A few weeks later he was given command of a Wellington bomber.

On the night of 15 September 1941 twelve aircraft were despatched on a raid of Hamburg. It was Ward's second mission as captain. Over the target his Wellington was trapped by searchlights. He made strenuous efforts to escape, but the bomber was passed from one cone of lights to the next as it made its bombing run. Anti-aircraft batteries hammered the Wellington and several fires broke out. This time there was no chance of putting them out. Ward held the crippled bomber at a safe height while his crew baled out. He was last seen at the controls and went down with the plane to his death. He was buried by the Germans in Ohlsdorf Cemetery, Hamburg.

Leonard Trent was luckier. He survived the war and remained in the RAF, commanding the first Vickers Valiant V-bomber squadron before being promoted to the rank of Group Captain. In 1943 he was No. 487 (New Zealand) Squadron Leader. On 3 May, 12 Venturas from the squadron took off for a daylight raid on the Amsterdam power station. It was Leonard Trent's role in this doomed attack which won him the third Victoria Cross awarded.

Venturer Courageous: Group Captain Leonard Trent V.C., D.F.C., A Biography
James Sanders

The sun's bright eye lit the spring day and all at Methwold that lay before its gaze was golden. When the station came astir and the tasks of the day were being made manifest, Wing Commander Grindell, CO of 487 Squadron, popped his head around the door of B Flight's office and said to Squadron Leader Trent: 'It's Flushing Docks today, Len. And I'll be leading your flight — and the squadron.'

'Hold hard, sir,' replied the flight commander. 'As you remember, I didn't cross the egg-line yesterday; and I've still got my bomb load hanging in my aircraft.' The egg-line, a whimsical datum point fixed by the squadron's aircrews, was an imaginary line a few miles out from the enemy coast. If this line was not crossed, the fliers were not entitled to the rare and prized bacon and egg meal on return.

A light-hearted argument took place. And it was won — or lost, according to one's point of view — by Trent. Later, as the crews gathered in the briefing room, they had the usual flutterings in their stomachs and the nervous visits to the latrines before they donned their flying gear.

Trent, whose battledress had gone to be dry-cleaned of battle sweat, was wearing his 'blues'. Within one hour before take-off for the sortie over the southernmost corner of the Dutch coast, he cleared the patch pockets of his tunic of any give-away memos, tickets or items which could provide clues to enemy intelligence should he be shot down; and, as was customary, he wrote to his wife.

He and his three crew members — Flight Lieutenant V. Phillips, Flying Officer R. D. C. Thomas and Sergeant W. Trenery — went to where their Ventura, V-Victor, stood at its dispersal point. All made their usual pre-flight checks of their respective departments.

But as the men of the attacking force were preparing to enter their machines there came a hail from an operations room messenger: 'Everybody back to the briefing room!'

When all the crews were seated Wing Commander Grindell announced there would be a new target. An early lunch would be prepared.

Fluttering butterflies took the fine edge off most appetites. And when the men returned to the briefing room they heard from Grindell that they were to bomb the electrical power house on the north-western edge of Amsterdam — a target and city which had never before been attacked by the RAF.

It was considered an important target; British Intelligence having reported that the Dutch workers in power houses throughout Holland were awaiting a lead from their Amsterdam colleagues to stage mass strikes because of British bombing. Intelligence placed much — probably too much — stock in what the Dutch underground had to say.

In the Methwold briefing room, on that bright morning of Monday 3 May 1943, the aircrews of the impending mission learned the facts of life (and death) concerning the operation which was to become known as 'Ramrod 17' — a Ramrod being a code name for a 'press on regardless' foray.

On the previous day, both Bostons of 107 Squadron and Venturas of 464 Squadron had unsuccessfully attacked the Royal Dutch Steel Works at Ijmuiden. The Bostons' bombs had overshot the target although the Australians' Venturas had damaged the coke factory, sulphate plant, benzole unit, the compressor and various store houses.

Now the plan was for six Bostons to make low-level attack on the steel works. In conjunction — and as a diversionary move — twelve Venturas of 487 Squadron would fly to Amsterdam and bomb the power house. Fourteen Ventura crews were to attend the briefing, thereby allowing for two aircraft on stand-by in case of last minute emergencies.

The Ventura force was to rendezvous with escort Spitfires from 118, 167 and 504 Squadrons over Coltishall airfield at 1700 hours, flying in two boxes each of six aircraft below 100 feet, the distance between each box not to exceed 500 yards. After crossing the English coast, bombers and escort Spitfires were to descend to sea level until 33 minutes out from Coltishall, flying at 190 mph indicated air speed. Then all aircraft were to climb rapidly at 165 mph to 10,000 feet, which was to be the bombing height.

For the second time that day, Squadron Leader Trent and his crew went to their aircraft. And at thirteen other dispersal points around the airfield at Methwold there was the same urgent pre-flight activity. Seated in V-Victor, Trent went through his cockpit checks, keeping a constant eye on time to ensure his force would be airborne to keep the appointment over Coltishall.

The lid of the auxiliary control box was open and he noted that his parking brake was on, the master switch was on and the undercart selector was down and locked; and its emergency bypass valve was wired open. Hydraulics, generator, gyropilot and cabin door were also checked.

He set his fuel selectors to the rear main tank, adjusted throttle, mixture, pitch, supercharger, air, gills and oil knobs and levers and signalled the waiting ground-crew airmen to turn each propeller by hand for a couple of revolutions before applying power.

Again, he looked at his watch. Good. Everything seemed to be on time. His engines were now running. As he waited for the oil to warm and course through the arteries of the Double-Wasps, he looked around the field as, one by one other machines of the force burst into life. Two of the Venturas had, however, developed faults which meant that the reserve aircraft would be brought into the breach.

At 1643 hours Squadron Leader Trent eased V-Victor off the grass runway. By 1647 hours all twelve of the aircraft were airborne and off on their first leg towards Amsterdam. Trent takes up the story in his own words:

Three miles from the Coltishall aerodrome we could see the Spitfires taking off and skimming the trees in a left-hand turn, to station themselves 100 yards on both sides of our formations. Right on time and all well, we set course for the Dutch coast.

We continued flying at 100 feet to avoid detection by the enemy radar. We hoped to get within 10 minutes of the Dutch coast before being plotted; at which point we'd do a full-power climb to bombing height. In this way we were usually able to complete the bombing run and be homeward bound before the German fighters could get into position to attack.

The enemy were not usually 'scrambled' until the bomber formations appeared on the radar screens. Then they had to take off and climb to get us within range of their forward-firing guns, keeping in mind our high and medium protective fighter cover. But only rarely did we catch a glimpse of this support. It was, however, a great comfort to feel that it was there, for the Ventura was cold meat for any aggressive German fighter. A dorsal turret with two .3s and the two .3s firing aft under the tail, was hardly a deterrent for a determined enemy in a cannon-firing fighter.

Admittedly, the Ventura pilot had two .5s and two .3s firing forward under the command of his thumb, but flying in formation and unable to manoeuvre quickly, he was not likely to have an opportunity to bring down any but the most naive German flyer — although, of course, such characters did exist (even if briefly) in air combats.

However, remembering the hard-luck story of Dibs Griffiths' gun button being in the 'off' position when a FW-190 crossed his track — and my own earlier and satisfactory experiences — my first move after leaving the English coast was to test my guns and leave the safety button on the 'fire' position.

Shortly after the formation had left the home coast and were skimming over the wavetops, Ventura Q-Queenie, piloted by Sergeant A. G. Barker, turned away and headed back to base. A hatch cover had come loose. This left only eleven bombers heading towards the target. As it was, Barker and his crew were to thank their lucky stars as the outcome of the mission unfolded. Already, the stage was being set for black tragedy.

The German Governor of Holland had decided to pay a state visit to the city of Haarlem that particular afternoon and fighter reinforcements of the Luftwaffe had been brought from as far as Norway and France to attend a convention of fighter pilots at Schiphol Aerodrome.

Some of the best of Goering's men in the Western Sector were there, ready and alert, as their lectures and discussions proceeded, to stand watch over their governor's welfare in Haarlem — a matter of 12 or 13 kilometres away to the north-west.

Both these functions — the Governor's visit to Haarlem and the convention of Luftwaffe fighter pilots — were unknown to British Intelligence when the Ventura bomber force was briefed. The track of the Amsterdam-bound aircraft was straight between the two hot spots.

By any other name, it was to prove as deadly a trap as any ill-defended gaggle of bombers could enter. To add to the hazard, there had been a bungle on the part of a wing of Spitfire IX fighters of 11 Group. Five support aircraft of 122 Squadron and eleven of 453 Squadron left Hornchurch at 1520–23 hours and, after refuelling at Martlesham, got airborne about 20 minutes too soon. Furthermore, instead of flying at low level until the Dutch coast was reached, they immediately began a climb to operational height. Obviously, the German radar picked up the Spitfires' movements and the Luftwaffe was on high alert. The seventy enemy fighters assembled at Schiphol had FW-190s to tackle the Spitfires and Me-109s for the bombers.

Realizing that his Spitfire IX squadrons were not only much too early but well north of their correct course — and so would not have enough petrol to complete their task — the Air Officer Commanding 11 Group had ordered their return.

Squadron Leader Trent led his bomber force into a climb at 1735 hours and soon, in clear weather, they reached the Dutch coast at 12,000 feet. But whereas the escorting Spitfires of 118 and 167 Squadrons were keeping close stations, the twelve Spitfires of 504 Squadron, which had come up from 10 Group for the operation, had found the climb to operational height too steep and at too low a speed. Unused to flying in formation in climbs of this type, they had dropped far behind.

Trent takes up the story again:

At 12,000 feet over the Dutch coast, and still not a shot fired, we were as yet unaware of the enemy fighters jockeying for position. Of course our escort knew all about the situation, but I was not informed.

Although my aircraft had recently been fitted with a VHF (Very High Frequency radio) set to be in contact with the escort, the Venturas around me had not had their magnetos and plugs properly screened (against electrical interference) and so my set was virtually useless. However, I was not unduly bothered, for we had an arranged signal that if for any reason the escort couldn't cope, a Spitfire was to fly across my bow, waggling his wings. I had never seen that signal and didn't ever expect to see it.

From a cloudless sky and in good visibility we could see Haarlem and just make out Amsterdam, 10 minutes' flying time from our position over the coast. We levelled off and, using the extra 2000 feet we had gained to increase our speed, I thought it was just a case of 'Look out, power house, here we come!'

My wireless operator had taken up his usual position in the astro dome. He was responsible for reporting the range and bearing of fighters attacking from beam and rear. The formations were reported in good order and I got the 'thumbs up' from my numbers two and three — the only Venturas I could see from my position in the cockpit.

It was at this point that more than twenty enemy fighters bore down on the Spitfires of 118 and 167 Squadrons which formed the close escort, while thirty other fighters attacked the Ventura force. The Spitfires of 504 Squadron, slow in their climb, were about 3 miles behind when the action began. The aggressive FW-190s charged in ahead of the RAF fighters, effectively cutting off assistance to the bombers. And although the close-escort Spitfires tried to stay with the Venturas, they were so occupied with fending off the attackers that they dropped well behind their charges.

> Suddenly, things began to happen. Flying Officer Thomas, from the astro dome, shouted: 'Here's a whole shower of fighters coming down on us out of the sun; they may be Spits — twenty, thirty, forty … Hell's teeth, they're 109s and 190s!' He called to Sergeant Trenery, our gunner: 'Watch 'em, Tren!' Then he began to give me his evasion patter: 'The 190s are engaging the close escort — so get ready to turn starboard, sir. And now the 109s are heading for us at 1000 yards … 800 … turn NOW … 600 yards. They're firing. Give 'em hell, Tren!'
>
> I could hear Tren's guns rattling away and smell the cordite. And as I turned to starboard I could see the enemy fighters coming in, one after another. About a dozen had attached themselves to my box and the same number were attacking Flight Lieutenant Duffill and his box.
>
> Duffill's aircraft was hit in the first attack and it went down with smoke pouring from its tail. I was to hear, long after, that he managed to regain control of the machine and get it back to base with two of his crew seriously injured — but not before one of the gunners had destroyed an enemy aircraft. Duffill's Ventura was the only one of the eleven to return from the mission.
>
> Next to fall victim was my number five, who was set on fire. And as the last of the attacking wave whipped by just underneath us, I straightened up and turned back on course, searching the sky

for signs of the top-cover Spitfires. I had caught a glimpse of a glorious dog fight going on miles back — so obviously the close escort was more than busy.

Our attackers could be seen pulling around underneath, with the obvious intent of attacking from the starboard again. By this time we were approaching Haarlem and, as I realized later, they were trying to make me turn away to the south — and to keep turning. This, to a certain extent, they were achieving. But I wasn't going to Haarlem, anyway. But about this time I had a horrible idea where I *was* going, if that top cover didn't arrive bloody soon.

Now there were only four Venturas left from the eleven that had crossed the Dutch coast. And the German attackers must have been exulting in their victories. Again they bored in on the valiant quartet.

As they came at us on our starboard, I turned slightly to the right to get the wings out of the way, thus giving our gunners a clear field of fire up to the last moment as the fighters ducked under our sadly depleted formation. It also presented a difficult full-deflection shot for the attackers.

As they came in this time I could see, at the ominous range of only 400 yards, the leading edges of the fighters' wings suddenly burst with flames as their wing guns poured shells at us. And this ugly sight continued until it seemed a collision was inevitable. Down, under and to the right again — and still we hadn't knocked one out of the sky.

But now my number four had been hit and had gone the way of number five. That had all happened in something like 2 minutes, which seemed like 2 years. I suppose the thought processes get speeded up a bit in such situations.

But I now realized that something had gone seriously wrong and that the top cover could not help us. What to do? We were on our own in a hopeless situation, obviously 'in for the chop' in a matter of minutes. We might as well be shot down over the land as come down in that rough sea — even supposing we managed to get back that far.

However, if the idea of turning for home ever occurred to the crew, they certainly didn't voice it. My navigator, Viv Phillips still had his eyes on Amsterdam and was urging me to keep turning left towards it whenever I could.

Again the enemy bored in from starboard bow and beam. From half up and 1500 yards they would wing over in loose pairs and come darting at our flanks like pack wolves ravaging the remnants of a buffalo herd. Once again at 400 yards I saw the twinkling flames from the leading edges of the fighters' wings and I pulled up slightly to put them off their aim and give them room to get underneath.

And then an encouraging thing happened. There was Leutnant von Prune himself! Instead of diving down, this merry goon pulled around level in front of our formation. I am sure he mis-judged our speed and obviously he had been studying anything but the armament of the Ventura. There he was, banked towards us — the range about 150 yards. He appeared very close. I scarcely had to move my aircraft. He was flying straight towards the centre of my old ring-and-bead sight.

There was a terrible vibration as I pressed the firing button. One … two … three seconds of assorted .5 and .3 ammunition ripped out to meet him. Got him! His wings suddenly rocked and he slowly turned upside-down and he was last seen going down on the starboard side, vertically and very fast.

Since the war a Dutch civilian, who was then a youth of eighteen, has written an account of what it all looked like from the ground. In his account he reports a Me-109 diving into the ground at this point. However, we didn't have time to follow the 109's progress to the ground — another pair of fighters was already shooting at us.

Frankly, I was amazed that we had lasted so long. Eight minutes had gone by and only now were we approaching the target. From hereon, it was a succession of attacks. Just then, my number three Terry Taylor pulled out with his port engine in flames.

And so Rusty Peryman, my number two was our only companion Ventura as the first of the Amsterdam built-up area appeared in the bombing sights. Suddenly, about a dozen or more black puffs appeared just ahead and slightly right. Good, I thought, at least the fighters will stand back a bit now. The black puffs started to appear all over the place and sometimes we could feel the bumps and hear the bangs. But, most surprising of all, the fighters took no notice of the ack-ack and appeared to redouble their efforts to get us before their own ack-ack batteries could claim kills.

My navigator now started to direct me onto the target and at the same time my fire controller was shouting the range and bearing of the attacking fighters and imploring me to turn right while my navigator was saying: 'Left, left … steady … 10 seconds to go … steady …' I had to tell my poor controller, Roy Thomas, to shut up. I never heard him say another word.

My mouth was as dry as a bone, for the ack-ack was everywhere and I wasn't able to take evasive action. 'Left, left … 5 seconds to go,' Viv Phillips chanted as I concentrated on keeping correct height and air speed. The bombing run was lasting a lifetime. And then there was a jubilant 'Bombs gone!' But when I looked up from my instrument panel I discovered my number two had gone. We were on our own.

As I reached for the lever to close the bomb doors, I thought: 'Down on the deck is our only chance.' But even as my hand came away from the lever there was a frightful bang and, horror of horrors, all the flying controls had been shot away. The engines were running perfectly so I waggled the stick and kicked the rudder pedals again to convince myself it was true. 'We've had it, chaps,' I called. 'No controls. Bale out. Abandon aircraft, quick!' My navigator dashed past me from the nose and disappeared into the main fuselage to collect his parachute and jettison the main door.

I struggled with the controls for perhaps another 5 seconds before the aeroplane suddenly reared right up. Although I throttled back, the nose wouldn't drop; and the Ventura fell off the top of the loop and promptly whipped into a spin. I jettisoned the hatch above my head and ripped off my safety straps and helmet. But in those few seconds the old aeroplane had really got wound up and the forces of the spin kept throwing me back into my seat. I started to fight like a madman to get out of that top hatch.

I remember getting my head in the breeze and my left foot on the throttle quadrant. Then there was a bang and I was outside. The aeroplane had broken up. A glance at the ground was sufficient to show that I was at about 7000 feet. And as I reached for the rip-cord I suddenly thought of the enemy fighters. I had always had a nightmare fear of being shot in my parachute descent and this was still uppermost in my mind, even in my present situation. I had seen a parachutist called Quiller do a delayed drop in New Zealand and it didn't look too bad, so I hung on to the rip-cord handle and kept an eye on the ground.

I fell surprisingly slowly — or so it seemed. 'I'll pull the cord at 3000 feet,' I thought. But just before that approximate height I found myself on my back, falling head first. I couldn't right myself quickly; and as I didn't like this attitude of fall, I gave the rip-cord a good smart jerk. The shroud lines came whipping out and one tangled in my right leg before I managed to kick free. And then I was pulled up with a jerk.

I was suspended in space at about 3000 feet. Then, to my consternation, several large hunks of aeroplane went fluttering by like large autumn leaves. But my luck held and nothing hit my canopy. I found myself still clutching the metal rip-cord handle like grim death. As I dropped it and watched it fall away, I saw myself heading towards a huge area of water. I thought of pulling on one side of the shroud lines, but decided to leave well alone. Then, as I glanced toward the target area I saw a large column of black smoke rising skywards. 'Good,' I thought. 'That'll be the bombs.' But, thinking it over afterwards, it was probably poor old V-Victor burning.

There was something wet on the back of my head and an investigation revealed lots of blood on my exploring hand. Fortunately, it happened to be a small scalp wound probably sustained after I had taken off my tin hat. Wearing the added head protection over the top of my leather flying helmet had become standard battle-station procedure with me, ever since my early sorties in Blenheims.

My left leg was aching and I had an awful feeling that I might have to land with a broken limb. By this time I was less than 100 feet from the ground and I saw with some relief that I was going to miss the large stretch of water by several hundred yards. Then I was over a ploughed field and the ground was suddenly rushing at me. Bang! I went end over end and finally came to rest half stunned and spitting out earth. Then the parachute canopy, caught by the wind, pulled me flat on my back and dragged me at high speed over the furrows before I banged the quick-release disc and came free of the harness. The silk 'chute billowed away with a Dutchman in pursuit.

Several people soon gathered. As I dusted myself down I looked ruefully at the left trouser leg of my 'number-one blue' uniform. It was rent from top to bottom. But, thank God, my leg was not broken — just badly bent. 'Escape is the first thing I do,' I thought. I looked at the dozen or more Dutch people standing near and asked: 'Do any of you people speak English?' No response. Then I heard a shout and, looking over my shoulder, I saw a German soldier with a gun, running towards me and waving frantically. I knew I was finished, so far as freedom was concerned, so I took my escape kit from my pocket and dropped it behind me so that the Dutch people could see. I kicked some earth over it and walked to meet my captor with my hands high.

As the war progressed and the reputation of New Zealand's pilots grew, various notables became increasingly keen to arrange opportunities where a little glory could rub off. Des Scott (who was now commanding 486 Squadron) describes one such encounter.

Typhoon Pilot
Desmond Scott DSO, OBE, DFC and Bar

As the volume of our offensive mounted, so did the interest of the Press, and a number of Air Ministry bulletins featuring our activities appeared in daily papers. This appeared to be a signal for some of the New Zealand 'chairborne' division at Halifax House in London to muscle in on our emergence from obscurity, and they suddenly started to invite themselves down to Tangmere.

Now although we were designated as a New Zealand Squadron, we were in fact under the jurisdiction of the RAF, and the only claim the New Zealand Air Force could have on us was that our pilots were New Zealanders, as were some of the wireless mechanics. The ground staff were predominantly RAF and in some of the so-called New Zealand squadrons a number of the air crew, including their commanding officers, were English.

I was determined to put a stop to these open visits forthwith. Writing to Halifax House, I requested that in future any staff wishing to visit my squadron must first apply in writing, stating the nature of their visit and the reason for it. I was not trying to be difficult, but some of these people were annoying my boys. Needless to say, the visits stopped, but there was one gentleman who arrived needing no invitation. He came at short notice on 17 April 1943, and was accompanied by our AOC, Air Marshal 'Ding' Saunders. He was our diminutive Minister of Defence, The Honourable Fred Jones, the Labour MP for Dunedin Central.

Many of our pilots were farmers and staunch Conservatives, and thus the Minister's welcome was somewhat tepid. He looked a comic figure, dressed in crumpled black suit, narrow trousers (in mourning at half-mast) and lace-up boots. He displayed an almost childlike desire to sit in my Typhoon, which stood directly outside the dispersal door. So we hoisted him up on to the wing of the aircraft and helped him into the cockpit, where he almost disappeared from sight. I spent ten minutes explaining the layout and function of the many instruments, and warned him to be careful of the undercarriage lever which, if released, would immediately set my aircraft on its belly. Then I let myself down to the ground so he could play about to his heart's content.

My pilots were dutifully looking towards the open cockpit when the Minister's small head appeared over the side. In a voice full of amazed wonder he called out: 'So this is a Spitfire. How many engines has it got?'

On that note, a few words about the aircraft themselves. Inanimate objects, for sure, but still able to inspire great affection. Consider Des Scott's descriptions of the Hawker Hurricane and Typhoon and also Bryan Young's lyrical account of a Spitfire landing.

One More Hour
Desmond Scott DSO, OBE, DFC and Bar

Saying goodbye to my Hurricane was like saying farewell to an old and trusted friend. As a young pilot I respected the Hurricane for its quiet manliness. Its ability to absorb punishment, and adapt itself to whatever was asked of it. The sweet purr of its Rolls Royce Merlin engine accepting every challenge without a note of complaint. Its wide, sturdy undercarriage setting you down safely on the roughest of airstrips and on the darkest of nights.

No, there was nothing temperamental about the Hurricane — in fact, I believe there were times when it forgave me my own shortcomings, and while so doing, saved me from an early grave. We had shared the dangerous skies for almost two years. It had flown me safely through flak storms that erupted like multi-coloured volcanoes above our targets in France, Belgium and Holland. The Hurricane may not have had the legs of a Spitfire, nor could it cope so well against the ME-109s in the higher reaches of the fighter pilot's domain. But its versatility more than compensated for its shortcomings. No skittish prima donna, it fulfilled its many roles with a quiet dignity and hard-hitting efficiency.

Operating from French bases, the Hurricane had bravely met the Luftwaffe head on during the battle for France, while a little time later they shot down more enemy aircraft than the Spitfires during the Battle of Britain. But that was not all. After the Battle of Britain the Hurricane had become a valuable jack of all trades, readily adapting itself to the dictates of a well-meaning, but often misguided RAF hierarchy. 'Put bombs on it' was the order from Air Ministry, and it became a 'Hurribomber', carrying a similar load to a twin-engined Blenheim bomber and with much greater accuracy and efficiency. Mounted on cargo ships it became known as a 'Hurricat', and was catapulted off short rusty decks to do battle against the German four-engined Condors that often shadowed our Atlantic convoys and the sea lanes that led to Murmansk. A brief moment of glory before parting with its pilot as they both headed down towards a deep and cruel sea.

In the Western Desert campaign, some Hurricanes acted as airborne artillery, mounting two anti-tank guns — one under each wing. And while the Spitfires slept on their native heath, the Hurricanes kept a night vigil above a bruised and battered Britain. But it was the arrival of the airborne rocket that gave the Hurricane its greatest punch. With its four 20 mm cannons and eight 60 pound rocket projectiles, it was to become the scourge of enemy shipping — particularly those vessels that tried to run the gauntlet of the English Channel or among the islands of the Scheldt.

The Hurricane's versatility therefore knew no bounds, yet as it became the recognised low-attack work horse of the RAF, its glamour as a fighter aircraft sadly diminished. The Spitfire on the other hand climbed higher and higher into the comparatively safe and less productive arena above Hitler's Europe.

I lowered myself into the cockpit of my Hurricane for the last time on the evening of 9 August 1942. Not to fly. Just to say a quiet thank you and goodbye. The north star was already shining as if lighting the way to the heavens where we had spent so much of our time. It was strangely quiet too, as I thanked God for delivering us both from the perils of the past. It was a difficult parting and I felt saddened as I let myself down to the ground and slowly made my way through the deepening twilight towards the noisy comfort of Manston's officers' mess.

Typhoon Pilot
Desmond Scott DSO, OBE, DFC and Bar

Whereas the Spitfire always behaved like a well-mannered thoroughbred, on first acquaintance the Typhoon reminded me of a half-draught: a low-bred cart horse, whose pedigree had received a sharp infusion of hot-headed sprinter's blood. It lacked finesse, and was a tiger to argue. Mastering it was akin to subduing the bully in a bar-room brawl. Once captured, you held a firm rein, for getting airborne was like riding the wild wind. One casual crack of the whip, and the jockey was almost left behind. But like the human race, the Typhoon had its good points too. In sharing the dangerous skies above Hitler's Europe I had good reason to respect its stout-hearted qualities. It gave no quarter; expected none. It carried me into the heart of the holocaust — and even when gravely wounded delivered me from its flames. As a young pilot I grew not only to respect the Typhoon, but also to trust — even to love — it.

After the death of the Third Reich, silently and unobtrusively the Typhoons flew off in obscurity. But I shall never forget them. Not ever, for by the war's end they had become part of me.

Beckoning Skies: A Pilot's Story
Flight Lieutenant Bryan E. Young

Not all landings are perfect, as any old pilot will tell you. There are bad landings and mediocre landings. Then there's the type of landing I pulled off in JF358 at touchdown on Habbaniyah. Once again I had to cope without the use of flaps and brakes but made one of those rare daisy-cutters that left me exalted, congratulating myself on being a pilot in a million. Although it is all over in a matter of seconds, that fraction of time, that moment of truth when the wings slowly lose all their lift and the weight of the aircraft is progressively transferred to the three points of the undercarriage at the precise moment it loses flying speed and touches the ground, is unconsciously and indelibly locked in memory. The mechanics of all the actions and reactions on the part of pilot and plane involved that day in that landing are easily recalled. The long and inviting stretch of broad black runway curving into line over the nose as the threshold is crossed, close the throttle right back and with clattering backfires from the engine, she gently flares only inches above the ground while the stick is eased back and the transition from air to ground is achieved with the delicacy of a flower unfolding. Three tons of Spitfire, weighing no more than fragile thistledown, tenderly kisses the runway; no need for brakes to keep straight, the rudder is still biting the decaying slipstream. At taxiing speed, ease off the strip as soon as possible to give the chap following behind a clear run. A wonderful, very personal experience, even esoteric and not given lightly to only a few in modern times.

Later in *Beckoning Skies* Bryan Young describes the awesome power of a
Burmese monsoon.

Let me now return to the rain-sodden paddy fields of Imphal. Out of fear
of things to come the peasant farmers were weeks behind in their rice
planting and it appeared that they had forsaken all hope of getting a crop
in that year. The sky's wrath of rain and wind descended regularly and
intermittently and the sun would blaze down from a clear sky with a
suffocating heat that dried off the runway in a few hours. We generally
flew every day despite the threat of monsoon conditions. These were a
greater danger to us than the Jap because sometimes an entry into cloud
could not be avoided and flying through a cumulonimbus could tear
your aircraft apart within seconds. I have known of Vengeance pilots taking
their aircraft into such cloud at 5000 ft and being spewed out into clear
air at over 20,000 ft, often suffering personal injury in the furious
buffeting if not having the aircraft broken into a thousand pieces around
them.

   In the cockpit one's hands are braced by a grip on the throttle handle,
one's feet against the rudder pedals, body tightly strapped into the seat.
However, you can't do anything about that part above the shoulders and
often, caught out in adverse weather, I would have to suffer my head
being thrashed around and dashed against the sides of the perspex
canopy, all the while being absolutely helpless to avoid it and straining
to keep the aircraft on an even keel. Flying in monsoon cloud could
be a terrifying ordeal; even flying beneath one I have had streaks of
lightning coming down from the cloudbase several feet above to the
ground below, flashing past between cockpit and wing-tip. Many were
the Vengeance crews, Hurricane and Spitfire pilots who met their end
during that season of bad weather. The Vengeance crews always had my
particular admiration. I was often a part of their escort and it was painful
to see these huge, lumbering aircraft plodding slowly, in tight formation,
to and from their targets. The Jap fighters could and did make
mincemeat of them. Without the necessary speed and agility to go round
and avoid a cloudmass, the Vengeance would often have to plow through
it to be further minced by turbulence within. The poor old Vengeance
was a kind of Cinderella in the aircraft world. The Americans who built

them had no use for them, nor had the French and Russians. It was only the RAF in Burma who put them to use for a short while until they were superseded by the Thunderbolt. They have had little if any mention in the chronicles of that service. The lack of recognition is a great pity as they and those who so gallantly flew them made a magnificent contribution to our efforts in Burma.

Having maligned the monsoon and described its nastiness in no uncertain terms, I am also bound to say that, as in all things, there is always another side to the coin. Although it was a physical burden, for those of us fortunate to get airborne at the right times it was sometimes aesthetically glorious. When up on an air test and taking the opportunity to indulge in the freedom of some aerial frivolities, those enormous black and white masses of cloud, with their concealed potential for instant disaster, would irresistibly beckon, enticing a closer look. On some such occasions caution would be thrown to the winds. I would approach a billowing cliffside of nebulous cottonwool searching for a secret entrance, a hallway into that fabulous castle. Through this passage I would sweep at a speed of perhaps 300 mph, suddenly real as overhangs and side tunnels in the fabric flashed by. The risk was great, for a dead end where there was no option but to be engulfed by the wall ahead and no room to turn back could and sometimes did call for a swift change of attention to the blind flying panel and a fast, if you were lucky, exodus to clear air.

On the other hand that tunnel of horrors frequently led to the centre and I would burst out to behold one of the most magnificent, instantly spellbinding sights ever seen by mortal man. Suddenly I would find myself inside a vast cathedral, many miles in length and breadth. In one corner it would have been able to fit St Paul's of London, St Peter's of Rome and all the pyramids of Egypt, without noticing their presence. A romantic soul would easily see the miles-long nave, the vast transept and the vaulted roof of Gothic arches soaring, thousands of feet into the heights above, the acre upon acre of stained-glass windows changing the diffused, ethereal light into colours that ranged from reds and purples to golds and greens. If ever one could put out his hand and touch the face of God it would be in the vastness of that magnificent amphitheatre. In no way could you compare these God-given temples in the sky with the puny structures erected by mankind on earth.

Words on wings

It was also in Burma that Bryan Young had an experience he still can't explain. Perhaps it was a hallucination, the trick of a tired mind ... or perhaps it was a truly supernatural event. Either way, it was an uncanny moment, introduced with an extract from the author's wartime diary.

They shot down nine Dakotas today. Scrambled at 7.30 this morning and six of us intercepted SW of Silchar. Four bombers and a gaggle of fighters. I was yellow two with an unserviceable radio and our section of two was the first into them. Bats Krohn my leader peeled off to the left and I followed. We were at 22,000 and the Japs at 12,000. We both went straight through them and up and I managed to get in a short burst.

Followed Bats in a second time and made a quarter attack. Lost Bats and made third attack alone and got one climbing vertically and at point blank range when he fell off the top of his climb and dropped away trailing gobs of black smoke. My number one was badly shot up about his tail but returned OK.

Although these two days were a little more intensive than usual, they serve to illustrate the build-up of action in our sphere of operations. A most curious little incident occurred after that dogfight when, out of ammunition and after finding another Spitfire to accompany me home, I went up alongside him and saw the FL code letters of 81 Squadron on his side. I was puzzled by the pilot who, staring all the time straight ahead, totally ignored me. He was steering a course directly opposite to the one which would take us back to base. By this time I was pretty close to him. In helmet, goggles and oxygen mask his features were unrecognisable but from the aircraft's individual code letter I could see that it was Captain Moon Collingwood (why he was called Moon I have no idea). He was one of our South African chaps, a mysterious character who was always regaling us with incredible stories of African witchdoctors that made our hair stand on end; he firmly believed in their powers.

To get back to the story. I had been close to him for about five minutes and as my radio wasn't working I couldn't get through to him to wake him from his doze. He then very slowly turned his head in my direction, stared fixedly at me for a few seconds, and began to gesticulate furiously at me, waving me away. My fuel was getting low by this time but

I still had enough to get me back home. His must have been about the same but he kept flying south and away from our base. My patience evaporated quickly and I sheared off toward home. I left him to run out of precious time and fuel; if that was the idiot's choice. On landing at Kumbhirgram, with only a few drops of fuel left, I found to my surprise that I was the last to land, the others had all returned at least fifteen minutes beforehand. At that moment no other 81 Squadron aircraft were in the air! I was astounded to learn that Moon had been one of the first to arrive. No way could I have been mistaken in identifying his aircraft. The code letters, red with a white outline, were at least two feet high on the side of the fuselage and as distinct as a bullseye. I had been within twenty feet of him. I thought it best to say nothing about the episode and kept my mouth shut. Whichever way I looked at it nothing about it added up or made any sense whatsoever. Metaphorically, it all had a rather fishy smell and that smell still hangs around today, even as I write about it!

Something similar happened to Des Scott in 1941 at a time when he and a young Aucklander, Bruce Hay, were flying Hurricanes.

One More Hour
Desmond Scott DSO, OBE, DFC and Bar

We were on night operations — working out of Manston, near Margate. Bruce was a tremendous all-round sportsman and we both loved rugby. He was the sort of fellow to have around when things got rough — in the sky or on the ground.

I had been flying myself almost to a standstill at about that time, and Wally the station doctor, a good friend to both Bruce and myself, ordered me on a week's leave. I went up to Edinburgh to visit a cousin, but after only three days a strong premonition that something dreadful was about to happen decided me to disregard the doctor's orders and return to my station. The train left Edinburgh at midnight, and after a few drinks with an old friend, Jock Blount, who served in the Argyll and Sutherland

Highlanders, and who also knew Bruce, I scrambled aboard the train just as it was departing.

I sat in a corner of one of the carriages and soon went to sleep. At precisely 1.30 in the morning, I was suddenly woken by a vivid nightmare. I heard Bruce call out, 'Boss, I'm on fire!' I could see him trying to leave his Hurricane as it plummeted towards the sea. I stumbled out into the passage and was promptly sick. I do not know what the rest of the passengers thought of me, but since they were nearly all soldiers and sailors, I guess they thought that I had become a little 'flak happy'.

Immediately the train pulled into London next morning I phoned Manston and asked Wally if anything had happened to Bruce the night before. There was a long pause, then he answered: 'I thought you would know. He was shot down off Dover at half-past-one this morning.'

Narrow escapes are a happy event whenever they occur, but especially in wartime. Here are two stories that must qualify as the closest of shaves — the second being an incident described in *Reader's Digest* as 'the world's first aerial hi-jack'.

Jump for It! Stories of the Caterpillar Club
Gerald Bowman

On the 8th October, 1940, Pilot Officer W. B. Parker, a twenty-five year-old officer of the Royal New Zealand Air Force, made what is probably the most terrifying parachute descent in the history of war — that is, if one excepts those fortunately rare incidents in which the parachute failed to open.

Parker was at that time one of the band of pioneer pilots who started flying over enemy territory and photographing German naval bases, ports, airfields and cities. At the outbreak of war it had been intended that this work should be done by Blenheim aircraft, but the unsung heroes who made the attempt soon found that their planes were no match for the German fighters, and they were often unable to get to the objectives marked down. The number of photographs obtained had to be accounted for against the heavy number of pilots who were never heard of again, or who finished up in German prison camps. Something, therefore, had to be done about it.

F. S. Cotton, a civilian air-survey operator working with the late Flight Lieutenant M. V. Longbottom, suggested that a really fast, high-flying aircraft like a Spitfire should be fitted with cameras, but with no guns. Without the weight of armament the Spitfire would be able to outclimb any enemy, and would be faster and quicker in manoeuvre. Moreover, if the Spitfire were fitted with long-range tanks which could be dropped, it would be able to penetrate deep into enemy territory, and it would still be in a position to lighten itself and run if it met unexpected attack.

As all Service people know now, this idea proved tremendously successful from the time it was first tried out by the courageous pioneers. It took courage to fly a totally unarmed aircraft deep into enemy territory, an operation which often included long and lonely crossings of hundreds of miles of ocean in any kind of weather. Later on in the war this special kind of work became the task of the famous P.R.U. (Photographic Reconnaissance Unit), the members of which were said to be the first to encounter gremlins.

Meanwhile, Parker and his companions had come to a base at Heston aerodrome and at St. Eval in Cornwall, to which they had returned after earlier operations in France, before the evacuation.

By October 1940 Parker had completed twenty-six operational flights with the Unit. He had already brought back a mass of valuable photographs of enemy territory and shipping dispositions in the Channel ports. During a routine flight from Cap Gris Nez to the Scheldt in the previous month he had spotted one of the first concentrations of barges and boats which the Germans were massing in preparation for the invasion. It was this, together with a mass of information, later brought back, which gave accurate targets for intensive bombing attacks and caused the Nazi High Command to think better of the whole idea.

The 8th of October was a bright clear morning of autumn with the sun warm and benign against a vast curtain of clear blue, and forming a death-trap for any airman without his eyes peeled. Parker knew its dangers well enough as he warmed up his engine, waved the chocks away, and then thundered off down the runway to do his job of photographing various Dutch and French ports. At that moment the sun-glare, and the enemy aircraft it might hide, did not worry him overmuch. He intended to put a good 30,000 feet beneath him before he set out to cross the Channel. Then, as he took his photographs through the tele-scopic lens of the camera, he would be far above the hornet's nest of enemy Messerschmitts and able to get well away in his unarmed aircraft if he saw any of them coming up for him. He would also be high enough to be reasonably safe from getting 'jumped' out of the sun-glare since no heavily-armed aircraft could hang about cruising at that altitude.

From Heston he climbed steadily, watching his altimeter until, when he was somewhere over the middle of Kent, he saw it registering 29,000 feet. Throughout the long climb his attention had automatically been on the sun side of the aircraft, but at this height he wasn't anticipating trouble.

Therefore, trouble chose just that moment to arrive, as is usual with airmen off guard.

Two Me.109 fighter aircraft suddenly materialized out of the glare and attacked him from the rear. Parker did not catch sight of them. He had no idea that there was any other machine within five miles of him when a burst of cannon shell ripped past him into his engine, which jarred violently and then stopped dead. He saw fire blasting back between his knees and realized that there was nothing to do but get out. He slammed his canopy back and began knocking undone his radio leads and safety harness; the aircraft shuddered as a second burst of cannon shells went home.

This burst penetrated one of the petrol tanks, so that he was blinded and sprayed with the fuel as he finally managed to kick himself clear and go head-first over the side.

At 29,000 feet a man cannot stay conscious for more than a minute or so without extra supplies of oxygen, and Parker knew it. He grabbed for the rip-cord of his parachute and jerked it almost as soon as he was clear of the blazing Spitfire. Then he passed out.

From that point until he reached around 10,000 feet, swinging beneath his parachute he was unable to take any interest in the pro-ceedings. When he regained consciousness he did it slowly and in the dazed, semi-detached manner peculiar to a victim of anoxia. He saw the vast green relief map of the earth swaying gently beneath him. He remembered the hectic moments of abandoning his aircraft and reflected that he was still alive, although of the shattered machine and his two enemies there was no sign.

Then, as his consciousness cleared, he smelled burning. He looked around and downwards, and realized for the first time that he himself was on fire. Already the petrol with which he had been sprayed had burned the lower part of his flying-suit. His hands were raw as he started frantically to beat out the flames. Petrol had soaked into the sponge rubber of his parachute cushion, which was well alight. He tore at it and wrenched away burning chunks of rubber; showers of sparks flew up, and his petrol-soaked flying-suit took fire around his face. He realized then that the whole of his equipment must have been soaked before he got out of the aircraft, because some of the rigging lines and part of his parachute canopy were alight in several places.

As a result, the speed of his descent increased and with the stronger draught the flames burned more fiercely.

It was at this point, when to observers below he looked more like a falling torch than a living man, that Parker gave up all hope. He decided that a quick death was much preferable to a slow one. He was in intense pain. The earth was still a long way below and his burning suit was flaming more briskly.

Parker reached for his chest, turned the quick-release gear of his parachute-harness, and banged it undone. In the next second he should have dropped clean away and fallen down the remaining thousands of feet to inevitable but mercifully-sudden death. As it was, only the two shoulder straps of his harness jerked away and left him. In the next second he had pitched upside-down, feeling the thigh straps of the harness slide down his legs and jam round his ankles.

There he hung helplessly while the flames of his burning flying-suit now played in the opposite direction. The fire travelled up his legs to his leather flying-boots and died away. Then one of the burning thigh straps parted in a small shower of sparks, and Parker was left hanging by one foot.

By this time, owing to the severe burns on his face, he was temporarily blind. Although he was now very close to the earth he could not see it. Therefore, hanging upside-down at the moment of landing, he could make no effort to protect himself when he dropped heavily into the middle of a ploughed field.

The thudding force of the impact broke one of his arms in two places and almost knocked him out. Yet he was still conscious when those who had been watching his descent dashed to pick him up. At which point Pilot Officer Parker remarked cheerfully, if a little hoarsely, 'Nice work, chaps — fielded on the first bounce!'

They Got Back: The Best Escape Stories from the 'R.A.F. Flying Review'
A. M. Feast

The air war came early to Malta and stayed late. By spring of 1942 the battle-scarred islands of Malta and Gozo were in a dire state of siege. No supply convoys had reached them since an ill-fated attempt in March of that year. Then, the three remaining merchantmen that had won through to Valetta's Grand Harbour were all sunk by enemy bombs before any substantial portion of their urgently needed cargoes had been unloaded. Before that, the last convoy to arrive reasonably intact had anchored in the harbour something like nine months earlier. By June, therefore, the plight of the island's inhabitants was indeed precarious, with the spectre of starvation and disease hovering over the 174 square miles that comprised Malta and Gozo.

The spirits of both soldiers and civilians, however, were buoyed up by Malta's increasing military activity. In spite of acute shortages, Malta not only increased the fierceness of its defence, but also launched a heavy air offensive against the enemy's airfields in Sicily, and against other enemy concentrations. As one fighter pilot phrased it, 'The tempo of life here is just indescribable. The morale of everyone is magnificent, but things are certainly tough. Bombing continues at intervals all day long. One lives here only to destroy the enemy or to hold him at bay. Everything else — living conditions, sleep, food, and all the ordinary standards of life — has gone by the board.'

In particular, British air and submarine attacks struck crippling blows at the enemy's shipping lines supplying Axis forces then rolling towards El Alamein.

These round-the-clock sorties against enemy shipping were at their height when Lieutenant Ted Strever, of the South African Air Force, arrived in Malta, in the early summer of 1942, to join 39 Squadron. Strever lost no time in being introduced to the air war over the Mediterranean.

He weathered his first strike successfully; and thus, by the rugged standards of those days, he was something in the nature of an old hand by the time his big day came.

July the twenty-eighth started out cloudless and hot. By nine o'clock the photographic reconnaissance Spitfire had finished its daily scan over enemy coastlines and water. Its successful sighting report had earlier been flashed back by radio; and, even before the Spitfire touched down on Luqa aerodrome, the Beauforts were rolling out of their dispersal pens, each with a torpedo tucked under its nacelle. Airborne, the nine Beauforts quickly formed up in fluid pairs formation, and began letting down on a south-east heading, low over the water. Their target was an Axis convoy off the Grecian island of Sapienza.

Three hours later the convoy, consisting of one tanker and three escorting destroyers, was intercepted in the shadow of Sapienza. At a mile and a half the destroyers' small-calibre armament spewed out a fine spray of Breda, Oerlikon, and Bofors that drifted lazily in coloured balls across the water, then suddenly whip-lashed by. The Beauforts fanned out and then banked into the attack.

Evasively skimming the water in air that seemed filled with tracer and flak bursts, Strever steadied his machine at sixty feet, laid off deflection and dropped his tin fish. Off his port wing Dawson's plane reared suddenly then crashed into the water. Strever opened her up to full throttle, and jinked in a wild caracole past the tanker — only to run into more flak from the escort vessel positioned off the other beam. He was well out when he skidded right into a string of Bofors that caught his port engine. Frantically he wound on right rudder trim as the aircraft was carried out of the fray under its own speed.

Almost automatically he began a slow turn to starboard — a turn towards base. But he knew without looking at his smoking port Wasp that it was useless. At best, his good engine might last twenty minutes before overheating. There was no choice but to ditch while he still had control.

The sea was freshening as he set her down; and, with gusts whipping the crests of the waves, Strever ran head-on through a trough. His Beaufort broke up fast.

But … no one was hurt. Scrambling out of the fast-sinking wreckage, the crew got the wing dinghy out intact, quickly inflated it, and pushed the little rubber boat away from the aircraft. All four men were aboard the dinghy when the Beaufort slipped beneath the waves.

When they got over the initial shock the four men took stock. The English navigator, Pilot Officer W. M. Dunsmore, was unhurt, apart from a few superficial scratches, as were the wireless operator, Sergeant J. A. Wilkinson, and the air gunner, Sergeant A. R. Brown — both New Zealanders. To the east lay Sapienza, with the larger bulk of the Peloponnesus and Messenia in the background.

They were adrift only a short time when they saw it. A dim speck appearing low on the horizon, and then the sound of aircraft engines, faint at first, then growing louder and louder, reached their ears. Their spirits sank when they saw the *fasces* markings of the Italian seaplane. The enemy aircraft flew overhead and past them.

Then … it turned round. The pilot had seen them; and flying back in their direction, the aircraft settled down on the water; and taxied over towards them.

The dinghy being virtually uncontrollable, Strever pulled off his jacket, tore his shoes off his feet, and dived over the side. A crewman slipped down on to the aircraft's float, uncoiling a line.

The Italian aircraft, a Cant, came up alongside, and soon the enemy aircrew hauled the dripping South African and his colleagues light-heartedly aboard.

At the aircraft's base on a nearby island, the bedraggled Beaufort boys were introduced to the Italian garrison officers' mess, where they were feted royally by their enemy hosts. They were plied with food in unlimited quantities, offered the best drinks in the bar, and were generally given the run of the place. Friendly Italian officers even gave up their own beds to make sure that the captives spent a comfortable night.

Thus it was almost with regret that Strever and his crew were forced to leave their generous hosts the following day. Before boarding the Cant Z.506-B, for a trip to Taranto and imprisonment, the four men posed for snapshot after snapshot, taken by the garrison officers.

The three-engined Cant had a crew of five: pilot, second pilot, engineer, a combined wireless-operator-navigator, and a corporal guard for the captives. Flying, they were soon to discover, was not one of the corporal's strong points.

Vibrating terribly, the aircraft lumbered through the air, sensitive to every up-current and down-draught. Before long the land-lubber corporal was looking considerably less happy. His sallow complexion turned progressively greener as the plane headed north-westwards towards Italy. Obviously, flying was not the corporal's *forte*.

It was Sergeant Wilkinson who first sensed that the time was right to turn the tables on their captors. Staring out of a nearby port-hole, Wilkinson pretended to be fascinated, then half-turned and gesticulated towards the unhappy corporal who was now holding his stomach and looking very ill and unhappy.

'Look out there!' Wilkinson's whole expression said. His inviting appearance distracted the air-sick groundling just long enough to allow for a solid right uppercut to the corporal's jaw. With a shout, Wilkinson gained his feet, and with the guard sprawled helplessly on the floor, the stockily-built New Zealander snatched his automatic pistol and tossed it to Strever.

The Lieutenant covered the aircrew, caught completely off their guard, up forward. He moved up the cabin, pointing the gun menacingly at them, ready to take over the aircraft. Suddenly, one of the crew reached for a tommy-gun lying just behind the pilot's seat. But he had scarcely put out his hand when Sergeant Brown scooped up and threw a convenient wrench almost the entire length of the fuselage, scoring a lucky hit that stifled the attack before it began. Strever, covered warily by Sergeant Wilkinson, stepped into the co-pilot's seat, took a firm grip on the controls, and turned the aircraft purposefully around on to a south-westerly heading.

Meanwhile, the mopping-up party of Pilot Officer Dunsmore and Sergeant Brown had subdued the other Italians in the cabin, and secured them. Then the four of them rested briefly and triumphantly, breathing hard, complete masters of the plane in which, only a few minutes before, they were being taken captive to Italy. For the next few minutes, Strever tried to hold the unwieldy aircraft reasonably straight and level, while his eyes took in the unfamiliar array of gauges and gadgets on the instrument panel.

'Maps! Get the maps!' he shouted to Pilot Officer Dunsmore. The English navigator searched the aircraft from nose to tail, but could find no maps or charts of any kind. Evidently the Italians were confirmed coastline crawlers. Using sign language Dunsmore persuaded his Italian opposite number to co-operate with him in drawing a map of the coastline from memory, and between them they produced a passable map of the relevant section of Italy and Sicily, from which Dunsmore was able to work out a rough course and estimated time of arrival for Malta.

The Italian pilot insisted in broken English, fright written all over his face, that they lacked sufficient fuel for the trip, but Malta was the nearest friendly base, and so they had no choice but to carry on. After some deliberation Strever and Dunsmore worked out an approximate course to Sicily, agreeing that a landing off the island's east coast, and a trust-to-luck inland dash, would be the only course of action.

Two hours later, to everyone's immense relief, land loomed up ahead; and after Dunsmore had identified the toe of Italy and Cape Spartivento, they pin-pointed their first reasonably good fix. The Cant was swung on to a more southerly heading and they flew on, all but clipping the wave tops. At one tense stage, a Junkers 88 formated on them briefly, and a pale-faced Strever exchanged friendly waves with the German pilot. After another hour the last headland of Sicily no longer lay on their right, and suddenly they realised that they might just possibly make Malta.

And now, with barely sixty miles to go, events were moving to a climax. The Italian pilot who had taken over the controls again, at Strever's order, in case of attack by R.A.F. fighters, tapped the fuel gauges excitedly.

'*Non c'e benzina! Non c'e benzina!*' he screeched. He flapped with reason. The gauges registered zero.

The panic-stricken Italian made a motion to land, but Strever gestured him on, while, for numberless times, a pale-faced Dunsmore strained his eyes to the horizon, looking for signs of land. The float-plane was still almost clipping the wave crests as they tried to slip in under the island's radar defences; and for a while it looked as if they might. It was a shout of joy from Sergeant Brown that registered the tip of Gozo far off the starboard quarter. But joy turned to wild alarm a second later as three Spitfires swooped out of the sun and boxed the Cant.

The next few minutes were terrifying ones. The Spits dived in from every quarter, raking the plodding, cumbersome float-plane with cannon and machine-gun fire. Brown frantically spun the rear guns to and fro as a gesture of peaceful intentions, while Wilkinson whipped off his white sheepskin vest and waved it wildly out of the hatch. Providentially, the marksmanship of the Spitfire pilots was ropy that day; for barring a hit in the starboard wing tip, the Cant came through unscathed. The reason for the fighter pilots' showing doubtless lay in their utterly unnerved state at finding a sitting duck right in their own back yard. Plainly it was too much for them.

A profusely perspiring Italian pilot-settled the Cant down on the water as the cannon crackled all around. Even before the landing run was completed, the three fuel-starved engines coughed out one by one, and the Cant began to weather-cock into wind. Overhead circled the three Spitfires, their guns now silent.

Within a few minutes an air-sea rescue launch came roaring out from Kalafrana Bay, to be greeted by the sight of Lieutenant Strever and crew sitting on the wing of an Italian seaplane, cordially drinking Greek wine with five late members of the Regia Aeronautica. (The Italian crew, it seems, were to have gone on leave after the end of their trip, and their suitcases were packed with wine.)

Lieutenant E. T. Strever and Pilot Officer M. Dunsmore were both awarded the Distinguished Flying Cross, while Sergeant J. A. Wilkinson and Sergeant A. R. Brown both received the Distinguished Flying Medal. The Spitfire pilots, regrettably, had a Commanding Officer not lightly swayed by the unusual circumstances, and the trio caught a sizzling rocket from him on the strength of their poor marksmanship.

Strategically, the D-Day landings were an enormous gamble. Logistically, they were a huge exercise that marked the beginning of the end for Hitler. By June 1944, the Luftwaffe was a depleted force, but there were other dangers lurking in skies over Normandy.

Green Kiwi versus German Eagle: The Journal of a New Zealand Spitfire Pilot
J. Norby King

We form up in the genesis of a beautiful morning and drone on. Other formations are flying in the distance — but in the same direction. Jerry will be backgrounded by the dawn glow. The Normandy beaches are jam-packed with men, movement and tank-landing craft. The man-made Mulberry harbour towed over here in sections seems busy with small ships. Warships on guard further out. All a picture of menace. It's a men-mincer scenario. Perhaps soldiers down there are thanking God they are not up here! A jungle of tethered anti-aircraft barrage balloons make evil the beachhead airspace. Down there, not far inland, below 5000 feet, there is a sputtering of Ack Ack's dirty dots, seeking Typhoons, probably, ranging with rockets for Jerry armour. The dots are fewer up here but swelling into bursting sacks of soot. Stone-coloured smoke puffs from Caen buildings, misting the town. It was another Dieppe, 9 miles inland, but we had tanks this time to tackle the ready-made fortress and take this road-centre. We needed it. And we need airfields down there, on the flat, to operate from in amongst it all.

Up and down off the coast, over it, our squadron patrols in three section of four. One or two 88s are creeping in low. The RT silence erupts into life with shouts. The calm voices of the Ops Controllers offer us no serious trade. Not yet! Those formations of black dots growing clearer through the perspex must be all ours. Jerry must be late getting up this morning. Or have the Fortresses bombed the Abbeville boys off the map? Time enough yet.

Hullo! Johnny Houlton is slipping away inland with the Blue section. That's weakened us. That's strictly against briefing instructions. Our job here is maintaining the fighter umbrella over the beachhead. He was muttering about getting them before they reach the beach. He'll get a rocket on return.

The debriefing adds to my education. I learn that the squadron hasn't seen a large formation of bandits for three months. Has Fighter Command been wearing the opposition down with the 9s over the past year? Also that Johnny and Bill Newenham got a 109 each while playing truant inland. I notice that a blind eye is turned towards the geography involved. Results justify risks but — splat — under a sleeveful of rings if things go wrong.

On the other hand our Top Brass support us when another service comes down on us with a heavy hand. Al Stead has been attacked while patrolling off the beachhead. Despite Al turning and turning to show our unmistakable elliptical wing-shape the Spitfire kept on attacking him.

And firing. In desperation Al called 'What shall I do?' He got some advice. 'Shoot the bastard down.' Probably from Chalky White, I reckon. A rough joker, he is. So Al did.

We gather round Spy. His news is that it was a navy Seafire — never been on ops before — and some admiral is jumping up and down. Al is all cut up, too. We can only wait, slumped down in the chairs.

Jack Yeatman fills the doorway. 'Is Bedstead here? Ah! Don't worry about it, Al. I've heard that Johnny Walker is going to shoot you off on leave. Group HQ is going to stand on its dignity and say that this is an internal matter of we blue jobs. Grapevine guff. Came from close to the telephone, though. Sounds right.'

With ground forces in France, constantly supported by the Tactical Air Force commanded by 'Maori' Coningham, there was one last objective: to depose Hitler, Germany had to be invaded. It was decided that RAF pilots would join Army units to liaise with aircraft circling overhead in 'the taxi rank', ready to attack enemy targets. The theory was that the best person to relay clear instructions to a pilot in the air was a pilot on the ground.

Of course, before that could happen, the troops had to get into Germany. Not easy, when the Rhine was a natural (and heavily defended) obstacle. The answer was an airborne invasion, using hundreds of gliders — and Bob Spurdle was in one of them.

The Blue Arena
Squadron Leader Bob Spurdle DFC and Bar

Op. Sortie 565 [my last] 'Up! Convoy moves off in thirty minutes. All up! Up!' And we stumbled out of our Nissen hut and into draughty three-tonners. It was 4 a.m., 24th March 1945. Shouted 'good lucks' to the other two teams, and last-minute hunting for gear, now marvellously misplaced, although we'd been checking and counter-checking for days…

At the 'drome we clambered into the big black glider. Glassy smiles and hollow laughs. The glider crew got in, cheerful as crickets. Thank God!

RAF types, a flight lieutenant and a sergeant pilot. Our jeep and trailer squatted in the darkness like fat brown toads. I didn't want to see the long towing cable being hooked to the glider's nose; the huge Stirling bombers trundling past with shining discs twirling vapour trails from spinning propeller tips. The whole air, the very ground trembled to the grumbling roar of hundreds upon hundreds of powerful motors. Our glider jerked, rumbled forward, was heaved bodily into the turbulent air. I'd seen enough and sat back behind the pilots' bulkhead and fiddled with my Sten gun, checked spare magazines: and thought.

How in hell had I got mixed up with all these brown jobs? I must have been mad! Outside the flimsy plywood skin of our glider, the huge armada took up formation; squadron after squadron from airfields scattered miles apart lumbered into the dawn sky, to rendezvous en route to the dropping zone. C47's in hundreds appeared, lugging pairs of the small American Waco gliders to join the huge Stirlings pulling the heavier Horsas and even bigger Hamilcars, which carried earthmoving gear, artillery and 'Tetrarch' light tanks. There were over forty thousand men in the air — glider borne infantry, parachutists, anti-tank men, machine gunners, mortar crews, Padres (no weapons), doctors, signallers — the lot. And, warming up on the airfields of England, France, Belgium and Holland were the fighter-bombers and anti-flak fighters to support this mighty air armada. Further back, light, medium and heavy bombers, the night machines of the RAF and the day machines of the US Airforce, took off to interdict and pound ground targets.

Already the Parachute regiments were well on their way — they were to be the first dropped and their job was to secure the landing areas for the glider-borne troops. Every one of the thousands of air-crew and the tens of thousands of soldiers knew his role in the operation. The huge airborne armada was to descend like a cloud of locusts onto the fields now being marked out with coloured smoke canisters placed by Pathfinder aircraft.

The whole thing was fantastic in its size and scope; over 6,000 aircraft were involved. This time it would be no Arnhem — the RAF were providing the radio links with our army and we had RAF fighter-bombers to give instant close support. And, in their trenches and flak posts, the Germans were ready. They may not have known exactly where the blow was to be struck, but every field was planted with heavy poles to wreck gliders, and mortar crews and gunners had every square yard zeroed for accurate defensive fire. And there were the minefields!

It took us two hours and forty minutes to cross the Channel, drone over the flooded Dutch fields and be cast adrift. Two hours and forty minutes of thinking, worrying, nail biting doubts, regrets and promises to be a better man in the future. If there was to be a future!

The whole glider vibrated, occasionally lurching violently when caught in the Stirling's slipstream. Creaks and groans of straining wood-work, the clinking of chain ends. It was bitterly cold: Dowlin and I sagged in our hard seats. It was too tiring to shout above the sound of rushing air. At the rear of the Horsa, behind the jeep and trailer, were Simpson and Holmes. After about an hour I heard crackings and splint-ering from the back. Good God, the bloody thing's falling apart! But it was only the two airmen hacking peep-holes in the plywood fuselage with their knives, to see what was going on outside.

They tell me some of the men slept quite soundly on the long flight, but for me the experience was too intense, too bizarre.

We were nearing the drop zone; I could see our Stirling tug slowly weaving as if checking its position. Others tended to drift across our path into some sort of approach pattern. Down below, by peering over the two pilots' shoulders, I saw gun flashes; sometimes rocket-firing Typhoons dived past and through the armada stream to engage German flak emplacements. They'd slide down in speeding curves, straighten up and then rockets streaked away, trailing grey gases, to burst in clouds of earth and smoke.

A thin brown haze smothered the whole area below; above, a brilliant blue sky. Hundreds, uncountable aircraft, Stirlings, Dakotas, Horsas, Wacos, Albemarles, Hamilcars, all in a huge multi-tiered unending stream. Fighters weaved above and on either side. We were in the vanguard group of the glider train, following immediately behind the paratroopers and combat infantry gliders.

Just in front of us a Stirling was hit by flak. A flash, a puff of smoke, then it staggered down to port, falling from the sky. Its glider slipped its towing-cable and veered sharply away to starboard; then the huge bomber ponderously reared up, up, up, to fall over backwards as we curved out of its path. A few parachutes and it was gone out of sight behind us. To our left a glider's tail plane broke off and the crippled machine was hastily cast adrift by its tug, the nylon rope whipping back in the slipstream. The fuselage and wings started to turn end over end at an astonishing speed of rotation; pieces broke off. The tail unit went. First soldiers, then a jeep, tumbled from the wreck. The men looked like puppets jerked by a madman — arms flailing — to fall thousands of feet to the waiting earth.

I'd seen enough and, with thoughts in a turmoil, clambered back to my seat. Now the sounds of battle were continuous — above the muted roar of motors and the hissing slipstream I could hear all manner of thuds and bangs. The thin plywood skin of the glider seemed to quiver with the din. Every now and again there would be a sharp whacking sound as the glider was hit by shrapnel or from flak bursting close by; the air was tainted with the smell of high explosives. I took off my tin hat and tried to squat in it. I guess I was trying to protect my genitals — an instinctive male reaction. Again I ventured a look ahead just in time to see our tow rope snake off and be towed away by our tug. I bet they were pleased to get shot of us. The glider crew were nattering away, the co-pilot gesticulating towards some feature below.

'I see it! Yes! That's it!' shouted the glider's captain, who promptly dropped his map and took over the controls. I clambered right forward and, clutching the pilots' seat frame, stared out at the fantastic sight.

Parachutes hung festooned from trees or lay spread across green fields like fallen washing. Already a few gliders were on the ground and more and more were diving steeply at incredible angles to level off and plough across the rough fields. Wreckage, more and more, was scattered by the planes hitting the German poles erected to deny safe landings.

'Hold on!' shouted the co-pilot, and I clutched harder as full flap was applied. The Horsa heaved up, then dropped, nosing over into a heart-stopping stoop towards the deck. The lift spoilers whistled in the slip-stream, tracer shells arced overhead. More and more bangs and thuds with the whole scene wreathed in smoke; men running or lying still on the fields.

Mortar bursts, black and ugly with cores of dull red flame, flickered at random across the earth, more running figures, dead animals, trees, ponds, a farmhouse and then — Crash! — and we were down.

Lurching and rumbling the big glider rocketed and heaved noisily over grassy farmland towards a belt of trees. Both pilots were swearing and cursing as they strained on rudder pedals to steer it clear of obstructions. Hanging on grimly I stared, fascinated, as bushes, fences and gates slid past. A loud bang! and the nose-wheel burst through the flooring — the glider skidded crazily on, lurching and bucketing in clouds of sods and dust. A last crunch and we were there.

The noise outside was frightening and we didn't want to leave the frail 'safety' of our plywood box. Bits of metal tore through the glider or clanged against the jeep and trailer. We had to get out — get the jeep and radio gear into action — but we didn't want to leave the shadows and face the sunlit exposure outside. Not outside into the noisy battle; the stink of cordite, and yells and cries of fighting men!

'Come on, you chaps! Let's get going and unload the jeep!' I shouted. I had to do it — such is the price of leadership.

The front of the glider was all jammed up. The sub-structure and nose wheel a shambles. We tried the tail unit and struggled to knock out the locking pins which were under stress. Although I shouted for them to stop and help, the two pilots sloped off, cheerful at a job well done, to join whatever unit they were supposed to fight alongside. Cursing and struggling, we toiled to swing the heavy tail unit out of the way, free the restraining chains from jeep and trailer and roll them out of the glider.

American parachutists went loping past — survivors from their group's unintended drop in the British sector. Fortuitous for us as it turned out — their error of judgment cost them dearly as they bore the brunt of German fire. But they were cocky bastards — one stole my binoculars left on a pile of gear waiting to be loaded on the jeep. They ambled off into the smoke and disappeared. Snagged white or camouflaged green-and-khaki parachutes hung down from the row of trees alongside us. And stretching them, the bodies of dead American paratroopers. Some lay in the fringing pond and some dangled, swaying slowly in the warm, still air.

We kept on struggling with our unloading; the ramp off the glider stuck at an awkward angle.

'Grab this bloke — we'll have to use him as a block.' We heaved the corpse under the ramp and backed the jeep out, coupled the trailer up and set off to find the farmhouse designated as Division HQ. The terrible sounds of battle were clearly diminishing — the heavy thudding of shells and mortars, the tearing chatter of machine gun fire, were dying down as

German resistance was overcome. Soldiers appeared and disappeared like wraiths in a nightmare as the pall of stinking smoke cut visibility to a few hundred yards. We drove past a large, fat woman lying dead on her back across a little path between farm buildings — she should have stayed under cover. A magnificent draught horse lay frozen in the rictus of death — locked rigid, head back and teeth bared, as when it tried to bite at a horrid wound in its shoulder.

Another wave of gliders swooped down, whistling and thrumming in the calm air. More bursts of multiple cannon fire sprayed up from a copse over to the left and one of the black gliders disintegrated crashing untidily and strewing men and equipment over an acre of ground. Rumbling, crashing into posts, the big aircraft rolled and skidded in all directions. The rending off of wooden wings, shouts, explosions — all a crazy backdrop of noise. Stretcher-bearers ran by followed by a Padre looking haunted and strained.

We located the site of our command post. Already a small glider-borne 'dozer was scooping out two holes to protect our jeep and trailer. Leaving the others to 'plant' the vehicles and get the radio ready for action I ran off to the HQ command post.

'You! Yes, you! Get after that tank — there are snipers in that wood. They're to be cleaned out!' The tall be-ribboned officer had on riding breeches and, for God's sake, carried a ruddy riding crop! He'd doffed his tin hat for the red beret and by his rank badges, was something more than a colonel.

'Sorry, sir, I'm RAF radio team.'

'I don't bloody well care what you are. Get cracking now and I mean *now*!'

Turning from me he rounded up a couple of passing soldiers and off we went at a shambling run after a light tank churning towards the wood. Clutching my Sten in sweaty hands, I had cause to regret the whole affair, and on rounding a hedgerow, kept on along it and around the far end to double back at a nonchalant walk. This was stupid! What if I were bumped off tracking down some lousy Hun and the army deprived of air support? Busily rationalising, I got to HQ and reported in.

HQ was a farmhouse surrounded by apple trees. One of the outhouses was on fire and a dozen Huns were running back and forth with buckets of water trying to douse the flames. They were soldiers without pants or boots. An incredibly funny sight.

'What gives?' I asked a wounded Tommy on a stretcher.

'They caught one of the bastards with a knife in his boot! Now they can't hide anything.'

Soon we were given targets and quickly got the Typhoons, Hurricanes and Spitfires diving into pockets of Hun resistance.

The hours flew by and then the day was done. The sun sank below the horizon in a red glow from the smoke haze and, as darkness fell, our radio team was stood-down until daybreak. We opened our tins of Spam and brewed up. But it was a grim night — 88's and mortars had the whole area bracketed and their explosions kept us awake. Worst of all was the vicious banging of Nebelwerfers, whose multiple crashes shook the earth. Hot, sweaty and tired to the bone our little crew couldn't sleep so we yarned away until completely exhausted by the day's excitement, we crawled into our bed-rolls and drifted off to sleep huddled on the sweet-smelling earth by our jeep.

My first day in Germany was over.

On the other side of the world, back in New Zealand, life was much more tranquil. Here, a pilot could concentrate on training, or familiarisation with some new type. Which is precisely what 20-year-old Miles King was doing when something very unexpected occurred ...

The Sky My Canvas: An Autobiography
Miles King

Crisp, calm, cold, and incredibly clear; as a camera shutter captures shattering glass at the instant of fracture, so is the clarity of that morning still indelibly inscribed in my mind.

Oxygen gently flowing into my mask, I relaxed in the cockpit of the F4U Corsair, delicately balanced upon the razor's edge of her absolute altitude, suspended between her mighty power, the little lift remaining from her wings, and the inevitable force of gravity. So delicate a balance that my gloved hand upon the control column could upset it with the minutest movement.

Far, far below, the Manukau resembled a puddle of mud. Southward, New Zealand lay flat as on a map, with the pinnacles of Ruapehu and Egmont piercing the matt, and far to the south, as on a canvas, the Alps backdropping, surmounted by the virginal sharpness of the snows of Mount Cook.

It was winter. The Royal New Zealand Air Force Fighter Command had commenced conversion from our tired but heroic P40 Kittyhawks to the great powerful new F4U Corsairs, the fourth fastest fighter in the world, and we were the first fighter squadron undergoing conversion at Ardmore. That morning the Officer Commanding, Grata Greig, had told me to take my machine and attempt to establish her absolute ceiling. We had estimates from the manufacturer, but so far none of us had attempted to establish the maximum. So for the past hour I had wound my way up into the sky, and sat now with the power plant gasping for air and the wings clawing to keep us airborne. My eyes scanned the panel and the rate of climb indicator was varying between plus 10 and minus 10 ft per minute. She was as high as she would go. The altimeter read 37,640 ft. The date was 9 August 1944.

It was time to return. I had achieved what I had set out to do. With a sigh I let the nose slide down, and we slipped down gaining speed, until once again she flew as she was meant to, the wings with lift, and the power plant breathing normally. Then, level again, I sat and gazed about me at the beauty and felt once more the feeling that was so familiar to me, and must be to all solo aviators: the affinity between our Maker, man, and his machine. The communion and peace that is beyond normal comprehension, so awesome, so totally divorced from earth-bound mortal problems. So, high as I was, and at such peace, there could be no room for a world distraught with newspaper headlines of disaster, distress and dastardly deeds.

How to return from my lofty place of peace? Simple. Slide the nose down a little to gain more speed. Then gracefully half roll and curl the nose down until we were vertical, nose centred on the heart of the Manukau.

And then we were slicing vertically down through the calm air, faster than I had ever travelled before. Air speed indicator jammed hard over against the stop. Altimeter unwinding so rapidly that it was impossible to read, and then I felt a tremble through the column, then a shake, and a split second later the column tore itself out of my grasp and thrashed wildly about the cockpit, hitting my legs as it went. So severe was this that I wore the bruises for a month. The whole airframe shook and twisted so violently that had it not been for the security of the harness, I would have undoubtedly been knocked unconscious against the side of the cockpit. It was no longer an aircraft. I felt as though I were hammering earthwards encased in a gigantic pneumatic drill. I could not grasp the controls, I could not bail out. I was helpless, committed to this brutal cacophony. There was nothing I could do. I had no time for fear. We hurtled earthwards out of control.

Then more quickly than it had commenced, suddenly it ceased. The tortuous buffeting stopped and the control column stilled, but still we sliced downwards through a silent but screaming stillness. It flashed through my mind that the blunt, winged monster had battered and bashed its body through a critical speed, and that we were now approaching the speed of sound. I grasped the column again and tried to ease it back, but no, it was solid and set as in cement.

So for about thirty seconds, stillness, then abruptly as we descended into denser air, the clamour and commotion broke out again. The aircraft shook and trembled and twisted, the control column thrashed uncontrollably about the cockpit, and still the nose was immovably fixed on the heart of the Manukau, enlarging in size much faster than ripples spread in a pool when a stone is dropped. I wound back the tail trim, but still we plunged down. Then, so slowly, the torture died to a tremble, the column gradually stilled, and still we plunged down.

Finally, the great long immovable nose moved. Half an inch, then an inch, then two, and reluctantly it started to move from the vertical and traverse the target of the Manukau. As it moved and started to lift towards the horizon a vast relief overtook me. We were not going to plunge unheralded and unsung into the muddy waters of the Manukau. The nose came up to the horizon and we were flying again. The altimeter read 3,000 ft. The air speed needle slowly slid back.

We streaked across the Manukau, the speed gradually falling off and I held the height at 3,000 ft. A great sigh of relief and I started to think about what damage we had sustained. What was my next move? The aircraft was undoubtedly damaged. Looking out of the cockpit I could see that the fabric on the ailerons at the wing tips had almost completely disappeared, so that almost the only control I had in that plane was in the little six inches by two inches of metal trim tab. I had no idea how much damage there was in the tail plane. Could I get the aircraft back on to the ground with safety? I eased the power back and at the same time wound the tail trim right back. When the speed was sufficiently reduced I dropped the undercarriage. The green light showed, so I had wheels, but I had to increase the power to hold the nose up. Obviously I had very little control.

I called the tower at Ardmore and advised them of my predicament, and continued experimenting. I found that by handling the aeroplane on the three trims alone, I could just manage to retain a degree of controllability. The Corsair had an approach speed of 78kts, but I found that with the undercarriage down I needed to maintain almost full power and 120kts to hold a gradual descent, and thus it was that I eventually lined up to the runway at Ardmore.

I came in on the approach almost at full power, 120kts across the fence and, more by good luck than management, the wheels touched the tarmac at the very commencement of the strip. We screamed along the strip tail high. To apply brake would have been fatal. I had to wait until the tail settled before I could gently apply a little brake, and then, as the far end of the runway loomed up. I was able to apply brake more and more heavily, and finally come to a complete stop perhaps ten metres from the end of the runway. I taxied back along the taxiway, accompanied by fire tenders and crash tenders, into the aircraft park. Ground staff put chocks beneath the wheels, and I closed the motor down with a great sense of relief.

For a few minutes I sat quietly, doing nothing, and then I clambered down from the cockpit on to the safety of the earth. I walked around the aeroplane. The ailerons were fabric as were the elevator and the rudder. The fabric had been almost completely stripped off all the control surfaces and the only means of control that I had were the little fabric remaining and the metal trim tabs on each one of those surfaces. I stood and looked, and thanked God.

The CO's jeep came across. The CO climbed out, walked around the aircraft, came back to me, and said, 'I hear you nearly left the squadron,

King.' I undid my parachute, slung it into the back of the jeep, climbed in and sat down with feeling.

To understand what had happened, it is necessary to have some understanding of the properties of air. The air consists of a mixture of oxygen and nitrogen, the molecules of which are moving in all directions very rapidly. The collision between these molecules generates pressure, which becomes less as the altitude increases because there are fewer molecules. The air is elastic in this respect and behaves like a spring. If you compress it in one place the wave of pressure you generate travels through the air like the ripples on a pond into which a stone has been thrown. The speed of such a wave at sea level is around 1200 km/hr. At 36,000 ft it drops to around 1000 km/hr.

When an aircraft's speed approaches the speed of sound in the air in which it is flying, pressure waves (or shock waves as they are generally called) are generated from those parts of it over which the air flows fastest. The waves travel away from the aircraft in all directions all the time, and because it is moving forwards, there is a cone of waves behind it forming a solid angle with the direction of flight. When the aircraft speed equals the speed of sound this angle is 45°. The Mach number — which is the ratio between the aircraft speed and the local speed of sound — is then equal to one.

A lot of things happen when shock waves are generated and the aircraft has entered the region of compressibility effects, as they are called. Usually compressibility effects begin to be significant above about $M = .85$. For example the air velocity over the top surface of the wing may reach sonic speed, whilst at the lower surface it is still sub-sonic. There will be sudden changes in pressure distribution, causing large trim changes, and making control extremely difficult. The control surfaces themselves may become ineffective or even have an effect of the reverse of normal — control reversal.

At the time I am speaking of, there were, of course, no such things as Mach meters on aircraft and nobody really knew how close to the speed of sound they were flying, because the speed of sound varies with temperature and altitude. Early tests of the Lockheed Lightning, in the United States in 1943, encountered compressibility problems during a prolonged dive which resulted in temporary loss of control and structural distortion. This particular dive was ultimately checked by the increased drag and reducing Mach number as the atmospheric density increased with reduced height.

John Derry, in testing the de Havilland 108, had the same experience of compressibility, and his report stated no means of recovery existed until the Mach number fell below that at which control was lost. This condition was realised at 30,000 ft in the DH108, when the Mach number fell below one and a slow recovery became possible. This was completed at 25,000 ft at Mach 0.95.

The world's first measured and controlled supersonic flight was recorded by Captain Charles Yeager of the United States Air Force in a rocket powered Bell X-1 in the autumn of 1947, when he achieved Mach 1.45. In the United Kingdom the late John Derry achieved Mach one in September 1948 in the de Havilland 108.

My Corsair flight was made after the Lockheed Lightning tests, but before Charles Yeager and John Derry officially exceeded Mach one. I would like to think that, in that brief period of stillness between two periods of compressibility on 9 August 1944, I preceded them both and could claim to be the first man to achieve supersonic flight.

The Corsair wasn't the only potent new aircraft operating in the Pacific at that time. Huge four-engined Boeing B-29 long-range bombers were beginning to replace earlier types such as the B-17 Flying Fortress and the B-24 Liberator. In their B-29s, US crews could attack targets on the Japanese mainland. James Bertram was one who saw their devastating effect.

A Rhodes Scholar at Oxford before the war, Bertram was working in Hong Kong as press attaché with the British Embassy when Japanese forces captured the territory. For a time he was a prisoner there, then was moved to a camp near Tokyo from where he watched the first mass raids on the city.

The Shadow of a War: A New Zealander in the Far East 1939–1946
James Bertram

Dead upon the stroke of noon, on a glorious late-summer day in 1944, the first B-29 appeared over Tokyo.

We had been standing around outside the Shiodome working hut waiting for the afternoon shunt. The blue sky was without a cloud, the whole yard lay muted and drowsy in the sunlight. When the long blast of a siren sounded in the distance, most of us took it for a noon factory whistle. Then the guns opened up.

An electric current of feeling — more wonder than fear — ran suddenly across the crowded platforms. Japanese railwaymen and lorry-drivers came padding out into the open, heads thrown back, sleeves raised in the inevitable gestures. And the first curious words soon became an excited chorus: '*Amerika shuoki! Niju-ku!*'

We all knew something about the latest American long-range bomber. For weeks the Japanese papers had been full of details of its construction — the four great engines, the wing-spread of fifty metres, the fabulous defensive armament. '*Niju-ku*' ('Twenty-nine') had already become a magic number; whenever we numbered off in Japanese at *tenko*, '*niju-ku!*' was always shouted a bit louder than '*niju-hachi!*' or '*san-ju!*' The Nips usually sniggered and thought it a good joke.

But the war was moving on; even in Tokyo it was about to catch up with us. Saipan had fallen to the U.S. Navy and Marines; the Second Front was joined at last in Europe. And Tojo, the man of destiny — Tojo, the iron General who had sworn he would never relax until he led his people to 'final victory' — Tojo was gone, like any other politician, not even pausing to take a bow as he made way for the venerable Admiral Suzuki and a new Japanese War Government.

The resignation of the hero of Pearl Harbour and that first year of smashing victories had shaken the populace of Tokyo in unexpected ways.

'It is like this,' Enomoto had told us in a quiet corner of the yard. 'The Japanese people cannot understand how Tojo could *resign*. He told them, and he told the Emperor, that Japan could win. If he was wrong, then he has betrayed his country. There is only one thing left to do: he must commit suicide. But instead, he just hands over the responsibility to another!'

The new Premier had not minced his words. 'Japan,' he declared, 'must now prepare to fight for her life. All our cities are in the front-line of battle.' And it was amazing how quickly a people deluded with tales of unbroken victory absorbed and digested the implications of the loss of Saipan. Perhaps it was because they had so long been familiar with the big display maps showing how vulnerable Japan was to bombing from Vladivostok. But now the danger-circles were drawn from a southern centre.

'*Ima,*' the Shiodome *hanchos* told us in the pidgin Japanese that was our *lingua franca* in the railway shed, '*Amerika skuoki takusan!*' ('Now many American bombers will come!')

Japan was being bombed all right, in that summer of 1944. China-based Super-Fortresses were hammering away at Kyushu, there were odd raids on Chushima. Sometimes long-range Alaska-based flights would visit the extreme north. But nothing, so far, had come to Tokyo — nothing since the recklessly plunging wings of Doolittle's fliers skimmed these roof-tops nearly two years before. 'Tokyo will never be bombed,' they had said before Doolittle's costly venture. And after it, 'Tokyo will never be bombed again!' Yet here, riding high with effortless ease over the enemy capital, was the first B-29.

'They might save that little lot!' said a P.O.W. gunner contemptuously, jerking a thumb at the white puffs in the blue. 'He's well over 30,000. And is he travelling!'

All we had seen at first was the soaring vapour trail, converging in an inexorable curve on the centre of the city. Then the white trail divided as the big plane closed in above us — we could pick out the vast, distinctive wing-spread of the new bomber, light-weighted for reconnaissance and climbing now at a speed no Japanese fighter at that altitude could match.

'Queer, ain't it, mate?' another prisoner mused aloud. 'Up there they're drinking coffee an' scoffin' chocolate. And we're down here working for these bastards. Well, happy landing, Yank!'

That was the first unlucky swallow of Tokyo's approaching winter. Before long the noon reconnaissance plane was to become a regular feature of our working-day; it was christened, of course, 'Photograph Joe', and became another prisoners' mascot, like 'the Witch'. Late that summer the raids on Tokyo began — small raids at first, by planes in strict formation, using every kind and type of bomb, but already favouring incendiaries. Fortunately for us, the first tentative bombing kept well to the east and north-east, in the main factory area of Tokyo.

Not once in those first weeks did we see a B-29 in serious trouble. The Japanese anti-aircraft fire was soon synchronised into a really impressive barrage; but the American planes kept high. They were regularly challenged by Japanese fighters, and Japanese fighter pilots were good. But in all the times we watched the fighters swoop down on the tight bombing formations we never saw a B-29 brought down; though often a fighter disintegrated in a puff of smoke and fire.

That was the first phase; and it was a remarkable demonstration of the fire-power of the new American plane. But then the bombing tactics changed. Raids were still made in daylight, but at much lower levels. The Japanese A.A. gunners were making better shooting; obviously, before long their luck would change. And we had noticed that the B-29's, which might come in from any point in the compass, had developed a habit of going out almost directly above Omori. A terrific cross-fire saw them off, from the massed batteries of Tokyo and Yokohama; on the canal bank a hundred yards from our camp, a light quick-firer usually got in the final bursts.

One autumn evening we had just got back from work as a raid was ending. A dark pall of smoke lay over the north-east of Tokyo; a few stray bombers were heading through it out to sea. Then one B-29 came over very low, just above Omori. Japanese fighters were trailing it, but keeping a respectful distance. Then the big bomber took a direct hit from anti-aircraft fire; smoke became fire, feathering out from the gaunt wings. Losing height rapidly, but still flying perfectly on automatic control, the doomed plane roared lower and lower across the camp.

By the guard-house the Nips were cheering themselves hoarse. '*Banzai, banzai!*'

We craned from every window of the barrack, gripped by a kind of tension we had not known in many months. Mick Cahalane, his boxer's hands clenched, his face working terribly, was pleading for us all: 'Come on, Yank, come on, me jewel! You can make it; come on, boy, come on.' For the impossible, the incredible, had happened: the champion was reeling from a knock-out blow. Dempsey was beaten at last. A B-29 had been hit.

Absurd, of course; but over these weeks we had almost come to feel that a B-29 was invulnerable. Now at last the Japanese fighters grew daring; they circled and stunted round the sinking ship, and we heard the dry rattle of their machine-guns with no answering fire. But if ever a bomber went down in majesty, it was this first B-29. And it rode to its doom on the hearts of six hundred Allied prisoners. No one baled out. Still on a steady keel, still proudly showing its markings to friend and foe alike, it dropped out of sight beyond the high walls of the camp. A moment later it crashed in the shallow bay.

A working-party detail went out from Omori next morning at low water to reconnoitre the wreck. The returning men reported that the bodies of the crew were still in the fuselage. We asked for permission to bring them in, or at least to hold a memorial service. But the wreck had been claimed as the property of the Imperial Japanese Navy (because it had come down at sea): the Army Staff at Omori were sternly warned off the scene. We made a mental note of the spot; and nearly a year later we remembered it, when American landing barges took us from Omori and across the bay to freedom....

'Alert!' Alf whispered from the blankets beside me. Out of the windy night a distant siren was wailing. Then, one by one, the nearer posts took up the mournful sound. We lay and shivered with anticipation; for the March night was bitterly cold, and soon we should be out in it.

'What time is it, picket?'

'Quarter past eleven. They're over all right — Nip fighters been going up for the last half-hour.'

It was pitch dark in the barrack; but night drill for an 'Alert' was routine by now at Omori. Get up and dress, boots on, all glass windows out; stand by for the 'Alarm'. Already the Japanese patrols were tramping outside. And at any moment 'the Bird' might be through with a flashlight. But what a night for a raid! ...

The famous 'long wind' of Tokyo was rising steadily, and in this gale nothing could control the fires. Groaning and cursing in the darkness, we struggled into our boots. The picket was already lifting out the glass windows that ran the full length of the barrack on either side, and finding it hard to hold them against the wind. With the windows gone, we were in a kind of pillared summer-house. Only it wasn't summer.

'Stand by while I check these rolls ... answer your numbers.' It was the voice of the barrack-commander, and he was reciting the numbers by heart. *'Barrack Fire Party. Stretcher Party. Cookhouse and Stores. Documents....'* The whole thing was quite scientifically organised; we had had plenty of practice.

'What are you on, Alf?'

'Not a thing! I'm staying right here, close to these blankets.'

I was on the 'Documents Party' — though, from what we knew of the Omori Camp Documents, few of them were worth preserving.

'*Alarm!*' The 'short blasts' were sounding now on the nearby siren in the Keihin Road. '*All parties fall in on the parade-ground!*'

Carried on the violent wind we could hear the characteristic throaty drone of Japanese night-fighters — a flight of four was climbing from Haneda with wing-lights on, in the favourite Nip formation of pairs in echelon. The night was clear, without a moon.

Our groups assembled without much fuss, though the Japanese guards were calling to each other round the camp walls, repeating orders in that wolf-like howl that had first pricked our scalps in the hills of Hong Kong. The Documents Party found their rendezvous outside the Camp Office. For the next half-hour we were busy enough to keep warm, lugging tin chests of papers out of the office and stowing them in water-logged shelters. To the north the guns had opened up; and the night was filled now with the smooth, high-pitched hum of American engines. It sounded like a big raid.

'Finished with those cases? All right, get back to your barracks.'

This was good; we should be in time to see something. Prisoners were supposed to stay inside the buildings, unless ordered into the shelters. But most of us preferred to find some vantage-point along the walls, where we could look through peep-holes carefully prepared for just this purpose. I slipped in behind the *benjo*, found the hole I knew, and looked out across the canal towards the city.

The whole eastern fringe of Tokyo was ablaze. Two enormous fires were raging beyond Shibaura; against the high bank of flame we could identify all the familiar landmarks — factory chimneys, warehouses, gasometers.

'They dropped a lot of flares at first,' said my fence neighbour. 'But now it's mostly fire-bombs. Low-level attack.'

The big bombers, coming in over the fires, were more clearly visible than in the beam of a searchlight: they looked to be no higher than 5,000 feet. Every gun in the Japanese capital was banging away at them, so that the barrage was continuous — an immense, shattering wall of noise across the night. It was the most daring night raid we had yet seen, and numbers of B-29s were hit as we watched. But steadily the fires were spreading. The two biggest ones had joined, and the wind was taking their golden plumes far out across the bay.

'Boy, are they going to town to-night!'

In fact, this raid of March 10th burnt out more than seven complete wards of the city of Tokyo (which has a total of thirty wards). In other words, nearly a quarter of one of the largest cities in the world was turning to smoke and ash before our eyes.

Modern war, most people will agree, is a barbarous and incredibly wasteful proceeding. But it does turn on some rare sights for the onlooker. I don't know how Tokyo looked that night from the air; from the ground it was the most tremendous thing I had ever seen. One could even understand Nero's exaltation — a burning city, seen from a safe distance, is the perfect distraction for a degenerate.

> 'At midnight on the Emperor's pavement flit
> Flames that no faggot feeds, nor steel has lit …'

But these flames a storm disturbed; and at least half the damage of that fatal night was not made by man, but by the elements.

Even on our little island the air was thick with flying ash and the reek of petrol fumes. Occasionally a twist of the great fires sent a wave of heat that smote our faces across miles of water. Beneath our wall, where we watched, we were in shadow. But behind us the barrack roofs, the wooden watch-tower with its Japanese sentry, the feathery tops of trees, were all clearly bathed in the glow, as in the light of some strange dawn. And faces looking towards the city were like the faces of Aztec priests in ritual around a temple pyre.

The fire-bombs were launched in containers that exploded and broke asunder in mid-air; from that moment the sky was filled with a golden dropping rain. Floating tinfoil and the vicious bursts of flak surrounded the target area — it seemed as though not a yard of that crowded space could be vacant. Yet still the bombers came roaring through smoke and cloud, plunging into the fierce light unwaveringly, and out again. How long had it all gone on? Two hours, four hours. We had lost count of time. My legs, when I moved at last from my peep-hole, were frozen and numb.

Now the planes were heading out to sea and home. We watched the Japanese fighters swooping down, saw the line of their tracers and the answering fire that leapt from the long bulk of the low-flying bombers.

Once more — as in the autumn before — a burning B-29 came right above Omori. This time we saw four parachutes drop from it like pink petals, to be lost again swiftly in the dark. The plane flew steadily on, white fire streaming from its wings — a burning dragon wheeling through the night — till at last it plunged and was quenched in the shadows of the bay.

A last brilliant shower of incandescent rockets, and the guns were silent: the raid was over. For the first time, now, we heard the roar and crackle of the fire. A few roaming searchlights strayed southwards. Japanese fighter patrols, their lights on again, were dropping back to their fields. And every fire-engine of Yokohama and Kawasaki was jangling along the Keihin Road towards the burning city. Little enough that they could do, until the wind dropped or until there was nothing left to burn....

'I was just wondering ...' said Alf's voice from the blankets again.

'What?'

'... if London was ever like that. In the blitz.'

I thought it over. 'Nothing so big, I think. There might be local fires as fierce; but they'd have a much better chance to control them. Everything here is wood. And that wind, of course....'

But what was in my own mind, with a sudden wave of thankfulness for a forgotten naval battle in the Coral Sea, was the wooden bungalows of towns and cities in my own New Zealand. A single raid on that scale would have wiped out the four largest cities of both islands. And it might so easily have been.

But it wasn't. On 8 May 1945 German representatives signed a document of unconditional surrender. Three months later, on the 6th and 9th of August, two new and lethal bombs were dropped on Japanese cities and the war in the Pacific was also over. In the towns and cities of New Zealand there were joyful celebrations. And for those who'd fought, there were memories ...

The Blue Arena
Squadron Leader Bob Spurdle DFC and Bar

On 2nd August 1945, I climbed out of a Mark XVI Spitfire after an air-to-air exercise, and walked back to dispersal.

I didn't especially savour this flight, but then how was I to know it was the last plane I'd ever fly?

On 6th August, the first atomic bomb was dropped on Japan and with it my dreams were gone forever.

To be repatriated back to New Zealand, I was required to join the RNZAF. I protested vehemently.

'No way! I want to remain RAF until demobilisation!'

But it was pointed out that priority was to RNZAF personnel and that my family and I might hang around in the UK for months. Thus, the handful of Short Service veterans — the few survivors of so very many volunteers — were forced to acquiesce and leave what was, in our eyes, the best service in the world.

Shirly, Anne and I embarked on, of all ships, the very one I'd left New Zealand on, six years before — the *Rangitata*! As we steamed over the oceans' lonely wastes, sheltering in the shade of canvas awnings, our motley collection of aircrew swapped experiences. Some were bizarre and, in their telling, maybe we were unconsciously healing ourselves of deep psychological traumas:

Of the mess barman who, getting impatient at the long drawn-out revelries, got tetchy and uncooperative. He was grabbed and rolled up in a carpet, stood in a corner and forgotten. Next morning the unfortunate chap was found suffocated and stiff as a board.

Of the dreadful sight of two Canadians who baled out of their burning B25 too late. They hit Volkel's runway to bounce and then roll along parcelling themselves neatly into two blood-soaked twitching cocoons of silk.

Of the two Typhoon pilots scrambled off from Manston's runway. One or the other had a tyre burst just before becoming airborne and they collided to end locked together in a ball of flame. When the fire engines had finished, all that was recovered were two torsos the size of small suitcases.

And on and on and on.

But there were humorous stories too. At Manston one of the chaps was practising instrument-flying when some Me 109's strafed the 'drome. All of a sudden the Link trainer went into a spin and the pupil thought the operator was playing tricks. Eventually, after trying every remedy he knew the 'pupil' lifted his canopy to find the airman long gone and the Link with bullet holes in its airbags. As far as I know he was the only chap ever shot down in a Link trainer!

The days drifted away and at long last clouds over our homeland's green hills showed snow white in front of our bow knifing through the long blue swells. We'd made it!

Back in New Zealand an old friend of mine, Squadron Leader Williams, came on board to interview all the more senior ranks before disembarkation. He was very embarrassed.

'Spud, believe me this isn't my idea! My job is to see all you chaps and advise you, tell you, not to apply for permanent commissions in the Air Force! You'll only be wasting everyone's time!'

It was clear enough to me but a severe blow to some of the others who had pinned hopes for their futures on an RNZAF career. The cunning ones were coming out of the woodwork now danger was all over and were protecting their own futures. And so, by this shabby, miserable means, a cadre of very experienced first-class air crew were shunted aside. Sure, there were a lot of experienced air crew amongst the Pacific RNZAF air crews but nothing like the calibre of those experienced in the European conflict. It didn't affect me as I had other plans.

One day, climbing down a steep ridge while deer-stalking, I found my nerve gone and I had to be assisted back to the track. Worse, I found I couldn't watch a plane flying overhead. I had to look at the ground — it was a year before I could look up at the sky again and learn to ignore the pictures in my mind.

Friends ask me why I never flew again. Fly? What for? To fly a runty light aircraft chained by authority to follow submitted flight plans?

Never!

To cut this red tape, I'd need a Sabre's power. To hear my lost comrades' voices again, to find again the wonder and glory of sunlit spires and the deep caverns of the clouds, I'd need the magic of a Merlin.

One More Hour

Desmond Scott DSO, OBE, DFC and Bar

As the war in Europe drew to a close, we had already entered the shadow of the Atom. It burst above the Japanese city of Hiroshima on the morning of 6 August 1945. It was an event that shattered the world.

Nine months later, as a member of Transport Command, I had occasion to visit Japan. It was a visit that may well have changed my direction. Dresden, Hamburg, Stuttgart, Berlin — none could be compared with this. Only Sir Winston Churchill could have given voice to my feelings as I stood alone in the centre of the wasteland of Hiroshima. 'What ought we to do? Which way shall we turn to save our lives and the future of the world? It does not matter so much to old people; they are going soon anyway, but I find it poignant to look at youth in all its activity and ardour and wonder what would lie before them if God wearied of mankind ...' But perhaps all this that I fear is mere melancholy and dismay and will fly away as the autumn leaves if I stand again by the old dispersal hut at Tangmere, Manston or Gilze Rijen and listen back through the wartime years. It cannot be that the spirit which once set our young blood afire has left our race forever. Nor could it have been subdued or cast forever in the nightmare of our dreams.

Here the autumn falls upon our River Avon, and near the bend, the tall trees of evening stand golden to the skies. Beneath the sycamores, the shadows deepen, and I am lulled into a sadness by the whispering of the leaves. It's a soft warm breeze that drifts down from the hills filling the world with pine forests, log fires and days of long ago. If only Smithy were here, Spike Umbers, Wally, Steve and Fitz. Just for one more day. Just for one more hour.

Perhaps if I close my eyes we might fly off again to those sunlit uplands where the clouds are full of laughter. Where good fellowship among all men is a never-ending thing.

PART FIVE

LATER
DAYS

The coming of peace meant a much smaller air force and an unsentimental melt-down for hundreds of surplus warbirds. Kiwis saw their first jet in 1946 when the RNZAF tested a Gloster Meteor at Hobsonville and Whenupai. With 'its twin motors roaring like a giant blow lamp' the Meteor took off 'at incredible speed and within a few seconds had climbed out of sight ... The screaming motors seemed to hurl the plane upward as fast as it could dive.'

Despite such glowing reports, the air force eventually purchased twin boom de Havilland Vampire fighters and, later, English Electric Canberra bombers. In fact, it was an RAF Canberra that landed first in the International Air Race of 1953. This race across the world from London to Christchurch was the last of its kind to be staged. With aircraft getting larger, faster and more reliable (and fares falling fast) long journeys became the rule rather than the exception and the great Air Race had done its dash.

Change too for New Zealand's domestic airlines, all of which were privately owned. Well, until December, 1945, when Peter Fraser's Labour government nationalised the lot. After decades of disinterest, politicians ruled the aerial roost with NAC officially beginning operations on 1 April 1946. By this time we'd also gone international with the establishment of Tasman Empire Airways — a joint venture with 50% of the shares owned by New Zealand, 30% by Australia and 20% by the U.K.

The arrival of the airline's first aircraft in 1939 was a grand event. 'Several thousand people had assembled and many more were sitting in cars along the waterfront road' when TEAL's 'great sea-bird ... alighted gracefully' in Auckland's harbour. Proudly named *Aotearoa*, the four-engined Short S30 Empire class boat touched down on 28 August 1939, less than a week before the start of World War Two.

During the war, *Aotearoa* and a second S30, the *Awarau*, flew a weekly trans-Tasman service, carrying 17 passengers in their 'roomy hulls ... Comfort is no longer confined to travel by land and sea. Within the air-conditioned interior, protected against noise and vibration, passengers may travel at nearly three miles a minute in the very lap of luxury.'

When peace returned, it was time for new aircraft. Some argued vehemently for American land-based airliners but Australian unease over competition (they still had a stake in TEAL) and energetic lobbying by anxious local exporters meant the Brits got the nod. Enter the Short Sandringham, a civil variant of the much more famous Sunderland, and frequently flown by somewhat phlegmatic ex-military pilots.

A Noble Chance: One Pilot's Life
Maurice McGreal

Auckland was a busy harbour in the mid year of the century and flying boats had often to contend with boat skippers who would hold blindly to their right-of-way in the fond belief that we always had everything under control.

On a sunny morning in July of 1949 the harbour was packed with overseas shipping, held up by some argument with the wharf labourers. It was near 8am and we had taxied from the pontoon bound for Sydney, planning to start the take-off from a point just east of the end of the breakwater, towards the west. I called Whenuapai for clearance through their traffic area.

Boat traffic on the harbour sector we were aiming for seemed clear and the Control launch gave us the OK for take-off. The skipper pushed the two outers to full power and then with a slight pull back on the port outer to correct for the torque, the boat started to rise on the step. He added the power of the two inboards. I placed my hands on the pedestal behind the four throttles to protect against any creep and now we were on the step and motoring with the speed building past 55 knots.

Slowly, behind the bows of one of the moored merchant ships, came the blunt flat bow of a vehicular ferry from Devonport. I gave a quick signal, pointing, and the skipper nodded. Then almost as I watched, from behind the breakwater came its twin from the city side terminal and the two were converging across our take-off path. A passenger ferry was crossing half a mile ahead. Our clear path was narrowing rapidly as the two vehicular vessels closed from left and from right.

The skipper's eyes were fixed ahead now. I could see that he had made his decision and had assessed his margins. We were set for 'GO' and his knuckles were white as he gripped the big black control wheel.

We flashed by the first ferry on the port side with a couple of hundred feet to spare. The city wharves lined the whole of the port side of our run and we edged fractionally towards them to try and win a bit more clearance from the second ferry coming in from the north. High on the step now and racing across the sunlit waters at near 80 knots; a long creaming wake settled back to the sea behind and on the starboard the ferry closed with an uncomfortable nearness.

The skipper suddenly poled back hard and the aircraft leaped a hundred feet into the air and then quickly levelled as we swept across the upturned faces of the passengers standing by their cars on the vehicular deck as the stream of water from our keel rained across their faces.

Climbing over the western harbour the skipper put the auto pilot into the engaged setting and with a grin he pulled at his moustache and said, 'Got a cigarette?' ...

Of course, with commercial flights then operating at much lower and more turbulent altitudes than those of today, take-off wasn't the only turbulent time.

At 9,500 feet, in the darkness of the cockpit the pilots watched ahead for the signals that would show the Front they knew to be there. The first sign, a strike of lightning across the ragged banners of cloud that now peeped above the horizon, and a quick calculation told that it was still probably an hour away. The Meteorologists briefed us saying that they expected the tops to be at 15,000 feet; but we knew they would more likely top 35,000 feet and we would have to search for some pathway in the sky to take our aircraft through in reasonable comfort, for we could not climb above them.

There were no aids, such as weather radar, to search out the heart spots of the violence that lurked within these frontal battalions. The rugged strength of the Solent and the skill and experience of the pilots backed by such rudimentary guidance as could be achieved using the radio compass, were all the aids we had. The radio compass tuned to some station within the sector ahead would respond with a quick swing of the needle when a lightning flash tore across the sky. We knew it had pointed to the core and the skipper would roll away to a heading twenty degrees off to the left or the right. Then tense eyes would watch again for the next flash to give another clue to the secret pathway we were trying to find.

The passengers sitting with their belts fastened in the cabins were either trying to sleep or to pretend the tension in their bellies was not fear but perhaps the sandwiches they had had for supper — or was it the glass of champagne they had sipped during the climb. Outside in the darkness the propeller tips were showing the first flickers of St Elmo's fire, the flickering glow of the static charges drawn from the droplets of water in the clouds and few would sleep a wink.

It was the night of 11 November; Armistice Day in the minds of those who recall the days of the first war and we had a full ship; forty-five passengers, one of whom was Sir Leonard Isitt our Chairman of Directors. Twelve house earlier and just fifty miles from the Sydney coastline we had flow westbound through this very new and very vigorous front which only then was beginning to get power into its loins. In the sudden fierce turbulence of the mid-afternoon the cabins were strewn with shards of costly bone china as the Royal Doulton tea service was flung about from floor to ceiling and now we knew that the front was still growing as we approached the great line of cloud castles.

Sir Leonard climbed the ladder to the flight deck to exchange a few words with the crew. He was a fine old gentleman with a lifetime of experience as an airman since the earliest days of flight. He parted the folds of the blackout curtain and pushed his head through, gripping the backs of out seats. Jim Kennedy was the skipper and Sir Leonard looked with interest at the signs of the fireworks ahead as he chatted.

There was not a great amount of activity; only an occasional strike across the clouds to detail with lurid clarity the power within. We passed through some thin middle layers of stratus and some ice started to build. The engineer turned the de-icers on.

Static electricity now outline, with flickering darts of fire, the edges of the wind screen and to right and left the propellers were spinning discs of green fire.

The wings now began to glow very brightly and the nose of the boat became suddenly brilliant with green fire. We were obviously very highly charged and at Jim's nod I turned the cockpit lights to maximum to offset the risk of glare from a discharge of static.

The front of the boat suddenly grew brilliant with intense light and without warning a great shaft of glowing fire that seemed the size of a telegraph pole shafted ahead from the bow into the cloud that was thick around. With a bang of thunder it exploded and just as suddenly disappeared. Our eyes were left seared with the continued memory of the light. I grabbed the Aldis lamp and triggered it so that it focussed on Jim's instrument panel. The powerful light was enough to show the blind flying panel clearly through the searing afterglow that still stayed in our eyes.

Jim disengaged the autopilot as the aircraft started to surge upwards like a lift in a skyscraper: 2000 feet per minute showed on the VSI and I put my hand across the bank of throttles and eased them back to keep the airspeed down to help reduce the effect of the sharp-edged gusts on the aircraft structure. About 1.5 times the stall speed was believed to be a good turbulence penetration speed.

The upward surge ceased and almost in an instant we were going down. Loose objects flew to hang weightless about the cockpit. I started to add some of the power that only a few minutes earlier I had pulled back and my teeth nearly came loose with the suddenness of the blow as we hit bottom of whatever it was we were in. My fingers started to tingle as I worked the throttle levers while Jim tenaciously flew attitude — keeping the wings level and the nose true to the horizon. Again my fingers tingled as they waited, ready, by the throttle levers and I glanced down. Small wriggling worms of greenish blue fire were running from their tips to the throttle levers and I realised that I, too, was charged and had started to discharge. Real weird, I thought.

The bottom of our world again fell away as we entered the big, fast, descending shaft of air somewhere on the outskirts of the thunder head we were flying through. I added some more boost and glanced across at Sir Leonard who was still standing semi crouched as he hung on for grim life to the structure of our seat pedestals.

Things remained still for a few minutes and he took the chance to say his thanks to the skipper and, turning to leave, he added: 'This is no place for Directors,' and the engineer helped him climb down through the hatchway.

Soon after they arrived in New Zealand, the Sandringhams were grounded for six months so that work could be done to rectify engine-overheating problems. It was George Bolt, then TEAL's chief engineer, who identified the reason — insufficient airflow caused by defects in the design of the engine nacelles. Initially, the manufacturers pooh-poohed this theory, but Bolt prevailed and his modifications became standard for the type. However, before those changes were made, the engine housing design fault did create some nasty moments.

To Fly a Desk
Noel Holmes

There were one or two incidents that caused misgivings, the most bizarre of which was perhaps 'The Affair of the Bishop's Robes'. This occurred at the end of 1947 when RMA *New Zealand* was forced to return to Sydney on three engines two hours after take off from Rose Bay. Because of squally weather the aircraft, with twenty-nine passengers and a crew of six on board, had been flying low when the engine failure took place. Air speed began to decrease and the machine continued to lose height until it was only fifty feet above the sea. In an effort to maintain height, 1600lbs of baggage, 600lbs of freight and 30lbs of company's stores were jettisoned. In the process a case fell open and a silk stocking twined around the arm of Betty Morton, one of the first six hostesses employed by the company. In the general panic she was seen frantically unwinding the stocking and throwing it out the door, apparently reasoning that it might be the final straw to break the camel's back if it stayed on board.

TEAL's board was in session at Auckland while the drama continued and was kept in touch with the situation by radio. Staid members were a little taken aback by a suggestion from an anonymous crew member that women and children should go next so that the breeding stock could be kept for posterity. As it turned out, the Sandringham held at a very low level and even gained a little altitude as fuel was consumed. Sydney was reached safely and the engine fault corrected in a few hours.

The affair was not forgotten in a hurry because of a claim by an American bishop whose robes were floating in the Tasman. TEAL gulped at the value put on the robes and there are those who swear the insurance company manager tried to jump out of his sixth storey window. There was general agreement that they were either dealing with a very rich bishop or a bishop whose diocese consisted of pious, warm-hearted millionaires.

Another frightening event occurred when Captain I.C. Patterson was taking off from Rose Bay towards the Sydney harbour bridge at about midnight. He heard an engine falter and, before he knew where he was, discovered all his engines were out. All were frothing oil. The event was without parallel but Patterson had no time to puzzle over causes. Completely without power he brought the plane round sharply and landed it in darkness, not up the flare path (there was no time for that) but across it. Drifting to a stop without hitting anything, he had his anchor put out and waited to be towed in.

While waiting Patterson had time to think and, with commendable acumen, insisted upon being taken to the oil dump immediately on stepping ashore. There, he had every drum sealed pending checks. Investigation showed the aircraft had been filled with tanning oil, not a lubricant to be recommended for high-powered radial engines. The exasperating aspect of the whole affair was that the drums were stamped with an Air Ministry specification.

Someone had blundered, a blunder that could have taken a high toll of lives. Who was to blame? The question was never answered.

Of course, TEAL's magnificent flying boats didn't only cross the Tasman.

New Zealand Aviation Yarns
Roy Sinclair

It was known as the Coral Route — an airline journey without peer through a South Pacific paradise offering passengers long stopovers at idyllic coral islands. TEAL inaugurated the Auckland to Tahiti service on 15 December 1951 in an effort to expand international routes, and create goodwill among people inhabiting islands spread through a vast area of the South Pacific. Initially the entire Coral Route was flown by Solent flying boats. Then, in May 1954, the Auckland to Fiji sector was taken over by DC-6 land planes. The Solent ZK-AMO *Aranui* was retained for the remainder of the flight to Papeete until September 1960 when it flew the world's last scheduled flying boat service on an international route. Dennis Marshall flew as chief steward with TEAL through the 1950s. He recalls the Coral Route, and one particularly bizarre incident that had him, two hostesses and 40 international passengers stranded for six days on a remote island. Paradise, as it turned out, was not found wanting.

'If we got up to Tahiti and back to Suva without incident it was unusual. On such rare occasions we gave ourselves a slap on the back. Very few New Zealanders flew the Coral Route. It was all one class — first-class — attracting well-off travellers, often captains of industry, who yearned for a South Pacific holiday experience. In those days Tahiti was an unusual destination to go to — it was the last, lovely, lonely place. It was thought of as a place where beautiful Tahitian girls were seen nude on the beach, but it was not quite like that. The people of Tahiti, and other South Pacific islands, were, however, the most gracious, fun-loving and natural people to be found everywhere. Today, I expect, Tahiti is more like metropolitan France.

'For people wanting to discover the South Pacific TEAL offered the one scheduled air service to Tahiti. Our Solent flying boat had a certain romance, and a certain glamorous style that was missing from other aircraft of the day. It flew at a maximum height of 8000 feet but most of the time we flew at 3000 or 4000 feet. With a speed of 200 mph — if we were lucky — passengers had a wonderful view. The Solent had two decks with a bar, and a galley where meals were freshly cooked on board. We even cooked bacon and eggs to order. Passengers faced each other across tables. It was like the luxurious club lounge of a millionaire's yacht.

'Once every week we took off from Lauthala Bay in Suva at about 9 am and arrived in Tahiti at 11 am next day. The first leg to Apia in Samoa — where we stayed in Aggie Grey's famous hotel until evening — took about three and a half hours. From there we would take off late at night when almost 2 miles of sea were lit by flares floating on little boats. It took an awful lot of runway — and power — to get a flying boat into the air. At dawn next morning we landed at the small atoll of Akaiami, in the Cooks, where the coral heads had been cleared to enable flying boats to land. The atoll is about 7 miles down the lagoon from the main island of Aitutaki. From Akaiami, it was another three and a half hours' or so flying time to Papeete in Tahiti where the whole town turned out to greet the flying boat.

'The lagoon at Akaiami was a very bright iridescent aqua that looked almost like a neon sign contrasting with the deeper blue of the sea. I have never seen anything like it anywhere; it was much brighter than most other island lagoons we knew. With palm trees growing to the water's edge, I thought Akaiami was the most fascinating place in the entire South Pacific. Our passengers could enjoy a swim during the two hours it took to hand-pump fuel into the Solent. We recommended they carry swimming costumes with their hand luggage. Akaiami was usually deserted (it was said to be the world's only deserted island with an international airport), but two native girls came over from the main island (Aitutaki) whenever we were scheduled to arrive. There was also an American named Harrington who supervised the refuelling from our large fuel barge. One of the girls, Dora, later married Harrington. TEAL had constructed a small hut from coconut tree posts, with fronds for the roof. A later addition was a short coral jetty for launches to come alongside — passengers did not appreciate getting their fashion shoes wet.

'On this particular occasion we were all back on the Solent at Akaiami when there was a defect in one of the four Bristol Hercules engines. Captain Joe Shephard got permission from New Zealand to take off with three engines and fly on to Tahiti for repairs. But the passengers had to be left behind — which was just as well because a fully laden Solent just got off with four engines. Take-offs were sometimes very exciting. We were unloaded with all the catering equipment, remaining food, blankets and pillows. And, there we were, castaways, spread among large piles of expensive alligator skin baggage as we watched the Solent take off and vanish. Captain Shephard had promised to be back in three or four days.

'Two of the passengers were New Zealanders. The others were from metropolitan centres of the United States and Europe. Among them was Martine Carol, the siren of the French movies. She was quite delightful. Unable to survive on the atoll — we had food for one day — we ferried the passengers up the lagoon in a small barge to the main inhabited island of Aitutaki. With no hotels, and very few buildings of any sort, we slept in huts built during the Second World War. They were rather austere, but the cold showers worked and we had ample wood for the stove. We purchased most of the food from the one small shop. Having little money, I had to take out my note book and pen to write IOUs — "TEAL will pay you so much ..."

'I recall learning to be philosophical during the enforced stay. The passengers were not equipped mentally or physically for such an adventure and I was afraid of being lynched as I prepared our first meal. I can still hear the German tourist complaining bitterly about his round-the-world bookings being wrecked, and demanding, "Venn do ve get off dis island?" But after a day of adjusting, they settled down to enjoy an island holiday. It was very hot, with temperatures up in the 80s and 90s in the shade. Martine amused us by always wanting to swim in the nude.

'In the plantations I killed my first pig, and we caught fish by throwing spears into the sea from the edge of the coral reef. But the flying boat's liquor, intended to provide for two flights, was running out. I was told by the natives that the administration's resident agent, a New Zealander, had an excellent wine cellar in his unoccupied house. To get permission to break it open we had to call the agent, who was visiting Rarotonga. Hearing of our dilemma, he told us to help ourselves. The wine was indeed very good, and TEAL later paid to restock the cellar.

'The children dressed up in their costumes as they would to entertain the Governor of New Zealand during his five-yearly regal visits. Using my IOUs, I asked them to make leis, and what the Tahitians call couronnes — rings of scented flowers — to sit on the women's heads. All the natives were magnificent people, even if they did not always understand us. One day I wrote my usual IOU for some tomatoes they said they had, only to find they were plants just 6 inches high and were to be ready by October!

'Each day, using a Morse code radio, I went to some considerable effort to find out about the flying boat's return from Tahiti, but each day the story was different. In the end I had to make up stories for my passengers, who seemed to take less and less interest — until the last day when our food was running short. Even then, I think there was some reluctance to return to Akaiami and the waiting flying boat. Looking back, it was really a most unusual time. Some of the passengers were still writing to me, and sending Christmas cards, many years afterwards.' ...

'When I started as a TEAL steward in 1950 I was on the trans-Tasman Solent flying boat service between Auckland and Sydney. There were certainly some unexpected and unusual happenings on this route. One such incident happened as we were departing from Sydney. I had gone through all the drills for taking off. They included fitting watertight doors

— actually pieces of the aisle — between the sections of cabin wall. The idea was, should we hit something, water would not rush through the entire lower deck. I then climbed up the ladder to tell the flight engineer that all doors were watertight. We were ready to go.

'Whenever a flying boat started its take-off run there was a sudden maelstrom of sea water swirling over the windows of the lower deck. The windows were quite large — about 2 feet by 2 feet 6 inches — and these were immersed in a deluge of water until the flying boat got up on its step and finally took off. On this particular occasion a young woman, sitting in E cabin towards the rear of the bottom deck, had an epileptic fit just as the sea swirled over her window. She must have given the window a sharp blow because next thing it was no longer there and the sea was pouring in. The first thing I heard was a rather alarmed hostess calling out. Once I realised what had happened I called to the engineer who told the pilot to abort the take-off. The flying boat quickly settled back into the sea as I helped the hostess attend the distressed woman until the epileptic fit passed. We were then able to commiserate with the other five passengers in E cabin who were, surprisingly, showing only slight agitation as they sat in about 3 feet of water.

'Fortunately, the water soon drained into the bilges, but we were delayed overnight while the cabin was dried out and a replacement window fitted. The large windows on the Solents also served as escape hatches, but it normally took some force to knock one of them out!'

'On another occasion, during a trans-Tasman yacht race — in 1952 or 1953 — we heard that one of the yachts had broached in heavy seas and among the bits and pieces lost overboard was the crew's one can opener. With all their food and water sealed in cans they were in quite a predicament. On a flight to Sydney we were asked if we could look out for the stricken yacht and drop off a replacement can opener.

'Well, as you can imagine, it was easy enough to suggest we do such a thing, but carrying it out was quite another matter. For a start, we did not even think we would find the yacht and, if we did, dropping a can opener anywhere near it seemed quite impossible to us. But as it turned out, we did find the yacht, and then the fun started. On that flight Ian Patterson was the captain, and Tom Brewer co-pilot.

'A plan had been devised for if, and when, we spotted the yacht. A can opener had been attached to half an old cork lifejacket we found back among the freight. Pieces of rag were cut up and tied on for streamers. We then lowered the small lift we used to send prepared meals between the flying boat's decks. With it wound down to the bottom deck it exposed a hatch we could open to store provisions at a terminal, only on this occasion we were to open the hatch to launch a can opener.

'Holding the package, Tom Brewer arranged himself face down across the flat surface of the lift while I sat on his feet and held his belt to prevent him from being sucked out. I could see through to the flight deck, and Ian, who was to signal at the appropriate moment. I was to tap Tom on the bottom, and he was to let the lifejacket go, hopefully landing the precious can opener just in front of the yacht. It must be remembered we had a full load of passengers while all this was going on.

'Ian positioned the Solent 500 feet above the water and, as he approached the yacht at about 170 mph, Tom opened the hatch and leaned out ready to make the drop. We were greeted by a tremendous rush of air, and a deafening roar from the two Bristol Hercules engines just a few feet away. It was all very unnerving, to say the least. Just then the top of the lift jolted. Unknown to us, it had not been completely wound down and our weight had pushed it the rest of the way, but we weren't to know that: we thought we were falling out. Tom's grip loosened on the lifejacket, involuntarily sending it overboard, as we put all our efforts into staying with the flying boat. Finally we got the hatch closed and sat there recovering from the ordeal as Ian took the Solent back up to its normal height.

'We had no doubts about our efforts being wasted — until next morning in Sydney. The *Sydney Morning Herald* claimed we had scored a bull's-eye, with the can opener landing in the rigging! Not long after that flight Tom Brewer decided to retire to run a motel in Rotorua. Whether the fright he got as he let the lifejacket go had any bearing on his decision, I will never know.'

To another airline now, this time a return flight to TACA, the Central American airline Lowell Yerex founded in 1931. Almost from the start, TACA roused the ire of rivals, especially Pan Am. The latter's links with the US Government put Lowell Yerex on a collision course with the politicians. The result was a long and bitter battle involving the British and American governments, Yerex and his commercial foes. Eventually, he lost control of TACA and decided to launch a new business in Argentina. In 1948, aged 53, he left New York in a Lockheed TI-72 (a commercial variant of the wartime Hudson). With nephew David Yerex on board, Lowell and his co-pilot flew to what was then British Honduras, then down to Panama.

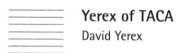

Yerex of TACA
David Yerex

From Panama to Lima, Peru, and then on south yet again. The skies were clear and once my duties as steward were done I could slip forward to the flight deck and wait patiently until someone suggested I take a seat.

Rod Boileau was always generous with the co-pilot's seat so now, south of Lima, as the tropical jungle gave way to the arid coastal belt of South America's west coast, I gazed out at the wide flat horizon of the sea on the starboard side; odd to think that out beyond that pale blue distinction between sea and sky lay the mountainous, twin-island country from which I had come … like Lowell, and Ted Scott, Keith Berry and Ian Cross.

When I turned back, Lowell nodded at the controls on my side and I gingerly took over, as I had done a few times previously. Lowell stood up in the space behind the throttle console and stretched his back, then leaned down and flipped a switch.

For some moments he stood there watching; the sky was cloudless and the coastline we were following to our next stop at Arica, just across the border into Chile, was plain to see.

'All you have to do is watch your air speed indicator and your artificial horizon and follow the coast.'

The plane seemed enormous to the occasional Tiger Moth and Harvard I had briefly trained in, but it ambled steadily along as though having no need of my hands on the controls. The land below, like an enormous relief map badly made, moved with leisurely slowness beneath our wings.

'You got a moment?'

I turned and found Raul in the doorway behind us, indicating some papers to Lowell. Quietly they began to discuss them, and I turned back to watching the scene below … the long straight road across the Tablazo de Ica, a small seaport down to the right with tiny boats clustered behind a breakwater.

When I glanced back, Lowell had turned back with Raul into the main cabin to carry on their discussions, away from the full throb of the engines.

The flying was easy; my confidence grew. Away on the left I could see the hazy outline of the Cordillera de los Andes, rising in a succession of crests beyond Cuzco and, further south, Lake Titicaca hidden under a mist-woven serape.

The coastline in the distance curved off seaward, but I had looked at the map and knew I could continue straight on and pick up the coast again when we … abruptly the engines coughed twice and died. TI-72 slumped in the air and then turned her blunt nose to the ground below. Without power, heavily overladen in typical TACA fashion, she seemed to me incapable of gliding. The ground leapt up at us like the hand of death.

In the main cabin there was chaos as the passengers tumbled from their seats. Cascading crockery sounded loud in the sudden silence of the stalled engines. Lowell, thrown forward, stumbled down the aisle and through the doorway to the flight deck, reached forward and turned on the fuel connection to the wing tanks … the tanks he had been waiting to change over to while the main tank drained, when Raul had interrupted him.

TI-72 spluttered like a swimmer surfacing and gradually lifted her head. Lowell clambered back into the pilot's seat and took over the controls.

'I think we broke a bottle of Coke,' he said. 'Maybe you'd better clear it up.'

At Arica we stopped the night and could have believed ourselves in Egypt … the surrounding desert, the palm trees, and the fishing boats in harbour with their strange triangular sails.

But next morning we flew south again, still following the coast. At

Santiago, Chile's capital, when TI-72 clumped onto the runway and finally shuddered to a stop off to one side of the main terminal, an elegant black Rolls Royce edged away from the shade of a hangar and moved sedately into position beside the plane.

'We'll be staying with friends of mine,' Lowell told me in his usual laconic manner.

So, while the rest of the Equimac team headed for a hotel, Lowell and I swept through the street of Santiago, down Calle McKenna, lounging back in ambassadorial comfort behind the chauffeur and the long bonnet with its famous symbol and a fluttering Union Jack.

'Not bad for a couple of Kiwis, eh?' said Lowell....

Three days later Lowell took TI-72 in a slow spiral over Santiago, gaining height for the risking crossing of the Andes that loomed so close along Chile's eastern border. Ahead of us, the pyramidal peak of Aconcagua rose to block the rising sun ... a black and white spire rising from the roof of the world.

'Close on 23,000 feet,' Lowell told me. 'The big airlines go straight over the top. But they're pressurised. We have to slip through the pass.' He pointed ahead to where the silhouette of the main range dipped at the very foot of Aconcagua itself. The pass, I noted, was still far above us.

'Keep her climbing at that,' Lowell said, and wandered back for a cup of coffee. I sat intent on the rate-of-climb needle and the artificial horizon, and every other dial I could see that might tell me something dreadful was about to happen. The memory of the switched tanks was still fresh....

So now I held with grim despair to the control column, terrified to pull too far back and stall the plane, but fearful too of the fact that our line of flight apparently would end several hundred feet below the approaching pass. Sluggishly TI-72 lifted on the upcurrent of air to the pass, but still it seemed to me we would never clear that massive rock wall.

Now I could see the tiny statue, the Christ of the Andes, that stood sentinel in the centre of the Pass of Aconcagua. I watched the reluctant altimeter needle in growing panic.

It was all very well having faith in Lowell, but by what stretch of the imagination did he have any right to place his faith in me? Perhaps he was getting old … perhaps the altitude was affecting him … perhaps today there was less air current than usual … perhaps the weight of cargo was …

Lowell slipped back into the pilot's seat and casually folded his arms.

'She's not coming up fast enough,' I whined. 'Is she?'

'Pass is 14 thousand,' said Lowell. I looked at the altimeter … 13,700. The statue was now a giant, facing us like awful destiny. The peak of Aconcagua was out of sight above us.

But gradually, painfully, TI-72 lifted herself on the upcurrent, and suddenly the Christ of the Andes was slipping beneath us as I unconsciously, ridiculously, tried to ease my weight off the seat and help the old plane over the barrier.

We were through. I had made it. There was nothing to it, and now I could take her down the long valley opening before us, a seasoned flier capable of …

'I'll take it,' said Lowell. 'It gets a bit slippery here.'

The air currents that had taken us up and over the Pass, now dropped TI-72 with sickening suddenness into the narrow gut of the valley just beyond the crossing. The Lodestar seemed to gasp for air, floundering wheezily as minarets of rock appeared to thrust straight up through the wings. My confidence vanished; my backside crawled in anticipation of the tearing, grinding, ghastly crash to come. I could not bear to look at the rock faces just beyond the window on my right, but I could not look away either. Could the wing tips be anything more than inches from them?

I looked ahead; the valley was widening and the downdraft had eased. Suddenly we were out, flying sedately down the centre of the wide valley between gentle hills that sprawled leisurely down to the vast pampas of Argentina, winging complacently out over Mendoza en route to Buenos Aires.

So at last we came to Lowell's new land …

There's a brief reference in *The New Zealand Book of Events* and four paragraphs in *The Super Men*, a history of New Zealand top-dressing. But, beyond that, nothing more about John Chaytor. And yet he did something pretty ingenious on his farm at Marshlands, near Blenheim. According to *The*

Super Men, John Chaytor '... could see the broad, fertile acres his father had won from the surrounding flax country.' But he could also see '... an area of swampy land which he knew could be developed. His problem was how to develop it.'

For a brief period in summer the land was dry enough to sow but the sun's heat meant '... any seed would be burnt ... before it had time to germinate. For the rest of the year, the land was too wet to work on. John Chaytor solved the problem by using a hot-air balloon tethered to the ground. Guided by ropes manipulated by brawny farmhands, the gas bag was steered systematically over the area while the seed drifted to earth. That was back in 1906. The operation was probably the first aerial agriculture ever carried out in the world.'

Then came the aeroplane, first put to agricultural use in America in 1921 when an infestation of pesky caterpillars was dispatched by a pilot sprinkling lead arsenate powder from a war-surplus trainer.

Twenty years later, the New Zealand authorities conducted aerial seed sowing trials along Ninety Mile Beach and followed those, in 1946, with experimental top-dressing runs in the Thames area. And that was the start of an industry.

In the early days, most top-dressing pilots flew Tiger Moths — because they were available. Austers were also used and then, late in 1954, the first American Fletcher FU24 — built to New Zealand specifications — arrived in the country. It was subsequently built here and in 2003, Pacific Aerospace in Hamilton flew their prototype PAC 750XL, a much-modified offspring of the old Fletcher, which can carry 10 passengers or 17 parachutists.

Mind you, it probably won't be using the rudimentary backcountry airstrips that come with the territory for a top-dressing pilot. New Zealand's weather and terrain — and human obstacles — mean top-dressing has always been a unique (and risky) business.

Flying Low
Roger Crow

Within five minutes I could not see it at all for, as we travelled inland, the overcast changed to a partial fog. In the Tiger I sat hunched against the cold, warily peering through my goggles as visibility closed down even more. Suddenly I realised I was almost brushing the tree tops, and the situation had become critical. I started to turn on to a reciprocal heading which would return me to the airport; a manoeuvre which should have been performed minutes earlier when the air was reasonably clear, for now I was definitely in a dangerous corner. I reduced the motor's power to slow the Tiger — visibility was now only a few hundred yards. My pulse rate leapt as a giant eucalyptus tree appeared out of the fog dead ahead and seemed to fill my whole vision. I slammed the throttle on to full power and rolled the aeroplane into its maximum rate of turn. The hideous monster slid harmlessly by. I levelled and crossed a paddock. It was too small to land in; they all were in this area. More trees! This time a plantation of pines. They appeared as an enormous blob of darkness in the fog at first, then materialised into feathery branches! Another bellowing turn. By now I was close to panic. This mad style of flying could only last a minute or less. With the trees I had been lucky, but there could be high-tension power wires in the area making an invisible trap. I had to get down, yet there was nowhere to land. I also forced myself to think logically, for panic resolves nothing. There appeared a certain lightness to the left. I banked towards this as a drowning man clutches at a straw and within five seconds the Tiger emerged from the mass of ground mist that had so nearly caused disaster.

I almost shouted with relief and climbed to a safe height of two hundred feet above the farms and their shelter belts of trees. The overcast was still present, its greyness merged with the horizon in every direction so that I seemed to be flying within the confines of a huge low-roofed cavern, whose floor was three miles of flat, colourless farmlands. In this restricted area I would leisurely inspect the paddocks for a suitable landing field. The fact that I was totally lost meant little. The main thing was to return to earth safely and gently. I could always wait for weather improvement and Brem should be impressed if I telephoned to say I was fogbound but intact. I started to look about, then spotted Pete Browne's Auster. It appeared as a miniature aeroplane at the other end of the cavern perhaps three miles away. It was turning and slowly settled into a paddock there. Indeed, fate was pushing kindness at me from every direction, for I was not lost and there was a landing field. I altered BNB's course towards it and sang 'I am as free as the breeze' at the top of my voice. The Tiger's throbbing snarl, the banshee wailing of the wind in the wires and the buffeting slipstream clawing into the cockpit destroyed most of my song and I soon gave in, for the machine was rapidly closing the intervening space between me and Pete's aeroplane now parked under the far trees of the field.

The medium-sized landing field was an oasis in a sea of small paddocks, all of which had tall poplars, pines and willows to shelter them. Several farmhouses were also dotted about. One of these was beside the distant parked Auster and in the morning's dullness, its kitchen window was yellow with light. I remembered the freezing air and thought fleetingly of the steaming pot of tea which would be there on the stove. Pete Browne had chosen this paddock well. I slowly eased back the throttle and the motor died with muffled backfires, the Tiger glided easily and speedily through the smooth air, its wires ceasing their intense wailing as the machine lost speed. A line of telephone poles and a road led to the house and I raised the nose to cross these at a safe height.

I did not see the power wires until they were a few yards in front of me. The tiger's long nose slid over the three gleaming cables in a second, and I was as fascinated with their sudden appearance as is a rabbit with an approaching snake. During the inevitable collision, the Tiger lurched in the air, and as the wheels gathered them together, the cables crackled and fused in a gigantic blue flash under the lower wing. It lit for an instant the fences and narrow road below, and a tall concrete power pole beside me began to crumble. What a fool I had been to be lulled into such a false sense of security for I had failed to see even one of those towers! And now I was a mere spectator involved in uncontrollable events. BNB stopped in midair, tipped forward as it shook itself free from the cables, and dropped forty feet vertically into the paddock beside the road. There was a bang similar to the sound of empty petrol tins being crushed, and a cloud of fertiliser dust puffed from the Tiger's every pore. Mortally wounded, she balanced on her nose for several seconds then slowly crashed upside down. She could not fall any further for she had reached the end. Then there was peace.

A few inches from my goggles lay the damp grass and some dried sheep manure. This and the turf smelt good. It was confusing to be sitting upside down hanging from safety straps. I dropped my hand down which would normally have been reaching up, to feel the damp grass. I was down at last, and at peace.

Then something thumped on to the ground beside me. I swivelled my head to observe the tip of a broken wooden propeller, which must have been hurled straight up and had only just returned. This was the Tiger's feeble finale and I frowned as somehow it reminded me of something unpleasant. Reality returned, and I sighed. Brem's wrath was going to be colossal over this episode. Then I stirred as the thought that I was sitting in a potential bomb entered my head, and I peered forward over the somewhat crumpled sleek lines to the nose where smoke was curling lazily from the motor. This was not serious as oil had splashed over the hot engine. None the less, this was not the place for a rest, and I gingerly levered myself out from underneath. All my joints were in working order. I stood up. Pete Browne, his driver, the farmer and his wife, were all running towards the wreck.

After the war, Miles King launched Rural Aviation, one of the first top-dressing companies in the country. He saw the industry grow, knew its characters and, years later, remembered them well.

The Sky My Canvas: An Autobiography
Miles King

Memories are made of these:

Of Goosegg (Don Erceg), with now over 20,000 hours of topdressing behind him, each hour equal in skill, risk, and concentration to five hours airline, making his first takeoff of the day from a valley bottom airstrip that opened on to the sea in North Taranaki. Crossing the beach as he climbed out, he found a Japanese mother ship, fish hold open, loading fish from fourteen dories less than a kilometre from the coast. Incensed, he instinctively hauled his 180 into a maximum rate climb, and at 1,5000 ft pointed down on to the mother ship, nose directly on the open hold. Like Jonathan Livingston he roared down like a bolt from the blue, and as he pulled out but a few feet from the ship, he dumped his three-quarters of a tonne, straight into the hold.

He then returned to the strip, climbed out of his aircraft, and reported to the farmer that he had wasted a load. The farmer asked him why.

Goosegg told him and the farmer, without hesitation, instructed him to repeat the performance. So away he went again, but this time as he crossed the coast the hold was closed, the dories aboard, and the mother ship was steaming for Japan with a hold full of contaminated and useless fish.

Or of topdressing some country immediately south of the Waitotara River, very early in 1950 when aerial topdressing was still relatively unknown. I was working the downhill side of a steep slope that ran down to the river, and a car came winding up the old road and around a bend, to be confronted with the horrific sight of a Tiger Moth curling down the hillside almost on top of him, spewing phosphate from its belly. The car, in total surprise, left the road and went through the fence and down the hill. I finished dropping my load and returned to the airstrip, where we sent out a party and discovered that nobody in the car had been injured.

Or Douglas Bader, immortal hero of the Battle of Britain, visiting Palmerston North where, in 1956, the Aviation Industry Association staged the first agricultural aviation show on earth. With sixty aircraft flying and many static displays, the show attracted huge crowds and the AIA organised and ran the entire three-day event, even replacing air traffic control normally run by the Department of Civil Aviation. In fact, the department's only appearance was in its communications DC3, which overshot its landing on greasy grass, ploughed through a fence into the public car park, and severely damaged some vehicles — and the aeroplane.

Douglas Bader watched the show, came over around our stand, and told me that he would like to see some aerial topdressing in action. With a little manoeuvring I bent his tin legs into a 180 and we took off for a live demonstration that we kept operating a few miles up the road during the show, lined the aircraft up to accept a load, and very quickly Douglas and I were topdressing. Seven runs we did, and returned to Palmerston North. I helped Douglas out of the 180, and his only comment was, 'I thought I'd done some dangerous flying, but never anything like that.'

Dangerous flying was commonplace for two southern men. Friends (and rivals) Ron Bush and Bill Hewett did just about everything with their planes: carrying passengers and cargo, instructing, delivering freight to remote parts — and top-dressing. They also raised numerous children — and did the occasional maternity run.

Famous New Zealand Aviators
John King

All the Bush children were born at the maternity hospital at Mosgiel, all attended by the same doctor and midwife. Number three, Stephen, had been an exciting arrival. Without a car to transport Pam once labour started, Ron Bush had the idea of putting her into the aero club's Tiger Moth, ZK-AKH, and taxiing across the aerodrome from the aero club, where they lived, to the hospital on the far side. Getting a heavily pregnant woman into the small cockpit of a Tiger Moth in the dark was no mean feat in itself, but together they managed.

In his rush, however, Bush set the throttle too far open, and when he swung the propeller the Tiger Moth started with a roar. Holding the wingtip of a rampant biplane and with nobody around to help so late at night, he was faced with the very real problem of how to stop things without injury to his wife or damage to the Tiger Moth — or club buildings or any of the other aircraft tied down nearby. In the end he resolved it all and delivered Pam to hospital, and Fred Ladd, stationed at the time at Taieri on the NAC Dominie run to Invercargill, became Stephen's godfather.

Bush was forever attracting the eye of the police and Air Department officials. On the occasion of the birth of another young Bush at Mosgiel he put on a particularly spectacular display of aerobatics directly overhead and, having landed, walked in to the hospital for a visit, covered from head to foot in white superphosphate dust. The police duly arrived to see about this highly dangerous event, but the hospital staff persuaded them that Mrs Bush would be terribly upset if they arrived inside in their uniforms and arrested her husband. Bush sneaked out the back entrance and into a taxi back to the aerodrome, but they were wise to that and were waiting for him when he landed back at Gore.

He offered some novel excuses for his misdemeanours, some of which were successful. On one occasion he was found to be carrying his loader driver in the hopper of the Tiger Moth with the lid shut, a common enough but illegal practice. Charged with the offence of having an unauthorised passenger in a single-seat aeroplane, he persuaded the magistrate that he was giving his crew member blind flying practice and was let off.

Years later Bush decided that Bill Hewett needed to be rescued from hospital, where he had been receiving treatment for a stomach ulcer. At that stage the new Dunedin airport at Momona was nearing the end of construction, so they thought it would be fitting if Hewett and Bush opened Momona. They landed there in the Cessna, but somebody took a note of the aircraft registration and Bush was once again up before magistrate J.D. Willis, a familiar sight.

His excuse this time was that they had made an emergency landing because one of the seat belts was hanging outside the door, flapping on the fuselage. That brought the suggestion that surely everyone needed to be strapped in securely before takeoff, but the persuasive Bush stated that NAC air hostesses were not strapped in for takeoff. Again he got off....

Hewett's own entry into aerial topdressing met with some initial reluctance from those he tried to persuade to back him, who had heard he was keen on chasing ducks up creeks. However, the company was formed at Mossburn, and Hewett started topdressing in August 1950 as Southland Aerial Fertilisers, the name changing after four years to Hewett Aviation. Tiger Moths were used at first, but in 1953 Hewett bought one of the first of the new Cessna 180s to come into New Zealand for aerial topdressing. ZK-BDD became his personal aeroplane, always his favourite, and most of the Cessna stories about Hewett concern that aeroplane, which is now based in the Awatere Valley, Marlborough, with Frank Prouting.

Throughout his topdressing career, which lasted until April 1967, Hewett continued his air freighting and other aerial work. In 1951 he bought from the RNZAF a Miles Aerovan, a tadpole-shaped aeroplane marginally powered by two 155 hp Cirrus Major engines. It was capable of carrying up to one tonne or six passengers, although Hewett was reported to have flown a complete rugby team to Harewood, Christchurch, without bothering about seats for anybody except the pilot.

He used it for his whitebait flights and flew fruit for Central Otago farmers, one day encountering a sudden down draught while carrying cases of tomatoes without lids, with the result that he found himself with a loose cargo of tomato purée. The Aerovan was also used for dropping supplies or parts of bush huts into remote areas, when the despatcher, secured by a line, would throw the material out the rear of the cavernous fuselage, the doors having been removed.

One day he was dropping material for a hut on a distant part of Nokomai Station in the Garvie Mountains with despatcher Malcolm Campbell. The Aerovan struggled up to a ridge in the down draught with marginal power, and once through that piece of turbulence Campbell came up to the cockpit with a very white face and said, 'I've been out.'

'Bull! What do you mean, out? How can you have been out when you're here now?'

But Campbell meant just that. For some reason he had no harness securing him to the aeroplane and a bump pitched him out the back into midair. But he managed to grab a small lip beside the door opening and, with the help of a bit of reverse flow of air, was able to haul himself back inside by his fingernails. They landed and Campbell went away to be sick, then said he had better go up again immediately or else he never would.

Like the other Aerovan, which was overturned by a gust of wind on the ground, Hewett's example ended its days inverted. Hewett had flown a group of hydro station workers from Roxburgh over to the West Coast on a scenic trip and, not being the sort who worried about weather forecasts, was caught out by a cold front when he arrived home at Mossburn at nightfall after depositing his passengers at Roxburgh. He turned round and headed north again and, with the lights of Ranfurly a welcome beacon in the gloom and murk, was landing in a paddock adjacent to the town when he went through a set of power wires. That was the end of the Aerovan.

It was far from the only aeroplane Hewett inverted on the ground. He took off from Te Waewae Bay in a Cessna 185 after a brief landing, as usual with one of his children in the back, behind the hopper, and a couple more in the front. Reaching back for his customary bottle of beer, he found it was wedged under the hopper and rather than face a dry flight back to Mossburn, he landed on the beach again.

But the Cessna was caught by a wave and flipped. 'The tide was coming in and we couldn't get it back on its feet, so a Land Rover came and tried to help us and it got bogged,' says Brian Hewett, one of the four sons in the family. 'A crawler tractor came out and got the Land Rover out and got the aircraft back on its feet, and we pushed it up onto the dry land.

'Mum was there so we drove home. Dad went back with a new propeller and new oil, changed the propeller and put new oil in the engine and flew home. He flew it home on his own over a route with landing areas the whole way. Upside down in salt water, the wings were a bit creased. I think it cost about £7000 to fix it.'

The West Coast is a fabled place that can rightly claim to be the home of New Zealand's first licensed, scheduled airline: Air Travel NZ, airborne in December 1934. Aircraft were an enormous boon to remote outposts like Haast, which, back then, had few links with the outside world. Air Travel flew on until 1947, when it became a nationalised part of NAC. The name may have changed but the same graceful little de Havilland biplanes kept flying.

Flightpath South Pacific
Ian H. Driscoll

The aircraft used were Fox Moths and Dominies, respectively single and twin-engined aircraft, though the latter had no performance on one motor. I remember boarding a Fox Moth at Hokitika with two other passengers. The freight was then loaded in our laps — bread, parcels, tyres, the lot. There was not much chance of getting out till the cargo and mail had been unloaded!

Scenically the route is one of great grandeur. The morning was brilliantly sunny. We had to buzz grazing sheep off the strip at Okuru before landing. There was a road linking Haast with Jackson Bay, but it had no contact with the rest of New Zealand. But standing on the airfield with the shining waters of the Tasman Sea on one side, untamed bush flowing away from it to the mouths of great valleys like the Landsborough which emerged from a hinterland of snow-capped Alps on the other, I was instantly struck by the fresh clearness of the air. There was an immense stillness and a soft peace about the place which I have never encountered elsewhere.

The south Westland route was a world of its own in every sense. At one stage we had a relieving traffic officer at Hokitika who stammered. He flew to Haast one day in a Dominie and during the flight one of his fellow passengers died. Gavin Grocott, assistant general manager of NAC today but Wellington district manager at the time with control of the service, instructed one of his traffic men to telephone Hokitika for details.

'What did you do, Bob?' he asked the relieving officer.

'W-w-well, th-th-there are n-no c-c-coff-coffins at 'Ahst,' he was told.

'I realise that, Bob. So what did you do?'

'I-I re-reticketed h-him, "M-M-Mr B-Bill S-S-S-Smith (d-d-de-deceased)" back t-to Oh-ki-t-t-tika a-an' t-turned him raund!' ...

There have been lighter moments, too, on the trunk route, such as the hostess of an aircraft flying off the Seaward Kaikoura mountains who, asked by a passenger with an historical bent whether Captain James Cook was the first European to sight the range, answered: 'Well, I don't really know, sir. I've never flown with him.'

Or the youthful first officer visiting the passenger cabin who found a woman suffering with her ears. He advised her to hold her nose and blow gently. But he couldn't leave well alone. 'It will clear your fallopian tubes,' he added.

A doctor sitting across the aisle leant over and said: 'Young man, if what you say is correct, you are on the verge of making a fortune!'

NAC didn't last long on the Coast. In the fifties they retired gracefully and West Coast Airways took over their routes. Things were pretty informal at West Coast Air — the rule was No Uniforms rather than No Smoking. Indeed, Paul Beauchamp Legg recalls a fellow pilot tapping his pipe on a Dominie's open window, expecting his plug of baccy to fall safely to earth. Instead, it wedged in equipment on the fuselage. With the plane's fabric highly flammable, disaster seemed imminent. Happily, the tobacco burnt out — crisis averted. But there always seemed to be something afoot at West Coast Airways.

West Coast Memories, Volume One
Paul Beauchamp Legg

On the 16th of December 1959 the Dominie was almost loaded at the Haast. Mrs Cron was sitting in the front left seat nursing her new baby when Shakey Joe clambered up the steps followed by his monstrous black dog Bulldozer slobbering good will and saliva as Shakey Joe worked his way to the rear seat. The lightest always had the rear seat.

Bulldozer, wagging his trunk-like tail, landed a big slobbery lick on Mrs Cron's legs and smiled up, expecting a friendly pat. Bulldozer loved all people, but Mrs Cron did not love all dogs.

'Take that dog off the aeroplane,' she shrieked clutching her baby before Bulldozer could express his love in that direction. 'It's not right. We can't have a dog in here with us,' she protested.

I had to agree. Mrs Cron was correct. Not all passengers with new babies appreciated a big slobbering dog.

Bulldozer was not only a friendly dog, he smelt. Not really bad; it was mainly his breath.

'Sorry Joe,' I said. 'Bulldozer will have to wait until I come back in the afternoon. Do you mind if I bring him then?'

'Aw. Yeah. O.K.,' agreed Shakey Joe easing himself out of his seat and past the other passengers. 'C'mon dog,' he said with a touch of emotion. 'Y've gotter wait.'

Bulldozer was tied in a warm sheltered place by the Haast reception shed, we could hardly call it a 'terminal building' although for us it was as good and important as the swankiest city airport terminal.

We flew to Hokitika in the normal hour and ten minutes.

At Hokitika Shakey Joe agreed to collect his dog that afternoon about 2.45 p.m.

Back at Haast with the second flight, I had no problem getting Bulldozer into the Dominie. He smiled happily and followed me inside the aircraft with his tail wagging and big friendly grin on his face. A pat or two and I had a friend for life.

I tied him in the back to stop him coming up front to help me fly the Dominie. He settled down and slept until we landed at Hokitika where he wagged his tail with delight and trotted down the steps to cock his leg at the nearest post. Thereafter he waited patiently, as did I, for Shakey Joe.

Came five o'clock but there was still no sign of Shakey Joe. He had disappeared, West Coast-style, into a pub, and forgotten all about time and dog.

I climbed onto my push-bike holding Bulldozer's rope. 'Come along boy. I'll take you home,' I said but Bulldozer had other ideas. As soon as he thought he knew the direction we wanted, he took over and towed me the half mile to my home in Hampden Street....

Next morning Bulldozer towed me back to work. He was surprisingly fit for such a fat dog. I liked him despite his appetite.

There was still no sign of Shakey Joe when I took off on the scheduled mail flight, nor when I returned, but that afternoon, right to the minute, but 24 hours late, Shakey Joe arrived completely oblivious to having lost a day.

'I couldn't have slept in,' he said. 'I haven't been to bed.'...

It was my weekend off. There being no Aero Club bookings for the afternoon I took my family to Lake Mahinapua, a popular picnic place a few miles south of Hokitika.

A picnic with my family was a rarity and today was warm with the sun shining. Lake Mahinapua was a lovely place fringed with bush and many picnic spots. There were small boats for hire, something we all had looked forward to enjoying. It would be a great family outing together.

We had only just arrived when a man came around frantically looking for me.

'Are you Paul Legg, the pilot?' he asked. I admitted that I was.

'You are wanted at the 'drome, there's been an accident at Haast,' he said with dramatic urgency.

That was the end of our picnic.

It was Ken Eden's turn on stand-by duty, but he had already been called out for a flight.

The information I received was somewhat vague. The messenger could not tell me more than that someone had been injured and needed to be flown out to Hokitika Hospital. On whose authority we did not know.

The injured man was neatly bandaged and lying in the Haast office.

He was wearing only a small pair of Jockey V underpants. It had been a hot day but by now it was cooling.

For want of a better name I will call him 'Jan.'

Early in the morning when out in the bush where the new Haast road was being built, 'Jan' had fallen from a pole and run a long spike up through his groin. There was some thought that he had lost his manhood.

The Ministry of Works team with whom he was working attended him as best they could and brought him out through the Haast river on an Abbey waggon, a big trailer affair fitted with tracks. At Haast he was attended by one of the ex-nurse wives, probably Mrs Buchanan.

When I arrived 'Jan' had no other clothes with him. Possibly his trousers had been removed to get at the wound. It was summer time and very warm when the accident occurred to him and it was common for men, particularly immigrants from Europe, to strip off when working in the warm sunshine. Seemingly 'Jan' had been brought out 'as is' but by late afternoon when I arrived the day was cooling. Once 'Jan' was loaded into the Dominie I stripped to my underwear and gave my clothes to keep him warm. I was decidedly cooler flying back to Hokitika than I had been going down to Haast. At Hokitika I had to go to the hospital in the ambulance to collect my clothes....

The engineer taxied the Dominie to the tarmac in front of our office. He had already removed some seats to make room for the mail, meat, groceries, paper, chain saws, and whatever other freight was wanted at the Haast or along the route.

This was a regular scheduled flight so I flight-planned as West Coast 691 going south and West Coast 692 returning North. Air Radio and Fire Service would be in attendance. The next flight along the same route, with perhaps the left over freight and passengers, would go as a charter and needed no official attendance. This I could never understand but back to West Coast 691.

Having loaded and signed the load sheet made out by our optimistic manager who figured that the pilot never really knew how much he was carrying, and a few extra pounds made it more profitable, I made a preflight inspection and loaded the passengers. 'Heaviest to the front. Lightest to the rear.' This method got the tail up quicker and kept the centre of gravity nearer its authorised limits.

I climbed in and checked the passengers' seat belts were done up as I squeezed my way up to my cockpit. At the cockpit doorway I heaved myself into the seat and sat comfortable in my own little world, a big window on each side; the nose pointing to the sky and the passengers strung out behind. The pilot of a Dominie was very conscious that in the event of any error on his part, he was the first to get it and possibly the last to get out. I know of no Dominie accidents on The Coast.

Back to this particular morning.

I had one passenger who had come for the round trip. He sat behind me so we could talk and I could show him the sights along the route.

We took off and made a low run across Lake Mahinapua, a pretty place with bush on either side. A slight turn to the right took us along the surf to the seals on the beach just north of the Wanganui river, then we turned slightly left again to see the white herons nesting. It was not then illegal to fly near the white herons. I kept a look-out for them and reported their arrival and nesting to some official person. Nowadays aeroplanes are not allowed within about five miles of the white herons.

After seeing the white herons we made a low run across the black water of Lake Okarito to see the Alps reflecting in the water, then I made a slight climb to maybe a thousand feet to cross the bush before landing on the grassed, banana-shaped, curving Franz Josef Airstrip. This airstrip was in an idyllic setting. Later it was metalled to give us a firmer surface. Sadly it has since been washed away by glacial floods and strewn with boulders.

As with Fox Glacier airstrip, irrespective of the wind, landings at Franz Josef were always done towards the mountains and take-off the other way. Fox Glacier had a 1 in 30 slope on it with an abrupt mountain at the end. There could be no second chance if you came in too high or too fast. Franz Josef, whilst less dramatic was in fact almost the same. (I did once have to sideslip a fully loaded Dominie into Franz from the other end, but that story will have to wait.)

Our regular scheduled flight operated to a strict timetable, West Coast Time. (When everybody was ready!)

I was supposed to pick up two men at Franz Josef. They had been three weeks driving the cattle from Haast. This was before the Haast road was built. Haast was part of the West Coast.

Cattle driving was a hazardous job. Rugged tracks wound through thick bush. There were bluffs and rivers to cross. Three weeks was the normal time to round up the cattle near Haast and travel about 60 miles. Some cattle were lost en-route.

There was no sign of my two passengers.

The scenic passenger and I walked up to the temporary bar which had been a temporary bar for many years after the hotel had burnt down.

Garth Graham, the manager, had not seen nor heard of my passengers.

I sent the scenic passenger off to have a look at the church with its lovely view of the glacier behind the altar while Garth put on a cup of tea.

'Maybe the telephone exchange lady knows where they are,' I suggested, knowing that one call on the coast party line phones alerted everyone. I phoned her.

'No. I haven't heard of them,' she said, then added helpful-like, 'Maybe they are over at the Fox.'

I was just contemplating flying over the 365 bends in the road to The Fox when the lady from the telephone exchange called back to say that she had just seen the Doctor go past.

'He's no use to me. It's not his week to go down,' I replied.

Doctor Hogg's dirt-splattered car shot past and down to the airstrip to reappear at the bar doorway a few minutes later.

'I've dropped your passengers off,' he said cheerily. 'Shall we go and get the dogs now?'

'Dogs?' I exclaimed.

'Yes. Dogs,' he replied as if I should have known better than to ask a silly question.

I climbed into his car. The scenic passenger returned and wandered back to the airstrip while Doctor Hogg and I went off to collect two large cattle dogs.

We pulled up beside the Dominie down at the airstrip. Three cattlemen were pleasantly drinking beer. They had a couple of sacks of beer bottles, it being easier and lighter to carry sacks than crates.

'Have a beer pilot,' they offered.

'Don't drink,' I replied eyeing the three new passengers, plus two cattle dogs.

'Well that's all five of us,' said Dr. Hogg unloading his medical bag from his car and walking to the Dominie.

'Three,' I bleated.

'No. Five,' he corrected moving towards the loading step.

'You haven't booked. It's not your week to go down Doc,' I almost pleaded.

'You know I didn't go down last week, so I'm going today,' he replied quite firmly as if I was a bigger fool than I seemed.

I turned to the cattlemen, still happily drinking.

'Look,' I said. Doc has to go down, would a couple of you mind waiting until this afternoon? I've a whitebait charter after the run.' I figured they were happy; the sun was shining and after three weeks away, what was another few hours to a Coaster with a sack of beer?

'To hell with that,' they replied.

'I haven't got enough seats,' I explained, thinking to win this time.

'Don't worry about that,' replied the first cattleman to poke his nose inside the Dominie. 'I'll sit on that keg of beer.'

'And I'll sit on the floor,' cheerfully called the other. There was a spare seat already piled with freight that could be rearranged to take number three cattleman.

'But I can't take it all,' I pleaded. 'We'll be too overloaded to get out of this strip.'

'Well leave the dogs behind and just take the beer,' concluded one of them.

I wrote down the names of the dogs in the hope that they would treat me with respect when I returned in the afternoon, tied a piece of rope to the rear of the Dominie for the extra passenger sitting on the floor to use as a safety belt and off we went. We were now about three-quarters of an hour behind schedule.

'Haast Radio. This is West Coast 691. Airborne Franz Josef. ETA Haast …' Then as an after-thought I added, 'I've got Doc on board. Over.'

Brownie acknowledged. He would phone two or three strategically placed Haast residents and soon everyone would know that Doctor Hogg was on his way down. Those needing his attention would arrange to meet him. All was well. Well one would think so.

I landed at Haast well behind schedule and unloaded.

'No mail this morning?' I said to Ivy Farmery, the lady who had the contract to bring the mail and freight from the Haast settlement three miles away. Ivy was also the official boatwoman to ferry people across the Haast River, but there is another story.

'I heard that you had the Doctor on board. You always wait for the Doctor,' she replied somewhat smugly with a smile on her face.

'Ivy, you know I have a whitebait charter this afternoon and have to catch the Christchurch train,' I replied.

Ivy shot off in her little Fordson van and I went for the obligatory morning tea with old Mrs Cron.

No matter what the circumstances, or how late, Mrs Cron expected the pilot to come in for morning tea. She cooked over an open fire, big black kettle hanging over the fire, or sitting on the hob. Her camp oven for cooking the five or six different cakes, scones and pikelets she had ready, snuggled down amongst the embers.

The Crons were one of the original pioneer families who made the Coast what it was.

Mrs Cron passed on the latest news and looked for news from the pilot. It was difficult to get away from one of Mrs Cron's morning or afternoon teas under half an hour. This morning was no exception.

I went back out to the aerodrome as Ivy shuddered to a stop with two passengers and their luggage. Ivy also had had a cuppa at the Haast.

'Light load of mail this morning,' I said wondering why there were no mail bags.

'Oh. Those horrible Ministry of Works men working on the bridge!' she exclaimed. 'They shouldn't be allowed there when I am going past. I was so concerned about them I forgot the mail.' With that she shuddered off again to the Haast post office.

I waited. Once Ivy had returned and the loadsheets were completed, we took off for Hokitika. We were now about an hour and a half behind schedule.

At Hokitika the Dominie was unloaded and refuelled then re-loaded with the balance of the freight and some mail that had accumulated. As many whitebait tins as practical were put in on top. I collected some sandwiches to eat on the way down to Franz Josef where I loaded the two dogs into the rear of the Dominie and took off for the Haast.

Once airborne at Franz, I advised Haast. Brownie asked for my estimated time of arrival for the load of whitebait. Knowing the whitebaiters to be notoriously late, I gave an ETA well ahead of the real estimated time of arrival, this meant that we should both arrive at the drome about the same time.

I landed at Haast to find that the dogs had been sick over the mail bags.

I had the bags and the aircraft reasonably clean when Des Nolan arrived with the whitebait.

The whitebaiters knew the load the Dominie could carry but in the hope that the pilot would not notice they brought out a few hundred pounds extra. It kept the price of the charter down.

All this meant some rapid mental arithmetic as the 4 gallon tins of whitebait were unloaded from the truck, lifted into the Dominie, then carried up into place. Half to three quarters of a ton of whitebait become progressively heavier as the plane is loaded.

Ultimately the whitebait was aboard and I was airborne. It took an hour and ten minutes to reach Hokitika. Long before then the train for Greymouth and Christchurch would have left but no one but me was worried.

I called our office on the radio. Merv advised the carrier. The carrier sent his fastest truck to meet me at the aerodrome.

I landed and rolled to a halt near the office. We quickly unloaded onto the truck. The truck shot off after the train. Abreast the train the truck driver hooted and waved and the engine driver stopped the train and took the load.

The whitebait reached Christchurch on time. Those were the days when service counted....

Myrtle [Cron] was a crack shot with a rifle. On one occasion she was attending to an American visitor in the West Coast Airways office on the Haast aerodrome when a stag walked out onto the runway. Myrtle reached over for her rifle, stood in the doorway and fired one shot which dropped the deer immediately. She placed her smoking rifle back in the corner, leaned on the counter, picked up her pen and continued ticketing the amazed American.

Another Dominie pilot, Brian Waugh, was born in Shropshire and flew with 75 (New Zealand) Squadron during World War Two. Hearing his Kiwi crewmates singing the praises of home roused his curiosity. In 1954, anxious to quit England, he decided to emigrate here, and subsequently worked for three small private airlines, South Island Airways, Trans Island Airways and West Coast Airways, on both sides of the Southern Alps.

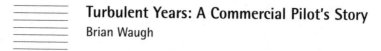

Turbulent Years: A Commercial Pilot's Story
Brian Waugh

With any job there are always flies in the ointment, and the nor'west wind soon became the bane of my life, especially during the summer months. One Saturday Chadwick and I took both Dominies to an air pageant at Timaru. It was very rough over the Canterbury Plains, and Chadwick had the misfortune to burst a tyre on landing. Johnny Neave, the chief flying instructor for the Canterbury Aero Club, came over to commiserate with us and I remember him explaining to me about the nor'wester.

'It's quite frightening,' he said. 'It sweeps across the Tasman, hits the Alps, deposits all the rain on the West Coast and belts across these plains as a hot, dry, föhn wind. It's a beaut.'

'Sounds just like the mistral in France,' I said.

'I wouldn't know about that,' he said, 'but the turbulence is quite fantastic. You won't enjoy it, I can assure you.'

Johnny was dead right, because the first time I flew to Nelson with Chadwick we struck the daddy of them all. Flying via the Kaikoura Peninsula, the Dominie took a hell of a battering, and I hoped the termites were holding hands in the main spars. At one stage a particularly savage gust blew the aircraft almost completely upside down. It was interesting to note the reactions of the passengers. Two old ladies looked as though their last moments had come, two gents were obviously experienced in New Zealand flying, and a boy of about twelve had his head permanently in a sick bag from Seddon to Nelson. Flying between Havelock and Nelson at about 9,000 feet, I experienced turbulence I had not known before in Europe. When we finally landed, Chadwick asked me what I thought about it. Always one not to muck about, I said, 'If that's the normal run of things, it beats me how we ever get any passengers.'

'Oh, Brian, that was hellish, I admit. I don't normally fly on — I wanted to show you things at their worst. I really should have landed at Woodbourne and pushed the passengers through by taxi.'

'But that would be fairly expensive, surely?'

'Oh, don't bother about the expense — it's not our money.'

'No, but you know as well as I do, Brian, it's the economics of the thing.'

He laughed. 'You're a typical bloody Pom — always thinking of the lolly.' ...

We flew back to Christchurch via Kaikoura and landed at Harewood in a 40 knot wind. (I don't need to say from what direction it was blowing.) Flying back to Oamaru it seemed just as turbulent, and the stench of sick bags wafting through the cabin just about flattened me. Landing at Timaru, two youths quickly disembarked, no doubt thankful to be down in one piece, and then one old chap, very distressed, holding two sick bags absolutely brimming over, managed to splutter out, 'What do I do with these?'

'The best thing to do is to go over there and sling them over the hedge,' I said. (This was before Timaru had a terminal building.)

'Not bloody likely,' said the old boy, 'my false teeth are in them.' ...

ZK AHS became my regular steed, and as it was lighter than the Dominie it provided a greater payload. It was a very good old aircraft, with a tight feel and smooth controls. However, I was always conscious of its being a 1938 model: after all we were now embarking on a new decade, the 1960s. Sometimes I even found myself wishing for a Lockheed Electra or Beech 18 or maybe even a new twin-engined Cessna. But for our small airline, that was out of the question for financial reasons, and we soldiered on for many more years with the reliable biplane De Havillands.

Although I possessed the aircraft engineer's licences, it was Des Wright our mechanic who carried out most of the routine maintenance work on the fleet, and he did a grand job keeping the aircraft in tiptop condition. Des owned his own Auster Autocrat ZK AUO which the company often hired to bring passengers from Greymouth to Hokitika, where they would embark on the Dominie for Haast. This Auster had a bad habit of giving an occasional cough and then picking up again. I asked Des if he had noticed this when he flew it, but he passed it off quite lightly: 'Yes, it's icing — nothing to worry about.'

Almost every time I flew it there was this cough, even when I had applied the hot air control. One particular morning, 21 March 1960, I landed the Auster at Greymouth to pick up Lew Tuck, a logging contractor who lived in Greymouth but had a work contract at Haast. Lew had spent most of the war in a prison camp in Germany. He was always a cheerful fellow and one of our most regular passengers.

'Gidday, yer old sod, how are you?' he shouted as he opened the fragile Auster door.

'Hop in, you old bugger, I won't stop the motor,' I replied.

He threw his bags into the back, heaved himself into the seat alongside me and slammed the door shut. We taxied to the far end of the field to take off over the town. He glanced up at the perfect blue sky: 'Looks like it might rain.'

'Does it,' I said, astonished, not knowing whether this was one of Lew's leg pulls.

'Sure,' he said. 'On the Coast it's either raining, just stopped raining, or looking as though it might rain, so on a day like this it looks like it might rain.' He roared at his own corny joke, and I couldn't help joining in: the warmth of his nature was infectious. I was ready to take off. 'I don't like these one-lung bastards,' he went on, pointing a blunt finger at the single engine. 'Always feel safer in the Dominie.'

I was soon airborne and was over Mackay Street at about 600 feet when without warning the engine coughed, and the prop hesitated and twitched to a stop. We looked at each other and said in unison: 'Bloody engine's failed!'

I hurriedly checked the switches and fuel. 'Land straight ahead if the engine fails,' was what they always said in the RAF. 'Never turn back to the field.'

We had little altitude, and there seemed nowhere to go — houses, or the Grey River. I turned to port and glided towards Blaketown beach. I knew we could never make it. Lew just sat there as tense as a bowstring.

'Never stretch a glide,' was another rule I'd been taught. But I'd have to stretch this one. Speed 45 mph. How I got onto that beach, I'll never know. There seemed to be some supernatural force that lifted the aeroplane at the last moment to enable flying speed to be maintained.

We touched down on the hard sand, made a remarkably smooth landing, and rolled to a stop. This was surely one of those occasions when there should have been an accident and there wasn't. I often wonder about such incidents — is it just luck?

For the first time since the engine had quit, Lew spoke: 'That's the last time you'll get me in this bloody thing.'

We both piled out. Without another word, Lew disappeared over the sand dunes towards town at a steady lope, taking his bags with him. I stared after him blankly.

There was nobody about, and a light warm breeze and the pounding surf helped me to relax — and also to ponder how close disaster had been. To satisfy my engineer's curiosity, I unfastened the cowlings, checked the plug leads and throttle linkage, and removed the petrol filter bowl. All OK. I'll start her up, I thought. I checked the beach and removed a few pieces of driftwood. It seemed like an acceptable take-off surface.

I swung the prop. To my amazement the engine started. I got in, checked for full power, switches OK — perfect. I'll take off. I got to 800 feet when — silence again! I landed straight ahead on South Beach as near the water as possible (the sand is always firmer there), and about opposite Baillie Neville's workshops. I carried out the same ritual as before, and everything was perfect again. But it was too risky to give it a third go. So I phoned Des Wright and Basil De Jong, our aircraft engineer, and they soon arrived on the scene.

'Hell's bells,' said Des upon arrival, 'it's never done that before. Where did Lew go?'

'Last thing I saw of him he was haring it hell for leather over the Blaketown beach.'

Further checks and adjustments were made, but nothing untoward was found. I took off again down the beach, and this time made it all the way to Hokitika. I knew about gremlins, but this particular gremlin had a distorted sense of humour, I thought. 'I'll put it in the hangar and we can give it a complete check,' said Des.

By the time I was ready to take off in the Dominie to go to Haast, Lew had arrived in a taxi from Greymouth. 'Where did you get to?' I asked.

'Five double whiskies at the Australasian pub, and here I am. Never mind, Brian, I'm running out of good stories, and this one will collect me a few free beers at Haast.' Lew was a great sport. For many years afterwards, whenever I saw him in Grey he always began the conversation with: 'Remember that Auster?'

We boarded the Dominie. I looked at the two engines. I don't give a damn about the statistics of failure, I thought, I prefer to have two donkeys. When you get to my age, you prefer to have braces as well as a belt.

Then again, there are times when it doesn't matter how many belts or braces you have.

The late afternoon sun was still warm and shining. Te Anau seemed a very pleasant place to be. It was hard to believe that not many miles away in Milford Sound the weather was reported as raining heavily with visibility severely restricted. I leaned back in my comfortable chair and looked through the smudged windows of the pilots' room.

My old Dominie ZK AKT with its smart red, white and silver paint-work and airline markings sat contentedly on the tarmac. I was waiting for two American tourists to arrive by taxi, before heading home to Queenstown. The last flight for the day. It seemed as if I'd been waiting for passengers all my life.

It was 15 April 1967. I was happily married with five lovely children and feeling pretty good about life. I was flying for Tourist Air Travel who were in the process of being taken over by the Mount Cook Company. I'd worked for quite a few airlines over the past twenty-five years. One day I would have to give up flying — every pilot has to live with that thought — but I felt I still had a long way to go.

In fact a new phase in my life had only just begun. A fortnight earlier I had transferred from Hokitika on the rugged West Coast to Tourist Air Travel's base in Queenstown. I'd spent seven years flying from 'Hoki', but economic factors had finally stopped the historic West Coast Airways service and I had to move south to retain my job. The new Haast Pass highway, widely acclaimed, was the cause of our demise, as it stole most of the freight work and most of our passengers. So a shift to Queenstown seemed the best thing to do.

Although I had flown the Milford–Te Anau–Queenstown route many times over the years, this particular flight was only my fourth since moving south. I had always enjoyed flying light aircraft because it is *real* flying. There is something exciting and vital about being at the controls of a small plane, even something as old as the 1930s vintage De Havilland Dominie. There is just not the same feeling in the cocooned and clinical interior of a jet.

I was of course well aware of the layman's aversion to small aircraft. For some reason people always think they are safer in a larger aircraft, but I don't think statistics bear that out. Certainly in social gatherings when people found out that I was a pilot everyone seemed to have an opinion and theory about the dangers of small aircraft. But the most annoying folk are those who boldly state that they dislike flying, and then upon further inquiry you find that they've never actually flown. It's funny how arrogance and ignorance often go hand in hand.

The approaching mud-splattered taxi interrupted my thoughts. I went out to meet my passengers. After introducing myself I said, 'I'm sorry you had to come out to Milford this way, but the airfield was socked in.' 'Brother,' said the man with a North American twang, 'you don't have to tell us — the rain simply poured down. I've never seen anything like it.' I laughed to myself. He introduced himself and his wife as Dr and Mrs Russell Sage from Indianapolis and said they were both private pilots back home in the States. I quipped, 'Good, then we'll be safe tonight with three pilots on board.'

After completing the formalities of the load sheet and filing the customary flight plan, I said goodbye to my fellow captains Ken Leahy and Gary Cruikshanks. I estimated a half-hour's run to Queenstown; soon I would be home with the family for tea. The Gipsy Queen engines were throttled to maximum revs and the biplane lumbered down the runway and lifted off casually in the fading light.

After fifteen minutes flying, without any warning, the port engine started to vibrate, and within seconds it stopped. Oil was spewing out over the cowling and onto the wing and spots were being flicked onto the side windows. I wasn't particularly worried by this as I had had several similar engine failures — including three in single-engined aircraft. It only confirmed my feeling that ZK AKT had never been my favourite Dominie. I had first flown it eight years before but it had never seemed to have the same smooth feel of the other Dominies or of my favourite Rapide ZK AHS.

Still, this engine failure was going to be no trouble as I had 3,000 feet of altitude and was lightly loaded. Altitude is like money in the bank, I always said, a nice comfortable feeling. I briefly talked to the Americans and then promptly sent out a mayday radio call to Peter Banks, the duty pilot at Queenstown. I told him I had an engine failure but was sure to make the aerodrome in one piece. He acknowledged my message. My friend Lyall Hood, operating out of the Hermitage in a Mount Cook Cessna, also picked up my message.

I trimmed the aircraft for the yaw, opened up the good engine and settled down to a comfortable single-engined flying speed of 95 mph. Passing over Queenstown and the dark waters of Lake Wakatipu, I suddenly sat bolt upright. The starboard engine was also vibrating! Blast it, I thought, I'll have to put the darn thing down in the lake. I eased the throttle back slightly. Those big wheel spats would not take kindly to a ditching. And the white caps on the water told me the westerly wind was quite strong. Ahead lay the airfield. No question of landing into the wind, I thought, get the darn thing onto the ground in one piece as quickly as possible downwind.

But I was too high and had no hope of getting down. The approach would be impossibly steep. There was no option but to go around. Thoughts raced through my mind. Will the old bus make it? Dearest Jean, [Waugh's wife] help me God! More power was needed, but the blasted starboard engine wouldn't take it — no power at all. I was sweating profusely. That left only the nearby Shotover River. I turned left, scraped some tree tops by a matter of inches, and headed for the riverbed. Never mind the dead engine, I thought, it's much shallower and safer near the bridge.

You just have to get away with it. Control the landing. He who controls the landing lives to fly another day. Don't let the aircraft crash you.

Control it … control it, put it down now before she graveyard spirals. Speed 65 mph. What a lovely bit of riverbed! It looks so welcomingly smooth. Watch those boulders, lift the port wing, look at those stones!

Down … down … we're down. Thank God, in one piece! Hell, what a bumpy ride. Damn, we've run out of gravel. Here's the water! What the … we're going over!

The terrible noise, the rending of metal and tearing of wood was deafening. The whole cockpit disappeared before my eyes. My harness snapped and I catapulted straight through the windscreen. This really isn't fair, I thought. The blood, hands torn, head and body bashed. My good foot hurt and I thought this is what it's like to die. Not bad really, but no — not under the water. Get the hell out of it quickly.

My head bobbed up into the air and I gulped some quick breaths. It was still and quiet, broken only by the sound of the quenched engines hissing with steam in the cold flowing Shotover River. Shotover into the Shotover, I mused.

It was all over in a matter of seconds. The engines very quickly hissed their last song. I struck out and found that I had surfaced some five yards ahead of the aircraft.

'How do we get out of here?' came a plaintive cry. The passengers, I thought. Such was the suddenness and shock of it all I had completely forgotten about them. 'Well you silly sods, open the door!' I shouted. It was not particularly good advice, as the aircraft was completely upside down in the water and the door was jammed tight.

I felt a helping hand around my shoulders. 'How many passengers, Brian?' I recognised one of the company mechanics, John Muir. 'Two,' I replied, suddenly aware of the effort to speak. 'Righto, I'll get them 'em out, I'll be back for you in a minute!'

He quickly returned. I could see a funny kind of red mist before my eyes, I was passing out. A hefty hand grabbed me and lifted me up. But I couldn't hold on and slithered back into the cold water. Oh, that foot! 'Come on, Brian, you're not going to die. I'll get yer out.'

Placed on the bank, I remember thinking how cold the Shotover is in April. I watched and shivered, my body shaking and aching all over. I felt a searing pain from my right foot. It stuck out at an odd bloody angle. The nearby bridge looked like a huge prehistoric monster, weird in the dark. Then blessed unconsciousness.

After the Shotover River accident, Brian Waugh had several operations to repair his shattered ankle. Eventually it was set in a fixed position, meaning his flying days were over. He and his wife bought a motel in Nelson and Brian Waugh, quite literally, settled down. Well, almost. 'On a clear sunny day, a common occurrence in Nelson, I often watch an aircraft droning across the sky. It makes me ache to fly the old Dominie once again through Arthur's Pass, over the glaciers of South Westland, to Haast, Queenstown and Milford.

That's *real* flying. At such times ... Jean will nudge me and say gently, "Time to close the hangar doors, Brian."'

A month before Brian Waugh's crash in 1967, another Tourist Air Travel pilot also made his last flight — one which, happily, ended without calamity. Fred Ladd was with Tourist Air Travel from the start, beginning as its pilot/manager in 1954. Over the next 13 years, Fred Ladd and his little Grumman Widgeon amphibian became Auckland icons, flying freight and passengers (some ill or injured) between the city and the islands of the Hauraki Gulf. A natural showman, Fred was renowned for japes such as riding down the ramp to his waiting Widgeon on a unicycle. But it was his take-off announcement that became a trademark — and the title of his biography.

A Shower of Spray and We're Away
Captain Fred Ladd, MBE with Ross Annabell

I had the audacity to do my first commercial flight into Kawau Island on 19 June 1955 — for the grand official opening of New Zealand Tourist Air Travel. I spent a busy few days acting as TAT's public relations officer, as well as its manager/pilot, and organised a fitting opening, complete with blue ribbon, a mayor to cut it, some newspapermen to watch, and a radio commentator, the inimitable Phil Stone, complete with tape-recorder.

A small crowd gathered to watch us at the old seaplane terminal at Mechanics Bay. His Worship the Mayor of Auckland, Mr J. H. Luxford, did the ribbon-cutting, and Johnny Veale and quite a large turnout of his 700 staff knocked off their various duties to make the crowd look larger and more enthusiastic. I'd cooked up quite a nifty little stunt to publicise our Widgeon's capabilities, by collaborating with the management at Kawau Island to put on a midday feast for the mayor. I'd promised Mayor Luxford that TAT could whisk him away from his mayoral duties at noon, fly him to Kawau for a grand inaugural flight dinner, and have him back in his office at 2 pm. It seemed an impossible feat at that time. The general manager of TEAL, Mr Geoff Roberts, was invited but was not available, and Johnny Veale represented him. We also invited two newspaper men as passengers, 'Pop' Shaw of the *Auckland Star*, and Jim Duncan of the *New Zealand Herald*. In that pre-hydrofoil era of launches and ferries, such a trip seemed almost impossible, and was shrewdly calculated to wake Auckland people up to the great value of a Widgeon service.

I was a bit despondent when I arrived at the wharf that morning and found a north-easterly wind getting up. A choppy sea was the last thing I wanted, because at that stage all my flying had been done in flat calm weather. I'd never landed or taken off with passengers in waves. Every half hour that morning I was out to check on the harbour, and to my horror the waves were getting bigger and bigger. Mabel said at least a couple of times that I ought to cancel the trip, but I didn't really see how we could without losing a lot of face. We'd put out so much publicity about what the amazing Widgeon could do that to call it off because of a bit of wind would have been a very poor show.

By 11.30 am there was a fifteen-knot wind blowing, and to my unaccustomed eyes it looked as though we were in the midst of a full-scale storm, coming straight out of the north-east, diagonally into Mechanics Bay. The official party duly arrived about the same time as various experts from TEAL were coming out to look at the harbour and saying: 'Are you going to fly in this?'

The speeches started, the tape-recorders were taping, and the blue ribbon fluttered in the gale, imploring to be cut. Mayor Luxford finally put it out of its misery, and we all boarded the Widgeon.

The mayor sat beside me in the co-pilot's seat, I started the motors and it was 'down the ramp into the damp' in capable Widgeon fashion, to the cheers of the crowd. Once in the sea I found I didn't like things very much at all. The waves were breaking right over us. I didn't know it, but my misgivings were shared by my wife, and by some of the TEAL people up on the wharf. I taxied out into the main stream to see how big the wave were, didn't like them any better, but felt I'd passed the point of no return. So I turned the Widgeon around and back to the shelter of the Jellicoe Wharf, turned into wind, finished off my cockpit drill, and fed in the power.

In those days I had not yet coined any magic words to counter the passengers' fright at the great showers of spray churned up by a skybound Widgeon on take off. My poor passengers just had to sit there and wonder as the waves came up right around the cockpit and all around the windows. Everybody, including me, was full of excited apprehension as she came up on the step, beginning to plane. Instead of taking off nicely, as all good Widgeons should, we bounced all the way across the harbour, leaping like a sea-horse from wave to wave. Back on the wharf all the official people and the 700 folk from TEAL were lining the wharves and Mabel told me later that every time I bounced, the TEAL boys cheered. They were just about hoarse with cheering by the time we laboriously clawed our way into the sky against a very strong wind. We made Kawau in twenty-five minutes, to find that the landing spot I'd planned to use was now impossible, with waves tearing over it like wild horses.

I found a place off a point at the tip of the Bon Accord, next to Mansion House in the main harbour. The only person feeling any apprehension as we came in was me, landing for the first time with my first load of VIPs on a virgin patch of ocean. But I made quite a creditable landing and felt very pleased as we taxied round the point and rolled up the beach at Mansion House. We were met by all the Mansion House staff and the owner, Alan Horsfall, and he and his merry men and women ushered us all into the diningroom and a very sumptuous luncheon. It was such a very good luncheon that when we got to the sweets the Mayor was beginning to relax completely, and regret all the chores and papers and things he had waiting for him back in his mayoral chambers.

He said what a good idea it would be if they could stay the whole afternoon, and the rest of the party dutifully agreed, as they are supposed to when mayors suggest things at official luncheons. I was the only one who didn't say 'Hear, Hear'. All through the soup and the entree and the fish I'd been listening to the wind getting up until, by the time the main course arrived, it was fairly howling around Mansion House. I kept wondering how on earth I was going to get airborne as I chomped my way through the roast chicken and sipped somewhat abstractedly at my coffee. When the sweets came on I piped up, addressing the mayor: 'I think, sir, that we'd better hurry the sweets and get away.'

His Worship was in no mood to hurry. 'Come on now, I'm enjoying this,' he said.

'Well, after all,' I said, 'the idea of this trip was to leave at noon and get back by two, and if we don't, the whole point of the thing is lost.'

'That might be so,' said His Worship, 'but I'm enjoying this, and I'm not in any great hurry to get back.'

I stuck to my guns. 'If you listen, there's quite a wind getting up. I think we'd better go,' I said.

The mayor agreed, rather reluctantly, I excused myself, and hurried down to warm the aircraft up. I was really apprehensive by this time. Mansion House Bay is a very small bay, and the Bon Accord Harbour is very long and big, and out there I could see the waves marching past in great style, bowled along by a twenty-five-knot wind. The party came out shortly after, and we waved our hosts goodbye and taxied out into the waves. I knew we couldn't get airborne in Mansion House Bay, so we had to taxi into the big stuff in Bon Accord Bay. When we got out there the old plane went up and down with the floats submerging on each side, wallowing around like a half-drowned duck.

It was terrible — impossible to get airborne in that stuff — and yet I knew that if I didn't get my guests back it was going to be goodnight to any favourable publicity for TAT's Kawau Island service.

I turned round and taxied back into the bay.

'Why are we gong back?' piped up the mayor, and it was fairly obvious that he didn't know much about seaplanes.

'Oh, I always come out here to have a look at the wave situation,' I said. 'Now we're going back into the bay, and when we get in by the beach we'll take off from there.'

'Great,' he said, still full of enthusiasm after the dinner, which was apparently sitting lighter on his stomach than my apprehension was on mine. One thing that came to me in the midst of all my worries was a conversation I'd had recently with George Bolt, the famous New Zealand aviation pioneer and seaplane expert, who'd given me a lot of tips on amphibian operation. He told me that if ever I was in a situation where I wanted to get airborne quickly, it helped to take the aircraft as close in as possible to shallow water, and then take off and allow the ground effect to take the aircraft up on the step much quicker than is usual in deeper water. I decided to try the George Bolt technique, and went right back in towards the beach before turning round. There were no magic words. I just said 'Ready?' and gave her the power.

We came up on the step very quickly and started to roar across Mansion House Bay, going fairly well because we were diagonally into wind. We were only just on the step when we reached the first big wave in Bon Accord Harbour, and the thumping that started then was terrific.

The book says that for a cross wind you hold the starboard wing down towards the wind and slowly let your nose come round, down wind. I tried to do that, though in my inexperience I exaggerated the manoeuvre, but I kept the power on and we kept dancing in until finally we got into the troughs and tore along parallel with the troughs in that awful gale. We began to bounce higher and higher, until what I thought were bounces were in fact 'porpoising', but I grimly kept the power on and somehow we came out of the last porpoise leap into the air — just in time to see the cliffs looming up on the other side of Bon Accord Harbour.

We were flying straight at those cliffs, and we just managed to scrape up enough height to clear them before we turned over the top and headed for Auckland.

The mayor was happily patting me on the back. 'Marvellous, Captain, marvellous,' he said. 'I didn't think you were ever going to get airborne, you're a great flyer.'

He apparently didn't realise how pale and sweating and trembling I was. We had the wind in our tail now, and we shot back to Auckland in twelve minutes, fairly racing in the storm — and it really was a storm by now. We landed at Mechanics Bay according to Hoyle, but as I went to taxi up the ramp I noticed something different about it, the concrete shoulders were sticking out at the bottom. The tide was out, and as I gave her the power to climb the ramp she merely went over on her nose. I pulled the power off, and the tail slapped back on the water with a thump, and there we were sitting at the bottom of the ramp, bogged in the Waitemata mud — an inglorious end to a very adventurous first official trip. It was my first low water spring tide, too.

I stopped my motors. Johnny Veale of TEAL looked around at every-body and said to the mayor in a loud voice: 'Well, Your Worship, exactly the same thing happens to us with our big flying boats. It's absolutely impossible at low tide, we've had a lot of trouble like this.'

I was very thankful to Johnny. I knew darned well that he never brought his big Sunderlands or his Solents up on low tide, he was just being kind to me. I signalled the TEAL mechanics and they brought a dinghy down the ramp, launched it, and ferried the mayor and Johnny Veale and the press boys ashore. As they reached dry land on the ramp I opened the cockpit window, and shouted: 'Whatever you do, gentlemen, when you get out of the dinghy you must walk up the *side* of the ramp.

Don't step up over the wall on to the ramp proper, or you'll slip and fall and possibly hurt yourselves. It's very, very slippery.'

The mayor got out first and did what I'd told him. Duncan and Johnny Veale followed, likewise obedient. But Pop Shaw, the *Star*'s man, decided to take a final picture of the Widgeon sitting ignominiously in the mud. He stepped off the correct path on to the ramp proper, which was covered in a very slippery marine moss. I was gazing horrified out of my window, knowing what was going to happen, but before I could get out a yell of warning Pop slipped, and I saw a pair of heels and soles flying straight at me. He was on his back, hurtling down the ramp, and his slide only stopped when he hit the bottom … knocked out. The TEAL chaps came carefully down and picked him up and they got him to the top and stood him up. He was groggy enough to be taken off to hospital for observation. I waved them all goodbye from the mud — a most depressing finale to New Zealand Tourist Air Travel's inaugural opening flight.

Staying with amphibians for a moment, the most famous of them all (in these parts, anyway) is the Catalina. Affectionately nicknamed 'Dumbo', the Catalina was a vital long-range anti-submarine reconnaissance and rescue aircraft during World War Two. Famous for their strength and endurance (the record is a 34-hour flight), Catalinas first served with the RNZAF in 1943, and remained on the inventory for ten years. All told, the air force operated 56 Catalinas and would probably welcome a few now, given there's not much else to fly.

Be that as it may, Ross Ewing was most certainly keen to see a Catalina back in southern skies. In 1992 he decided to make his fantasy fly and started canvassing other enthusiasts. Eventually, the syndicate bought a suitable Catalina they'd found in Arkansas and set out to fly it back to New Zealand. Having reached Hawaii, the next leg was a flight to Tahiti. All went well, until the Catalina's crew received a fateful message.

Catalina Dreaming
Ross Ewing

Ten hours after take off we were about four degrees of latitude north of the equator when air traffic control in Honolulu asked us to climb to 5,000ft if we wished to continue with air traffic clearances, or stay at 1,000ft and accept 'flight following' services only.

We decided to climb. It was a fateful decision. During the climb we went through some rain showers but then levelled off in clear air. There were stars above, but no moon.

Kirk started looking for the Southern Cross. I peered over his shoulder — the distinctive group of stars should be up ahead, somewhere. I thought I had just seen it — it looked different from this new perspective — when it happened ...

The left engine backfired, giving off a seemingly huge belt of fiery orange/red flame, then ran smoothly again.

What could have caused that?

A keen discussion developed on the intercom — carburettor icing? Spark plug fouling? Water in the magneto system?

Kirk had brought back the boost on No 1 engine, and now eased it up again — okay, so far. But then, BANG, it backfired again. Power back again. No problems, running smoothly.

Gingerly, Kirk increased the boost again but the backfiring increased in frequency again until, gradually, less and less boost became available. With reduced power we were forced to descend. Kirk set METO power on the No 2 engine. We were coming down, at about 100ft per minute.

I dialled the Christmas Island co-ordinates into the GPS and hit the 'go to' button. It gave an instant position: 300nm to the right at about two o'clock by the clock code. We headed towards it. On Kirk's instructions I put out a Pan call on the HF. Honolulu came back immediately asking what assistance we would like? I said we didn't know but we'd let them know.

Our speed was now back to the minimum control speed of 83–85kts. Kirk still had the other engine on METO power. N5404J was wallowing markedly, Kirk fighting to keep control at times.

After a brief discussion with Kirk I put out a Mayday call. The transmission, we were told later, was heard in the Solomon Islands. It was also heard by two Air New Zealand Boeing 747 jumbos en route to the US. On board one of the jets was Sir Tim Wallis. The captain called him to the flight deck and he was able to listen in to our plight.

The voice of the lady on Honolulu Radio never wavered — she remained very cool and calm. She was very professional, very reassuring and very helpful. I told her of our predicament. I had to repeat several times to her that we were a flying boat.

Kirk was still struggling to keep N5404J under control. Outside it was pitch black. I kept myself busy talking to and encouraging him and keeping Honolulu in the picture. Honolulu called and said a US Coast Guard C-130 was on the way, then that a ship, the *Direct Kookaburra* was 13 hours away. This ship information, I learned later, came from a US coastguard system called AMVeR which stands for Automatic Merchant Vessel Report. I gave our GPS position, several times, just to be sure.

By now the No 1 engine was throttled back almost to idle power. Any slight increase in power would cause it to backfire. The windmilling propeller was causing a lot of drag which compounded our problem. Kirk then tried to 'feather' the propeller but announced that it wouldn't stay in 'feather'.

That was it. I knew we were now in real trouble. We were going down. There was now no way that we would make it to Christmas Island. We couldn't jettison fuel — the six drums were empty and the Cat had no fuel dump system. We couldn't lighten our load in any other safe or practical way.

I called Honolulu again and told them we were 'going down' and would be force-landing shortly on the open sea. The lady, as calm as ever, said, 'Roger, the ship is now due in 12 hours.'

It was about 4.45 am. It remained pitch black outside. We turned the landing lights on but they were blinding. This may have been because there was a light sea fog. We had to turn them off again. We didn't have an accurate QNH pressure setting for our altimeters and therefore had no way of knowing how high we were above the ocean. Things were looking decidedly grim.

I deliberately avoided thinking too deeply about the trouble we were in. 'Denial,' in times of great stress, is said to be an effective coping mechanism and I found this to be so!

I remembered back to when I was analysing the risk of taking a

Catalina across the Pacific and had come up with I suppose what was the ultimate in rationalisation — 'Oh well, if it all turns to a can of worms at least we could land on the water'! This is precisely what we were about to do. But at this weight, and at night? It didn't bear thinking about further.

Nobody panicked. I made sure everyone was strapped in in the forward cabin. I looked back into the front cockpit — the altimeter read 300 ft. My mouth became *very* dry. It was hard to realise this was really happening.

I turned around and signalled to the others that we were about to 'go down'. I strapped in to my seat on the right side of the forward cabin and, with the starboard engine still roaring away in my ear, we all waited. I remember thinking, 'This is it.'

Kirk fought for control, keeping the Catalina in as near to landing attitude as he could on instruments and in total darkness. With one engine set to full power and the other throttled back to 'idle', the aircraft was trying to continually tear itself free of his grasp and enter a steep diving turn to the left.

I tried not to think of the consequences of a heavy water landing. I knew that if we hit too nose-low the nose wheel doors would burst open, inviting a huge tube of water into the main cabin which would sink us within seconds. I also knew that a heavy crash would most likely tear off the wing at its pylon mountings and cause it to immediately pitch forward into the main cabin area ...

I'll never forget the noise N5404J made as it struck the water a glancing blow. We had clipped one of the 5 to 6 foot swells which we later saw were running at the time. It was a sudden, very loud 'metal on metal' graunching sound — it lasted about half a second — then — silence.

The ocean was as hard as concrete.

I remembered somewhere from my air force training days that this silence is *not* the signal to relax, undo your seat belt and walk around but instead to 'brace' for the next impact. I was about to call out 'brace!', but was checked by the thought that 'brace' was not really a good clear term for our present company and situation, so instead I called out 'HOLD ON!'

Shortly thereafter we hit again — not quite so hard — then again, and again, each time the impact of touching becoming less and less ...

Then, quite suddenly — miraculously — there was silence. We were on the water, and all in one piece. The feeling of relief was immense.

We all whooped and cheered but our yelling quickly turned to alarm as the Catalina lurched sideways, the left wing-tip entering the water. The landing floats were not down!

Tony Butcher yelled 'FLOATS!' to Kirk, who immediately hit the float switch and we soon took on an even keel. More feelings of great relief.

I grabbed the HF radio microphone and radioed to Honolulu, 'We are on the water and we are okay.' I gave our GPS position. It was: 3 degrees, 11 minutes and 43 seconds North: 154 degrees, 45 minutes and 01 seconds West. The controller acknowledged.

Kirk shut down both engines. For a moment we sat in silence. But straight away we knew we had been mortally damaged — water could be heard trickling in under the floor boards.

Kirk looked distraught and was visibly upset at what he had been through, and no doubt at what he had done with 'our' Catalina.

I grabbed his hand and shook it, congratulating him for saving our lives, which he undoubtedly had done. We could have so easily all been killed. Straight away I made him promise me that he would never take the aircraft's loss as a personal failure on his part. He agreed.

Within minutes Tony Butcher had the cabin floor panels ripped up and he and John Strutman were designing a way of containing the leaks. The bilge pumps were switched on, but soon both had failed. Tony modified the electric fuel transfer pump to a bailing pump but it too failed rapidly. He tried the manual fuel pump but it, too, soon became blocked. Why?

Tony tore the electric pump apart — inside was the answer — paint flakes. The impact must have loosened a large amount of old paint flakes which would soon clog up any pump.

PK screws, chewing gum, we tried the lot but it soon became obvious that we were not winning. In desperation we started bailing — setting up a chain gang with buckets and chilly bins — at one stage we thought we might have the leaks under control but the physical effort was not sustainable and after an hour it became clear that we would have to abandon our Catalina. We sat around, feeling exhausted.

Meanwhile dawn had broken. Good communications continued with Honolulu and we told them what we were up to. They wanted us to change frequency but I told them, no, this was our lifeline and we were hanging onto it. They confirmed that a Coastguard C-130 would be there in about five hours.

I sat on the large cargo door exit ledge with the door locked wide open above, and looked outside and up towards the for'd fuselage. Peta Carey was looking out the navigator's hatch and into the far distant horizon, her mind obviously elsewhere. I caught her eye. She looked slightly embarrassed.

'It's alright,' she said, 'I've just had an incredible deja vu.' I understood, and nodded.

Our situation regarding saving N5404J had become hopeless. Our Catalina was sinking. Kirk gave the order to abandon ship. Tony Butcher pulled out the liferaft and pulled the air bottle lanyard. Thankfully, the liferaft inflated. The raft was of generous size and could hold 42 persons. We threw our bags into it and one by one leapt aboard.

As we let go from the Catalina the feeling was one of general insecurity. Awkwardly, the raft immediately drifted under the port wing and became stuck there. It was an unpleasant moment as the rubber raft scraped against the wing struts and floats. I hoped there were no sharp protrusions. Kirk suddenly hit on the idea of making a sail with a blanket. He and John Strutman held up a blanket and Tom Neave and I followed suit. Soon we were sailing out from under the wing and free from further trouble.

Tony Butcher immediately burst into song — 'Sailing, sailing, over the water blue....' We all laughed.

The sun was now coming up. There was little wind and very little cloud. It was a fine mid-Pacific day …

Once things had settled down aboard the liferaft I became consumed by a strong passion to be with my wife and family. For years we had jokingly said that flying was my first love and that my wife, Raewyn, was my second … She certainly let me have a free rein as far as flying was concerned and for that I was always grateful. But the ditching of N5404J in mid-Pacific had somehow seemed to have changed all that.

There is nothing like being stranded in the middle of the Pacific Ocean in a liferaft to help focus one's priorities in life. I reached for my notebook and wrote a letter, not knowing whether I would ever see Raewyn or the rest of my family again, let alone how a puny letter penned thousands of miles from anywhere would ever get to them were we to perish out here. The letter read:

> Saturday 15 January '94
> In liferaft
> 200 miles from Christmas Island
> Dear Raewyn, Lee, Scott, Charlotte, Grant and Madison,
> This is to let you know that I am thinking about you all at this time.
> Our Catalina was so full of fuel that, when the left engine began to fail we could not maintain height and slowly began to descend in the pitch black night towards the sea — which we finally hit — in a more or less controlled crash.
>
> Kirk our pilot did an excellent job — his first night landing in a Catalina! But we had sprung a few holes in the hull and despite our best efforts at pumping and bailing we started to sink, so had to 'abandon ship' — and so here we are in our dinghy.
>
> A US Coast Guard Hercules is due here soon and a ship is due in about 10 hours.
>
> So here's hoping.
>
> I don't feel afraid, only deeply disappointed.
>
> However, that's the way things go.
> With Love,
> Ross

It was a traumatic time. The desolation of being in the mid-Pacific is hard to describe. With the weather being so fine it was hard to appreciate this and not to be ridiculous and say — well, who's for a dip, then?

We chatted amongst ourselves as the Catalina drifted away, conversation flagging at times. N5404J gradually adopted a nose down, tail high attitude as she filled with water. I found it difficult to look at her.

We were all in a state of shock. So much so that none of us even thought of turning on our survival beacon! Tony had the beacon. I had the aerial in my bag.

One of our group began hyperventilating badly, another became violently seasick.

Periodically, we would all become quiet and withdrawn. In response to this Tony Butcher, recognising that things were getting a little serious would — from time to time — crack a joke. We would all laugh, and feel better. That happened a number of times. Tony, who is amongst other things a professional yachtsman by trade, was a tower of strength on the liferaft. Peta Carey is also a 'yachtie' who has sailed the Pacific.

We drifted and chatted for five hours, conserving water, putting up the liferaft roof to protect us from the now fiercely blazing sun. Still no one thought of turning on our survival beacon.

We could barely see N5404J now, its tail sticking up almost vertically in the air like a mortally wounded whale.

My emotions were mixed — I was crestfallen that we had lost N5404J after all the hard work and the syndicate and other money that had been poured into the project — and the public support we had generated — and yet I was so *elated* to be alive! I felt like singing at the top of my voice.

We all heard it together — the C-130 of the US Coast Guard — 'Mother Goose' as Peta Carey so aptly dubbed it. It had found us using our last GPS position, and nothing else — no help from our emergency beacon! As it roared low over our bright yellow raft I waved out.

We had John Strutman's hand-held VHF two-way radio on board and Kirk soon made good radio contact with the Hercules captain. He was such a cool dude all of us could have hugged him.

The C-130 dropped a smoke float. It landed some distance away and began drifting towards us. It started to look as if it would come dangerously close to our liferaft. How ironic, I remember thinking, to be found by our searchers and then sunk by one of their flares! The flare, still burning furiously, eventually drifted harmlessly by, about 10 feet from our rubber craft. The Hercules then dropped three small liferafts, another radio and some more water which we gathered up.

After circling for a while the C-130 captain — our hero — radioed that he was going to Christmas Island to refuel and would be back about 4.30pm to guide the *Direct Kookaburra* to us. The ship was due at about 5.30 — just before nightfall. Brilliant!

We said cheerio and told him our liferaft's course and speed from the two hand-held GPSs which we had on board. Westerly 2 to 3 knots. I sat back in the liferaft, feeling somewhat happier now about the prospects of coming out of our ditching alive.

We chatted and dozed again, Tony occasionally bringing us 'round with one of his wisecrack jokes.

'I hate it when this happens,' he would say.

I knew we were not out of the woods yet, but I considered our chances were now better than even.

The C-130 was back right on cue at 4.30 and was soon setting up and flying a race track pattern — overflying the ship and then heading straight for us. The weather remained incredibly good. I began to feel quite positive.

Because of the earth's curvature and because we were so low on the water we did not see the ship until it steamed into view about 12 miles away. It grew in size quickly and then came to a stop about 300 yards from us. It then manoeuvred more closely and a crewman threw out a line. Tony grabbed it and we swung in alongside the ship which was now moving at about 1 to 2 knots.

The ship towered above us, its layers of deck-borne containers making it look even more huge. Several of our crew looked in dismay at the 80-foot rope ladder we had to climb. Personally, I was glad to take *any* ladder!

'One limb at a time,' called the crew as we climbed up the rope ladder.

As we boarded the ship someone said they saw a shark …

The ditching didn't dampen Ross Ewing's enthusiasm. He continued to search for a Catalina and finally found one for sale on the Nile. It now flies regularly down under, visiting air shows and carnivals around the country.

Time for one last flight over water. Well, no, not one but two — from Oz to New Zealand and back, plus a couple of islands en route. Even now, a VFR flight across the Tasman in a single-engined plane is a risky business. But worth doing when you want to fly home and see Mum.

Journeying with Aviators in New Zealand
Roy Sinclair

Whenever Cathy Pegg drives onto one of Melbourne's freeways she heads her 1970 Mercedes Benz towards the fast lane. The classic car, named 'The Baroness May' — after her mother — is the second love of her life. Her first love, and one that involves all her passion, is flying. But Cathy Pegg is no ordinary aviator. She is Maori. And, as a symbol of wealth, the distinctive Mercedes Benz emblem on her $4000 car is as phoney as are the bullet-hole transfers on the rear windscreen. Her upbringing in New Zealand is the sort of stuff that inspired Alan Duff's *Once Were Warriors*. Her father, 'a violent woman-ising alcoholic', was far from being the ideal family man. She grew up on the Whenuapai Air Force Base where her father was stationed, and eventually graduated from Auckland's Massey High School, near Henderson, as the 'school dope'.

Aged 21, and a trained nurse, in 1977 she escaped to Australia and subsequently, at the Point Cook Royal Australian Air Force Base, met and married her former husband, Phillip. Late in December 1993 she returned to New Zealand for the first time to see her much-loved mother. She flew herself in a single-engine Cessna 182 via Lord Howe and Norfolk islands, becoming he first Maori woman to fly the Tasman Sea solo. The momentous event, which also included a return flight to Australia, went almost unnoticed in New Zealand. In Australia, Dick Smith, the entrepreneurial aviator and founder of *Australian Geographic* magazine, claimed Cathy's trans-Tasman flights as a double aviation record.

In May 1997 I met Cathy Pegg and the Baroness May, simultaneously. I am not sure which I noticed first. Both were impressive. Both promised to be as forthright as they were intriguing. I had tracked Cathy Pegg through *Australian Geographic* and the generous help from the editorial staff. I wrote to her and in return she phoned me. We spoke for five minutes and I knew a deserving story was ready for the telling. Next day I was peering through travel agents' windows, checking out the best deal for a return flight to Melbourne. A few short weeks later and I had arrived, unprepared for the verbal adventure of my life.

'I love this old classic car. When I am driving it my friends call me Cathy von Pegg,' she told me as we pounded along in the fast lane. Yet, despite an inherent fear of fast cars, I felt secure with the Baroness May and it exuberant driver. And we did spend a lot of time criss-crossing Melbourne's freeways during the three days it took to build my story. We were either on our way to Point Cook airfield, located near Melbourne's outer western suburbs, or to Moorabbin's general aviation airport south-east of the city. And for three days we talked, and talked, and talked — mostly about flying and aviators. Cathy had taken a week's leave from her job in the publications department at the Laverton RAAF Base.

We were two excited Kiwis inextricably caught up in a huge Australian metropolis, one in which we had a shared affinity. In 20 years Melbourne, unlike her native New Zealand, had shown Cathy Pegg some kindness. Perhaps, too, I could make a similar claim. Many years previously, in Melbourne, I had met the first love of my life when, as an uninformed and clownish 23-year-old, I made my first tentative venture away from home.

Cathy, I soon discovered, had experienced 42 years of life that had been as poignant as it was quirky. She does not know self-pity. The greater part of her conversation is spiced with genuine laughter, although she unashamedly admits to knowing how to cry — a lot — whenever she is overwhelmed. Her more poignant experiences include a temptation, as an unemployed young nurse, to street walk in St Kilda's notorious red light district when money ran sort. 'I was desperate and prepared to do anything — well almost. But I couldn't imagine allowing any man to be up to his nuts in my guts for any amount of money.' She then landed a fortuitous job at the Freemasons Hospital in South Melbourne.

On another occasion, a tragedy at Point Cook had her hauling a young Air Force flying officer from a burning Tiger Moth that had spun into the ground. At the time she was learning to fly herself. Most student pilots would have given flying away after such a shocking experience. For myself, I had wimpishly let young and short-tempered flying instructors destroy my ambition to fly. I told Cathy about it, and I felt ashamed.

'Flying to New Zealand was an ambition for as long as I can remember,' she said. 'It was something I used to fantasise about. I remember telling someone that I was going to be the first Maori woman to fly the Tasman Sea. I never thought it would happen. It did happen. I was successful. It nearly killed me.'

Cathy Pegg was looking for an adventure, one that would give a private pilot the same sort of satisfaction a mountaineer gets on the summit of Mt Everest. It all started when, as a dutiful wife, she joined her husband clay-pigeon shooting at the RAAF Point Cook Gun Club.

'Phil never forced me to do anything I didn't like to do but I thought it was my duty to accompany him rigorously every weekend, a spare Browning slung over my shoulder, as he skilfully blew multitudes of clay birds out of the sky, one after the other. God, how I hated the place. It was a wasted weekend but anything was better than the routine of housework and mowing lawns, the weekend highlight of every woman married to the defence force.'

To escape the talk of the other women comparing the unlikely qualities of their respective offspring, Cathy would walk the short distance to the perimeter of the airfield. From there she loved the sight of Melbourne's skyscrapers, but more enthralling were the small aeroplanes and their student pilots attempting to take off and land with some degree of precision.

'I could watch them for hours. I imagined I was one of those pilots, swooping down over the gun club, my left fist jammed hard against the windscreen and my middle finger jutting upright at those pathetic clay targets that made me play second fiddle to my marriage.

'Then, late one Saturday afternoon, on our way home, we passed the RAAF Flying Club as one of the aircraft was taxiing in after a late flight. I shouted for Phillip to stop the car. I got out and went so close I could almost touch the Piper Tomahawk. I was so excited I thought I might have an orgasm — that was something that didn't happen to many women at the gun club. Phillip had sensed my excitement and suggested I visit the flying club, which I did.'

Cathy's first flying lesson was booked for 28 June 1987. From her first step into the flying club she had found a home away from home, and new friends. Her instructor was Jack Edwards, a man in his mid-sixties who had flown several years for Ansett Australia following his discharge from RAAF flying.

'He was incredibly patient, and he had to be, the poor man. Aged 32 I was known as old "Flossy", and I was about the worst student pilot God had ever put on this earth. They told me that flying was the most fun anyone could have outside the bedroom, but what they neglected to tell me was I had to learn how to fly the bloody aeroplane first.

'That first lesson was a nightmare. We lined up on runway 17 and gave the Tomahawk VH-PCF full throttle. No matter how hard I tried I could not keep that aeroplane straight on the runway. That day, I really made Ding Duck look like Tom Cruise flying those fighter jets in the *Top*

Gun movie. As I struggled to battle the rudder pedals, my fantasies of giving the gun club the majestic royal salute soon went out the window.'

Cathy persevered. Jack was always encouraging and she got to love VH-PCF. Then one day, while waiting for her flying lesson, she heard a May Day call on the clubhouse radio. PCF had had an engine failure. The instructor and his student pilot were looking for a paddock to land in. The paddock had long grass that hid a large hump. The Tomahawk flipped. The occupants, lucky not to have been killed, suffered back and neck injuries. WH-PCF was wrecked, beyond repair.

She continued her lessons in other two-seat trainers: a Cessna 150 Aerobat and a Beechcraft Skipper. Both were almost the same to fly as the Tomahawk. On 6 September, when she had 8 hours in her logbook, Jack decided Cathy would fly solo.

'Jack had booked the Skipper VH-UMX for my lesson. There was nothing out of the ordinary to suggest what Jack had in mind. We took off from runway 17, flew three circuits, and on the third touchdown Jack told me to pull over to the edge of the runway. I thought I had done something very wrong and braced myself to have my poor bum kicked from one end of the runway to the other.

'Jack calmly undid his safety belt, opened the door, and stepped out onto the wing. He then shouted, "THIS IS YOUR FIRST SOLO. GOOD LUCK AND DON'T FORGET TO FLARE." With that, he slammed the door.

'And I said, "SHIT!"

'I lined up and pushed the throttle forward. UMX started to roll and, without Jack's added weight, the Skipper leapt into the air. Did I feel great!!! — until I turned to base and realised I had to land the thing all on my own. I trimmed the aeroplane to 65 knots, put on flap as required and did my utmost to stop myself from pooing my pants. As I got closer to the ground, I pulled the throttle to idle, eased the control column back, held my breath and waited for the crash. To my great astonishment UMX greased onto the runway and came to a gentle stop on the centre line. Jack was smiling as he waited for me. I was shaking in disbelief.'

In 1993 Cathy Pegg was saving for an airline ticket to New Zealand. She phoned a friend from the flying club and told him about it. Unimpressed, he said, 'You're saving money to give to an airline? You are a pilot, Cathy. Why don't you bloody well fly yourself?'

'I thought that was really not such a bad idea so I phoned a friend in Queensland. He was Jimmy Connor and he had a Sud-Horizon, a small French aeroplane. I came straight to the point and said, "Jimmy, how would you feel about lending me your aeroplane to fly to New Zealand?"

'He said, "Cath, what a damn great idea."

'It would have eased my mind had he said, "No." The idea of flying over all that ocean terrified me. I knew I would be carrying a spare pair of undies, and they wouldn't be lace!'

Cathy was told to make preparations for a late December departure. But just a few weeks later the aeroplane was wrecked by a hailstorm. The huge hailstones demolished all the control surfaces — the ailerons, flaps, and elevators.

'I thought it was all over. Then John Laming, a retired RAAF squadron leader, told me he knew of a little twin-engine Piper Comanche that would be available for my flight. John even offered to see me through my twin endorsement, for nothing. Just a few weeks before departure the owner asked for $10,000 up front. I was truly pissed off. I didn't have that sort of money. I was annoyed that he had waited until I had made most of my preparations before dropping the bombshell. I used ample choice expletives to express my utter disappointment. I was almost ready to throw all my charts in the rubbish. I had had two chances, and you don't usually get a third.

'Then, John rang again. He said, "I have found the perfect aeroplane for you."

'I thought, "Doesn't this man ever give up?" To John, I said, "Shit. It's on again. I've got to face that bloody ocean."'

The use of the Cessna 182, Victor Hotel-Echo Kilo Foxtrot was going to cost Cathy $4000. That was her budget. Fuel would cost another $1000, which could go on her Mastercard. The Cessna was overhauled thoroughly two weeks prior to takeoff. The engine was found to have a cracked cylinder.

'An overhaul was not due for some time, and a cracked cylinder might not normally be found in the usual 100-hour check. Had I flown as it was I think the cylinder would have cracked wide open and I would have ditched.'

The RAAF helped with providing survival equipment, which included a Uvic flying suit, one-man life raft, life jacket, emergency locator beacon, flares, and two-way radio. Cathy was taken on RAAF training courses and shown how to use the equipment. She was put into a decompression chamber and taken up to a simulated 28,000 feet to demonstrate the effects of hypoxia.

'The RAAF was very supportive, and it was great fun jumping off Zodiacs and being rescued by helicopters. I was also shown how to use the Pyrotechnics, the flares and sea dyes. I even spoke to Squadron Leader Garley, in Canberra. He is also a pilot. "If we can help a fellow aviator, we will," he told me. That was brilliant, and it was very encouraging to be backed by a defence force.'

Members of the Australian Women's Pilots Association wrote letters seeking sponsorship. *Australian Geographic* publisher, Dick Smith, was impressed by Cathy's meticulous preparation and her commitment to the RAAF safety training. He donated $1000 to foot the fuel bill.

Later, Dick Smith, under the heading 'Landing an aviation record' wrote:

> In a feat worthy of more experienced pilots, Cathy flew from the Royal Australian Air Force base at Point Cook, south-west of Melbourne, to her homeland, via Lord Howe and Norfolk islands ... with just 200 hours' flying experience and only two other cross-country flights under her belt ... All the support and planning in the world, including advice from around-the-world fliers Gabby Kennard and Peter Norvill, could not prepare her for the dramas she experienced.

She was island-hopping, starting across the ocean from Coffs Harbour in New South Wales. It was the reverse of Francis Chichester's trans-Tasman Gipsy Moth seaplane flight of 1931. Cathy even studied Chichester's basic navigation involving wind drift calculations that had enabled him, with very basic instruments, to pinpoint a small island in a vast ocean.

'I was not a bit confident when I set out. I was certainly scared. I had thought I was well prepared but I did not have an instrument or night rating, and that almost cost me my life. I was reluctant to have anyone with me. I didn't know what was going to happen. My biggest fear was surviving a serious mishap, and my passenger dying. I could not have coped with that. I did take my Teddy bear given to me for the trip by my half-brother, William. I call the bear Charles Kingsford Smith.'

Cathy started flying across the ocean on 30 December. Her Cessna, laden with fuel, tie-down kit, and emergency gear, struggled into the air and flew with its tail down and nose up. With standard fuel tanks, the Cessna could hang in the air for five and a half hours.

'I kept looking down at the water thinking that I would soon be down there and it would all be over. But after half an hour I relaxed. You become very accepting of what might happen. I stopped worrying and just flew.'

The flight of 316 nautical miles took two and a half hours, and was almost uneventful. The Cessna had twin ADF (Automatic Direction Finder) needles on the one dial to assist with the compass navigation. Once the ADF needles were aligned to the appropriate NDB (Non Directional Beacon) ground stations, at the departure and arrival points, it was easy to see if the aeroplane drifted off its course. Cathy also had her own Global Positioning System, a Garmin 55. It plugged into the cigarette lighter.

'At the time there were no GPS units approved for Australian aviation. Some pilots say a GPS is cheating on navigation because it is so simple to use. If it is set correctly it will always take you right above where you want to be. A GPS will also count down the nautical miles remaining to fly, and will indicate the aeroplane's ground speed. When flying over a featureless ocean I believe a pilot should use every means of navigation available. My GPS is a very basic model, with nothing outrageous about it. It got me all the way to New Zealand and back.

'You always feel confident you will find your island. But it was a beautiful sight seeing those twin peaks of Lord Howe framed in my windscreen. I was 100 miles out, flying at 8000 feet. As I got closer the picture just got bigger and bigger.'

That night Cathy stayed on the island. She had a dream: Her older sister was flying a Cessna across an ocean. Cathy pleaded with her, 'Mary, don't do this. This is a very silly thing to do.' But she went off and did fly over the ocean — and the Cessna crashed.

'It did worry me. Then I thought, if I go back, or if I go on, either way I will be flying over water. I could have caught a commercial flight back to Australia. Someone would have had to fly the Cessna back to Australia and I would have spent the rest of my life not knowing how my flight might have turned out. So I decided to fly on. I had made my Last Will and Testament before I left home, and not told anyone. Phillip would have found it when he went through my things.'

Next morning Cathy was preparing for the 480 nautical mile flight to Norfolk, expected to take just over three and a half hours. Dressed in her all-weather flying gear, which included thermal underwear in case she ditched, she was sweltering. The Cessna had been refuelled the previous day so all she had to do, apart from the usual pre-flight checks, was to recheck the fuel levels and do a fuel drain in case condensation overnight had left water in the tanks.

'I did a drain of the wing tanks and they were OK. The Cessna 182 also has an engine drain but that one is very awkward. I couldn't reach the top of the fuel drain inside the engine cowling, pull a little plug, and also put a drain under the engine. My arms, encased as they were in all the flying gear, just weren't long enough. I thought I wouldn't bother. Then I realised I'd better do it properly if I was going to fly over water. Perhaps I was still thinking about that disturbing dream?

'I managed to jam a jar under the cowling flap, and pulled off two full jars of water. Had I not done the drain properly, the engine might have got me into the air over the sea and stopped. I would have landed in the sea, and it would simply have been bad airmanship.'

Flying on from Lord Howe becomes an international flight, involving clearing customs and filing an International Flight Plan. She had been told that she could fly on an Australian domestic flight plan but New Zealand said, 'No.'

'I said, "Shit!" The flight plan was important should a search and rescue be necessary. While I was contemplating what to do, a Dash 8 landed. I ran out waving my arms crying, "Please, Captain. Have you got an International Flight Plan?"

'He gave me one but I didn't even know how to fill it in. I felt a dork. He showed me, and he even filed it in to Auckland International Airport. He took off for Norfolk half an hour after me, and much later passed me again on his way back to Lord Howe, long before I arrived at my next island.'

Again, the ocean flight had been almost uneventful apart from minor visibility problems caused by the smoke from New South Wales bush fires drifting out across the Tasman. She was met at Norfolk by Bonnie

Quintal, one of the island's characters and an accomplished private pilot who has flown around Australia, twice, in a Cessna 172, and around America in a Piper Warrior.

'Bonnie told me she had cancelled my motel booking and I was to stay with her. She is a beautifully spoken woman aged in her sixties. We went up to her house where she had afternoon tea set out ready. The cake was all cut up nicely. She sat me down at the kitchen table. I was feeling sweaty and uncomfortable in all my flying gear. She passed me a cup and said, "Here you are dear. Have a nice cup of tea and a piece of cake. After that long flight you must be feeling fucked." My jaw dropped.

'That night Bonnie took me out on the town and I soon discovered she was prepared to make the most of it. I was decidedly worse for wear when we arrived home. Bonnie then said, "Do you want another spot, dear?"

'Having been out on the town with Bonnie, it seemed I was known by everyone on the island as I prepared to take off next morning. The customs officer even knew me. Knowing Bonnie certainly made all the formalities simple. Bonnie helped with my flight plan and had it faxed to Auckland. She has since been one of my best friends in aviation.'

Norfolk to Kerikeri is almost the same distance as from Lord Howe to Norfolk, but not quite so difficult. The last part of the flight, for some distance, is over land. Much of Cathy's flight was made at 8000 feet, but during the first hour she climbed to 11,000 and suffered from milk hypoxia.

'Auckland wanted to know where I was along that leg. They couldn't contact me, nor I them. I was operating VHF radio. So they asked a Qantas jet to try and contact me, which they did on the 1215 emergency frequency I was operating on while flying over the ocean.

'The Qantas captain called Echo Kilo Foxtrot, twice. Suddenly I realised that was me. Who wants me in the middle of all this ocean? He then told me Auckland wanted to know my arrival time into Kerikeri. I was operating on Melbourne's Eastern Standard Time. Flying operates on UTC Greenwich time. For me, the time is one hour forward, and evening becomes morning. But I couldn't work it out. It didn't occur to me something was wrong.

'Then the Qantas captain said, "For Christ's sake, just tell them how long it's going to take you."

'And I said, "Three hours."

'He was very nice even though he must have thought I was a bit dopey. He then said to me, "You are a woman; in a single-engine aeroplane, flying from Australia to New Zealand?"

'I said, "That's right." He said, "Bloody women. Only a woman would want to do that!" To this day, I have no idea who he was.

'Realising I was probably suffering from hypoxia I double-checked my compass alignment. I knew then that something was desperately wrong. If what I was seeing was correct, I should have been flying backwards. I was reading my compass backwards! I thought, ever so casually, I had better do something about this. I descended to a lower, and safer, altitude but it didn't make much difference. I continued to read my compass backwards.'

Approaching New Zealand, Cathy saw one of the most dramatic sights she had ever seen from the air. Beyond the tip of Northland the sea appeared to be boiling where the Tasman Sea and Pacific Ocean met. 'It went on for miles and miles. It was magnificent; I just wanted to keep on looking down at the angry currents.'

Another beautiful sight was the cloud covering the land. 'I knew then why New Zealand was always known to Maori people as Aotearoa, the Land of the Long White Cloud. I wanted to fly above the cloud but I knew I had to descend and fly under it, and fly south to my arrival at Kerikeri.'

Cathy Pegg landed at Kerikeri early in the afternoon. It was New Year's Day, 1994. 'I was pleased I had arrived within my estimated time. I thought I might have felt more emotional. But I was just feeling an overwhelming relief at making it.

'The inside of the aeroplane had to be sprayed before the customs people would allow me out. They handed me the spray can and told me to do it myself. I didn't know how much to use so I must have almost emptied the entire can. I was coughing and spluttering, and they were tapping on the window saying, "That's enough, that's enough!" I was almost suffocated.'

Maria Stewart, a friend with whom she had done her nursing training in Auckland, was there to meet her. They celebrated with a huge meal of seafood. But the seafood was too rich for Cathy's stomach and next day she was unwell.

'After all my years in Australia I was unable to cope with New Zealand seafood. I wanted to stay in Kerikeri until I felt better, rather than fly on to my home in Gisborne. But I had been led to believe that my mother was in hospital and she was anxious to see me. I had come all this way and was worried about my mother, so I decided to fly on.

'As I flew over the ranges from Opotiki I was feeling worse and knew I should not have been flying on my own. I was flying in a space of about 50 feet between the mountains and the cloud base, and that was a really silly thing to be doing.

'When I called inbound to the Gisborne tower I was told to hold west of the field because of a helicopter operating over the vicinity. I was feeling so sick I never really heard what I had just been asked to do. It went in one ear, and out the other. I just wanted to be down on the ground. I flew over the airfield and almost collided with the bloody helicopter.

'Realising I was about to be killed, along with those in the helicopter, I instinctively put the Cessna into an initial and pitch manoeuvre which banks the aeroplane into the circuit at top speed. With the wings almost 90 degrees to the ground, you bank cross-wind and then bank in the opposite direction to fly down wind. Having the wings at 90 degrees means the speed washes off as the aeroplane turns. It was one of those little tricks Mark Thoresen had taught me and I did not think a Cessna would do it. I was chronically ill but in aviation you take responsibility for whatever you do.

'Once I was on the ground the controller called me to the tower. I knew I was going to have my arse kicked red raw from one end of the strip to the other. We both knew I was in the wrong. When he saw my RAAF flying suit with the flying officer's stripe on the shoulder all he said was, "In civilian street we do what the air traffic controller tells us to do." I was let off very lightly.

'My mother was there to meet me. She had not been in hospital at all. My father was there, too, but he walked away without speaking to me. I tied the plane down, threw up and went home to my mother's house where I passed out, and slept, for the next 24 hours.'

Further proof that aviation's never been just for blokes comes from Hanafi Hayes, a popular NZBC television presenter during the 1970s. One of the happiest shows he made featured veteran athletes and sportspeople. It was called *Life Begins at Seventy.*

Hayes over New Zealand
Hanafi Hayes

The last person to appear in this scenario for super-annuitants certainly kept her feet off the ground. At least she did when she was performing her thing. Indeed when the camera first picked her up for the opening sequence of her bit in the film, she was 800 feet above the ground. Standing on the wing of an aeroplane.

Hilda Alexander, who didn't know that I knew she was seventy-eight at the time, was the oldest woman ever to fly in this fashion in the world. She still is, I believe.

Incidentally, this was one episode where I really did dread having to participate in the same manner as the 'old expert'. But when it came to it, I had no choice. Dear Hilda challenged me before we actually started filming.

'If I can do it at my age, then so can you … if you're man enough.'

Let me explain exactly what she meant. She was inviting me to stand, just as she did, upright, on top, outside that is, on the wing of a Tiger Moth single-engine biplane. While it was in the air.

Now, I'm not really a coward. I mean, I was prepared to risk my life trying to stand vertically on ice skates. I have, in my time walked across rope bridges over deep gulleys. Well ten feet deep. I've even walked unaccompanied down Queen Street in Auckland after dark. On late shopping nights. But stand on the wing of a plane?

At that time, I had a phobia about even flying inside of one. It's true. I was the archetypal 'fear of flying' passenger. Every time I boarded one of those unnatural vehicles, it was like the first time for me. The thought uppermost in my mind was not so much will it crash, as when will it crash?

Maybe I would have been happier if I'd studied aerodynamics. As it was, my mind was always full of terror. Every noise was a warning of impending disaster. Every movement of the craft, a sign of dreadful trouble. And there was no way that I would either look out of a window or leave my seat during a flight. However long it took. I wouldn't even unbuckle my safety belt when the little illuminated sign came on with a 'ding'. In truth, I always assumed the 'ding' was another passenger in strife, calling for a stewardess. As for getting up and going to the toilet at the rear of the plane, there just was no chance of that. I would rather sit in kidney-bursting agony than risk being in that compartment when it fell away from the rest of the aircraft. Hard to believe that I was once in the RAF isn't it?

Thank goodness, I've since conquered the problem. Simply by having to fly to so many places. However, at the time we were filming Hilda Alexander, that, quite truthfully, was my state of mind on the subject of flying.

But what could I do? This fantastic great-grandmother had thrown down the gauntlet. I had to pick it up. All I could hope for was that I might pass out as we took off so that I would know nothing about it. Meanwhile we had to film the little troublemaker.

Mrs Alexander was a wee ball of a woman, especially in her flying rig. Gingery frizzy hair escaped here and there from under her leather helmet and she wore a great big fur coat on backwards to keep out the cold. She was a feminine person for all her antics in the air. For the camera she'd managed a touch of rouge here and a touch of lipstick there and her shiny brown eyes showed she enjoyed a good joke.

First we filmed her way up and over the Cathedral in the centre of Christchurch. That was done with the cameraman shooting from another light plane a few feet away. When they all got through that dice with death, I talked to her down on the ground.

Why on earth, to use a singularly inappropriate phrase, did she do it? Well in the first place, because her son had done it. She actually didn't take up this sport or madness, call it what you will, until she was seventy-three. I never found out what she did for kicks before that date. Bullfighting on ice I shouldn't wonder.

The spirit of thrill-seeking was certainly in the family. Even her grandchildren had been up there, she told me. That was another reason for my soon-to-happen act of bravado. They were all there watching us.

It had all started though, as Hilda said, with her son, Ron. He was the first man to fly across Cook Strait on the wing of a plane. Now they've got ferry boats of course. Usually though Ron performed at the controls. Indeed he was one of New Zealand's best-known aerobatic flyers. Over the previous five years, Mum and son had appeared with the act at pageants and air shows all over the place. Ron in the cockpit and Hilda standing outside alone, waving her hankie to the crowds from upwards of 500 feet. And if that wasn't enough, this incredible lady specially requested her son to tilt the plane from time to time. She got bored just following a steady course.

However, she was still human. When I asked her if anything unusual had ever happened whilst she was up there, she answered: 'Well, no, not so far. I wouldn't like anything unusual to happen.'

My sentiments entirely, I thought, when my turn came.

The little climb up the ladder was like mounting the platform for the guillotine. Madame Lafarge Alexander was standing at the side knitting and grinning away below me. Actually just getting onto the wing was enough for me. Already I was a good ten feet off the ground. But once I was up there, I couldn't turn back. They'd taken the ladder away, and started up the engine.

In the centre of the plane, an upright stand was bolted into the top of the wing. It was set at a slight angle and served as a back rest, supported by a pole each side. The idea was to stand against this and be harnessed to it with safety belts across the chest. Once I was strapped in, I pulled my goggles down, tried to control my legs from shaking violently and gave a weak smile to the camera below. We were off.

At first, I couldn't believe what was happening. Had I really agreed to this? Was this really me standing on top of a Tiger Moth with a whirring lethal propeller a yard or less from my feet? What would happen if I fell? If the harness were to snap and I fell forward onto the blade?

They say that just before you die, you see your whole life in rapid playback form in a matter of seconds. Well I didn't exactly experience that. But I did manage to conjure up just about every permutation of horror that could possibly happen to me in that situation. And we weren't off the ground yet.

Maybe Ron won't actually take off. That was my next thought. Perhaps he'll have realised how absolutely terrified I really was. Sure, he could just fake it for the camera and then pull up at the end of the runway and turn around. He didn't though. We were off the ground. My God. Yes, that's right, that's all I could think of, My God. Please save me.

If ever you're having a problem with your belief in a Creator, any doubts at all as to where you naturally turn when faced with certain death, then fly on the wing of a plane. I prayed incessantly. Over trees, over fields, over buildings. The whole of north Christchurch has been prayed over. And then we were back on the ground. My faith was intact. My prayers had been answered.

All the same, I wouldn't be doing that again. Challenge or no challenge, that was for sure. Hilda Alexander was more than welcome to that particular thrill.

As things turned out, though, the little flying great-grandmother never went up on the wing again. I found that out some time later, in 1977.

I happened to bump into her in a small coffee shop in Riccarton. The old dare-devil didn't recognise me at first, but then she came over. She'd aged slightly I noticed, but was still full of fight at eighty-odd. She wanted me to do her a favour. Would I write to an address she gave me, to verify that she had flown on the Tiger Moth wing, on the day that we filmed. Apparently the information was being sent off for entry in the Guinness Book of Records. Her name would be recorded officially at last as the oldest lady to fly on the wing of a plane.

If standing on the wing of an aircraft sounds perilous, trying jumping out of one in flight to 'bull-dog' a fleeing deer ...

High-fenced deer farms are now a common sight in the New Zealand countryside, but it wasn't always so. For years, deer were simply seen as pests and shot, sometimes from light aircraft. When venison prices rose in the 1960s and 1970s, helicopters entered the fray. With licensed hunters, poachers and cowboy operators all competing for the kill — and the cash — things got pretty stroppy.

For a time, the Wild West atmosphere of the 'Deer Wars' prevailed, until a few astute souls noticed deer grew much quicker in captivity than they did in the wild. So the cullers became captors, initially jumping on their cervine targets before sensibly switching to tranquilliser guns and nets.

Sir Tim Wallis, who established the Alpine Fighter Collection and also launched the internationally recognised Warbirds Over Wanaka airshow in 1988, was an early deer hunter, flying fixed-wing aircraft as well as helicopters. He was one of the first people to see the agricultural potential of deer and helped set up the country's first deer farm.

Equally, Sir Tim would be the first to insist he was not alone. There was many another wild pilot and hunter also doing the business. Their adventures, misadventures — and truly awesome feats — read like a ripping yarn, a work of the most fantastical fiction. But what they did out in the bush and up in the peaks was real, and it's a story that's unique to New Zealand.

The first person to chronicle aviation's last frontier was Mike Bennett.

The Venison Hunters
Mike Bennett

Our helicopter always carried a round 22-litre tin with emergency provisions, two sleeping-bags and a light tent in case of accident or being caught in fog. There was also a spare rifle aboard for the pilot as the shooter always took his rifle with him on getting out of the helicopter if it was going to be away for any length of time. We also carried pencil flares, which were next to useless, but at least we tried to the best of our resources at that time to reduce the hazards of our job as much as possible.

We spent a night wrapped in heavy, damp fog in Limbo Creek at the head of the Cascade. It was a narrow hanging valley, incredibly difficult of foot access and possibly one of the remotest spots in all the mountains, seen or traversed by very few. It was an unreal night in the fog in that place of infinite solitude; the rest of the world might never have been, such is the atmosphere of the elements and the remote, secret places of the mountains. In the dawn, such as it was, we decided to try our luck as despite our wordly toughness we could not quite shake the feeling of absolute trespass in hostile ground where there was almost a tangible aura of malevolence.

The exit from the valley was an incredibly narrow slot leading to a steep gorge bestrewn with enormous boulders and strange pinnacle rocks like great splinters flung from the heights. The pilot was almost blind in the whiteout, the bubble an opaque mass of perspex covered in fog droplets. We crept out into the unknown in not much more than a hover, holding a deal of power, with myself hanging half out of the machine with feet on the skids. I will never forget a vast finger of schist looming out of the fog immediately ahead of the helicopter. The pilot stopped at my shout and we gently waltzed away to the left. It had been very close. Despite the near-freezing temperature, Ben was sweating profusely: despite the help of my eyes, the responsibility was all his. When we broke through the fog layer we landed on the first flat below the gorge and, not speaking, walked away the tension we had both felt.

A couple of days later we went back in clear air up the gorge of Limbo Creek to pick up our kill and when we saw what we had not seen in the fog, realised just how luck and/or foolish we had been.

Then there was the day when it really was too turbulent for flying above the bushline, so of course we were up there. I had just climbed aboard the hovering machine, which was fortunately facing out from the face, when a great gust came down from the crags and literally hurled us down the slope. I was still only half in the cockpit and was clinging to the loose end of a seat-belt with frantic strength. My head was level with the control console and with the speedometer only inches from my eyes was amazed to see us go from a hover to over 100 miles per hour in seconds. We cleared the lip of the shelf with not much to spare, as the machine was being forced closer to the ground all the time in the massive downdraught. Over the yawning valley below the machine plunged and bucketed as the pilot fought for control, with the blades clattering madly as they too fought for air to grip. It was a nasty moment. It was time to go home.

Helicopters handle turbulence better than fixed-wing aircraft, but they do have their limits, although the skill and/or desperation of the pilot in command has much to do with it. A salutary experience is to be a passenger in a chopper which has 80 mph indicated on the clock and is actually being blown backwards. Now it never ceases to amaze me how much trust we used to blindly put in such complicated pieces of fragile machinery....

I shot a monster stag in the Upper Cascade flats one time and the animal staggered into the bush before collapsing. That particular section of the flats was no more than small boulder beaches covered in sizable rocks where the helicopter could not land without difficulty. I jumped out, ran under the trees and gutted the animal, but as it was well over 135 kilos could not drag it through the mud and out of the bush.

The Canadian pilot could see I was having a struggle, and next thing I knew he was hovering right on the treeline, having a good look. I could hardly believe it when he just touched the skids down and walked the machine in under the trees. The blades were cutting off all the hanging branches of beech with all the precision of an oldfashioned barber with a pudding-basin. The stag was hooked on to the machine with a strop and the pilot backed out. How he did it I will never know, but he did....

The stags in the early days would often stand and dispute their ground with the closing helicopter. During the Roar it was not uncommon for a stag to roar at the machine as his hinds were being shot about him. This often gave the hardworking helicopter pilot a rare chance for a shot — if he could find a place to land. It was frustrating for these men to see hundreds of animals shot and never get the chance themselves at a particularly good trophy animal which, if one was generous enough to disregard that it had been shot with the aid of a machine, would look well on the wall of the old homestead in little old Noo York or some such. This hankering for something to hang the Stetson on had an unfortunate sequel for me at a later date, but in that instance the pilot had wanted a good trophy chamois, an animal which we didn't often shoot in those days when there were so many heavier deer to choose from.

The basin below the small receding glacier in the head of the Donald River, a tributary of the Waiatoto, was teeming with chamois and it was nothing unusual to see well over 500 animals in that small hanging valley. It was also a place where the helicopter could be landed close to the creek, and as Jim Davies the pilot was due to go back to Canada shortly, this we duly did. There aren't many places you can land a helicopter and be within range of chamois, but you certainly could in the Donald. We landed and started glassing through the telescope sights for a decent head. At 300 metres Jim lay down and shot three or four to better his chances of a head and he did get a good one. The chamois carcases were picked up and ferried down to a previously made up load of deer on a small terrace below the waterfall, where I jumped out to restrop the two loads together.

The wind had started to gust and I had no sooner hooked the load to the machine when Jim could hold his hover no longer and he lifted off without warning. The trouble was, he lifted off with me as well. As the load jerked I tried to jump back, but a curved chamois horn hooked neatly between the tongue and the laces of one boot.

Everything happened so quickly that I was almost unaware that it was actually happening. One moment I had been standing on top of a load of animals, the next I had been plucked effortlessly away to find myself hanging upside down under the load of an airborne chopper. It was, to say the least, an unusual sensation. It was so unreal I had no time to feel fear. Instead I was aware of a hard pain in one foot and feared that the chamois horn was buried in it. Fortunately Jim Davies, as he had lifted off to regain stability, had instinctively looked to sight me as he lifted off, and as he gained height and could see the entire terrace that I should have been on he realised that I couldn't just have disappeared into thin air.

In this case the unbelievable had happened and I *had* disappeared into thin air, and this worried Jim's sense of the fitness of things. He gently banked the machine and very gently lowered the load back to the same place and as the weight came off the machine he pressed the release and let go the chain before backing off to see what he had caught on his hook with 300-odd kilos of carcases as bait.

My head had touched down first, then I was buried by the load of animals. As I struggled to get clear, two metres of released chain hit me on the head and neck and I began to feel that it was not my day. It was almost the last straw. I still felt a sense of unreality, I was also cold, angry and in pain at both ends — and then, I believe, shock took over. Managing at last to get clear of the load I found myself aboard the helicopter. Jim didn't argue when I told him, 'Just get me off this bloody mountain.'

Various accounts of this lucky escape were reported in articles ranging from Australia to the States and Great Britain, with various heights under my tartan tam o'shanter ranging from 100 to 1000 metres. We were working the edge of a terrace so that as soon as the machine lifted the load Jim had had possibly seventy metres below him where the Donald meandered through a scrub-and-boulder flat before the hanging valley ended and plunged straight down to the Waiatoto 750 metres below. Jim's turn with the load might have taken him over this edge as it was only a couple of hundred metres away; neither was the operation seen by

Ian Rendell, who was working on another load with his back turned, farther along the same terrace. The height below at any one time was hardly important either then or now, as I am under the impression that it is not the extent of a fall that does the damage, but the sudden stop at the end of it.

And that's it. Touchdown; the end of this particular 'three-century-flight'. It would've been good to stay up a bit longer — indeed there was going to be one last rollicking tale but alas, it got shot down at the very last minute.

So it's back to the hangar with, let's hope, a greater sense of what it means to take a risk and a greater willingness to take a few yourself — both in the air and on the ground. At a time when our media and legislators seem utterly obsessed with victims and the many perils of existence; when they appear totally determined to give us a world so strictured and structured and safe as to stupefy the most timid of souls, the authors of the words on wings you've just read serve to remind us that the unknown is still the best place to go.

All of us are heirs to this adventure and might do well to remember what took these Kiwis into the skies. It wasn't rules; it wasn't regulations; and it certainly wasn't the fretful mood of a fearful age. Perhaps we could all do with some of their aviation spirit in our tanks.